Multivariate Morphometrics

2nd Edition

R. A. Reyment

*Department of Historical Geology and Paleontology,
Paleontological Institute, Uppsala, Sweden*

R. E. Blackith

*Department of Zoology,
Trinity College, Dublin, Ireland*

N. Λ. Campbell

*Division of Mathematics and Statistics,
CSIRO, Wembley, Australia*

1984

ACADEMIC PRESS

(*Harcourt Brace Jovanovich, Publishers*)

London   Orlando   San Diego   San Francisco   New York
Toronto   Montreal   Sydney   Tokyo   São Paulo

ACADEMIC PRESS INC. (LONDON) LTD.
24–28 Oval Road
London NW1

United States Edition Published by
ACADEMIC PRESS INC.
(Harcourt Brace Jovanovich, Inc.)
Orlando, Florida 32887

British Library Cataloguing in Publication Data

Reyment, R. A.
   Multivariate morphometrics.—2nd ed.
   1. Morphology—Mathematics   2. Multivariate
   analysis
   I. Title   II. Blackith, R. E.
   III. Campbell, N. A.   IV. Blackith, R. E.
   Multivariate morphometrics
   574.4'028   QH351

   ISBN 0-12-586970-3
   LCCCN 83-71980

Photoset by Paston Press, Norwich and printed in Great Britain
by St Edmundsbury Press, Bury St Edmunds

Preface

The range of topics that could reasonably be expected in this book is now so great that having three authors instead of two goes only part of the way to enable us to cover the ground. The need to condense the text was a further constraint.

We now realise that some of the most fundamental problems of morphometric analysis are inadequately resolved. It is not even clear what steps should be taken to resolve some of them, and bringing the reader face to face with these problems has been a preoccupation.

The response to the first edition has strengthened our belief that the book should serve as a source of practical examples, drawn from a wide range of disciplines, though some improvements are offered for the analyses in the first edition. Twelve years ago the relevance of multivariate theory, as it then was, to our concerns seemed tenuous; more recent work has brought theory and practice closer together.

We have removed virtually all aspects of numerical taxonomy, cladistics and related topics, partly because other matters seemed more urgent and partly because that debate has been soured to a point we find most unattractive, and is likely to remain inconclusive, at least as regards the central theme of what we consider as morphometrics. We have also deleted the glossary and several chapters on topics which have not developed as we thought they might do.

Inevitably, the bibliography has grown with the expansion of the subject, and we are painfully aware that work which would meet any reasonable criterion for admission has had to be omitted.

A quarter of a century ago, a distinguished biologist, responding to a talk by one of us, dismissed multivariate morphometrics as "just youthful enthusiasm" and predicted that within three years it would be forgotten. We are happy to report that the subject is patently thriving, our enthusiasm remains undimmed, and our admiration for others working in this field grows with their achievements. We note with satisfaction that morphometrical work is now playing a decisive role in the rapidly expanding fields of evolutionary biology and quantitative genetics.

The preparation of this largely rewritten edition of Multivariate Morphometrics has been undertaken in several intensive episodes in Uppsala and Melbourne. We express our thanks to the numerous bodies providing us with funding and working conditions: Uppsala University, Trinity College, Dublin, Swedish Natural Science Research Council, DMS-CSIRO, the Institute of Special Education, Victoria, Australia and Melbourne University.

January 1984   *R. A. Reyment*
                                                       R. E. Blackith
                                                       N. A. Campbell

Acknowledgements

We are indebted to A. W. Davis, S. D. Hopper, D. Ratcliff and L. G. Veitch for their helpful comments on various chapters.

Contents

1
The Historical Background to Morphometric Studies

Although the history of morphometrics is, inevitably, entangled with the much wider issue of the history of attempts to classify animals and plants, the subject antedates classificatory attempts. For there is an area of thought in the consideration of the form of living things which is much more a question of art than of science in the modern senses of these words. The desire to abstract from the great variety of living organisms forms which are essentially harmonious, aesthetically superior to other forms, or else capable of abstraction as the archetypical form of some wider category of shapes and sizes, is very deep-seated in human behaviour. Moreover, it is ignored at the risk of serious misunderstandings, because there is evidence that at least a part of the current difficulties of communication between leading schools of thought in taxonomy stems from a lack of historical insight into the motives of taxonomists, and into the extent to which different practices are likely to be aesthetically satisfying.

In the earliest studies of the forms of living organisms of which we have any record, there is evidence that classification, aesthetics, and what we should now call functional analysis, are interlocked to an extent which makes any attempt to separate them futile. For the Babylonians, back into the beginnings of the second millenium BC and for the Greeks before Plato and Aristotle, that is to say before about 350 BC, there was a strong conviction that form, as much in music as in anatomy, was capable of numerical expression; more, that the very essence of musical harmony and of aesthetic harmony resided in that theory of numbers which came to be one of the focal points of the thought of the Pythagorean school of philosophers in the Greek-speaking, and at least partly, ethnically Greek city–states of southern Italy from around 600 BC to around 300 BC. As Stapleton (1956, 1958) has put

1

it, the mathematics of Pythagoras and his successors must be regarded as the offspring of both music and mysticism. Pythagoras (580–497 BC) himself, the son of a Phoenician jewel engraver by a Greek mother, and reputed to have been an Olympic boxer, knew of the numerical relations between the lengths of strings (and pipes) emitting concordant notes (Levy, 1926). His immediate successors probably knew most of what we should now call Euclid book II and part of book VII (van der Waerden, 1963).

Activities which a modern numerical taxonomist would instantly recognize were part of the teaching of the Pythagorean school; for instance, the essential quality of an animal or plant was "captured" by drawing an outline sketch, almost a caricature, in the sand, and then "recording" this essence by counting the number of junctions between the lines of the sketch, keeping this number of lines to the minimum required to evoke the image of the animal. It is of interest to record that elements of this approach are contained in the Thompsonian attitude to the measurement of shape and size promulgated by Bookstein (1978).

Both Theophrastos and Aristotle describe how Eurytos, at the end of the fifth century BC, made a low relief sketch of a man in thick limewash on a wall, sticking coloured pebbles into the junctions of the lines. A later account by the pseudo-Alexandrian commentator adds that Eurytos would finish his sketched-in representation of man with pebbles equal in number to the units which in his view defined a man (Guthrie, 1962): the "number" of man was 250 and of a plant was 360. No doubt a numerical taxonomist would part company with this activity at the point where the mystical interpretation of the numbers began, as combinations of the male and female principles of the universe (3 and 2) or as symbols of the hand of the Creator (1 and 4). Nevertheless, no one who has followed the development of Palaeolithic art can fail to recognize the deep-seated need to "capture" the essence of certain animals of the hunt by sketching them, perhaps on the walls of caves, perhaps by making bone or stone models. It may, indeed, be that the constant practice of this ritual sketching led to the extraordinary disparity between the skill with which animals of the hunts were depicted by Palaeolithic man and his crude and fumbling attempts to portray the human form, a disparity which becomes ever more striking as more and more artefacts of many periods are discovered.

Nor can anyone who has watched a skilled cartoonist portray the essential qualities of a person or of an animal, with great economy of line, fail to appreciate how quickly the human mind responds to visual clues which evoke an emotional response.

The way in which Greek morphometrics began is not necessarily, or even probably, true of other civilizations. The classification of animals in India, for example, was based, so far as we know, from about 600 BC to about 200

AD, on the mode of reproduction of the animal (real or supposed), or towards the end of this period, on the number of senses that it was supposed to possess (Sinha and Shankarnarayan, 1955).

The early Greek attempts to discover the essence of form, and hence, as a secondary aim, to classify organisms, were cast in a geometrical rather than an algebraic mould, and although with hindsight we can see how the highly sophisticated branch of geometry known as topology would have enabled them to continue their trend of thought, it is clear that, in the state of knowledge then prevailing, geometry was at a dead end: the kind of problem soluble by geometrical techniques and epitomized by Euclidean geometry was of very limited use in describing the forms of organisms. By the fourth century BC the time was ripe for a new approach, based on the logical syllogism of Aristotle (384–322 BC), systematizing and amplifying the thought of Plato and Socrates.

This school was opposed to the Pythagoreans because, to quote Aristotle "they construct the whole universe out of numbers, not however strictly monadic numbers, for they supposed the units to have magnitude". To some extent there seems to have been an inadequate distinction, natural enough in the fifth century BC, between the unit of number *per se* and that geometric point whose motion generated lines, which in turn generated planes, and they again generated solid structures. But it is also possible to argue that much of the disagreement stemmed from Aristotle's concept of number as an essentially scalar quantity, with ratios and proportions to represent form, whereas the Pythagoreans had an almost vectorial interpretation of numbers, which were conceived as "flowing". Their system of representing numbers by rows of pebbles (hence "calculus") arranged in regular patterns (gnomons) instead of numerical symbols would have encouraged a way of thinking about the system of natural numbers not far removed from some of the simplest ways of thinking about a vector plane. Aristotle's inability to understand the Pythagoreans "flowing" numbers rankled, and he became quite petulant on the subject.

Few phenomena are more curious than the erection of Aristotelian ways of thinking into the epitome of classical teaching; a way of thinking about Aristotle which is still widespread. For he was in many ways a destroyer of the old Pythagorean habits of mind as well as an innovator of genius himself. As Bertrand Russell has commented, he was inept at quantitative work, and was thus forced to think in terms which were qualitative, and although he made tremendous advances in what we should now describe as taxonomic theory, these advances were made at the high cost of destroying the concern for what underlay the external form so characteristic of Pythagoras' school. This change was made necessary, in the conditions of Greece in the fourth century BC, partly for historical reasons stemming from the growth of scepticism and partly because the Pythagorean philosophy was non-operational in the sense that one could applaud its teachings with the heart but fail to see how the reason could act on its precepts. At this period too the Pythagorean ideas of the heliocentric (or at least fire-centred) solar system, the spherical earth, the atomic nature of matter, propounded by Pythagoras' contemporary Democritos in the Macedonian city of Abdera, and indeed the wide-spread practice of partial democracy faded into the mists from which they were to re-emerge only in comparatively recent times. Pythagorean thought was eclipsed in the intellectual revolution.

Although Aristotle was willing (indeed he had little option in the matter) to abandon to the straight-jacket of the syllogism the descriptive function of classification, he was also the heir, and of even more concern to us, the principal instrument for bequeathing to later generations the concept of pure form. In the immediate past, Plato had discussed the idea that behind the everyday appearance of things there was an ideal form which possessed, unsullied by the imperfections of ordinary life, the quintessential qualities of the things studied; that immanent in the bulls of the fields around Athens, one could create in the mind a concept of a bull perfect in its power, its virility, its aesthetic proportions. This idea was to be carried still further by Aristotle, mainly by way of his influence on Plotinus whose transcendental philosophy at once diminished the operational nature and hence the emotional impact of Aristotelian biological theory and increased the chances that Aristotelian philosophy would survive the harrowing by the mediaeval Church. This survival was a near thing. Even by the time of the foundation of the University of Paris (1215), the fear of Aristotle was so great as to induce the insertion into the statutes of a ban on carrying Aristotle's work on pain of excommunication. Gradually, as classical learning revived, so did the syllogistic framework of classification, and the ideal of pure form which, to varying extents, influenced the taxonomic concept of a type in its formative stages.

We need only to look at two very different workers in the biological field, Linnaeus and Goethe. Linnaeus was taught syllogistic logic at school, and it would have seemed to him natural enough to employ Aristotle's syllogisms to form the intellectual framework of the Linnean classification. Moreover, operating within a strictly logical and theocentric framework, it would also have seemed natural to Linnaeus to find one good character by which species could be differentiated, for if each species is thought of as having been sent by the Creator for man's benefit, it would be attributing some defects to the logic of the Creation were man not to be in a position to see the distinguishing marks. We owe much to Cain's historical research for our understanding of the main-springs of Linnaeus' thought (Cain, 1958, 1962).

With the genesis of the dichotomous key in syllogistic logic, and that of the monothetic classification in the theology of Linnaeus' times, we turn to Goethe.

In some far from trivial ways, Goethe's conception of form was nearer that of Aristotle than that of Plotinus (Wilkinson, 1951). He distrusted Plotinus' idea that form is quintessential, perfect only when it does not descend into matter; but if his mental image of the ideal form transcended the outward manifestations of an organism, it also embraced its outward form, since he was deeply concerned with function in relation thereto. To a considerable extent Goethe was a link between the Linnean approach and that of the taxonomists of the school now known as Adansonian, for his concern with "gestalt" (the whole organism) was much more in the line of Pythagorean ideals.

The general consequences of the introduction of evolutionary theory were surprisingly unhelpful to taxonomy. Linnean systematics accommodated itself so easily to evolutionary modes of thought that almost no changes of practice were necessary, only the description of that practice needed alteration. This fact itself is witness to the superficial nature of biology at the classificatory level. What was worse, the ready acceptance of, and indeed demand for, phylogenetic speculation lowered the standards of classificatory work. Agnes Arber has commented, "The whole attitude of many post-Darwinian botanists . . . has been distorted through trying to compel the study of form to observe phylogenetic ends" (Arber, 1950, p.7). Now that standards in conventional as well as in quantitative taxonomy are sharply rising, we are entitled to insist that the one great, generalizing principle in biology, evolutionary theory, should not be prostituted to this end. Very little serious work has been done on the inheritance of form and Grafius' work in this field stands out as a promising line of study (Grafius, 1965).

Pythagorean ideals, if not ideas, emerged strongly in the work of D'Arcy Wentworth Thompson (1942). It is a measure of the problems facing the earlier Greek philosophers in this field that Thompson was, in his own way, as remote from an operational solution to the comparison of forms as they were.

The analysis of shape-variation and growth has turned out to be a more difficult problem than one might have thought a decade ago. There are now several approaches to the subject although none of these can be unanimously claimed to be the long sought-for final solution. Mosimann (1970) has attempted a specific solution to the problem and Bookstein (1978), who, not without just cause, is highly critical of all previous work on the subject, except that of Thompson (1942) and Oxnard (1973), has attempted a Thompsonian solution to the problem which, however, although of interest and a valuable advance, is still not the last word. A useful "state-of-the-art" paper on the subject is that of Sprent (1972).

Thompson seems to have thought in terms of gradients of growth whose end-product was the final form of the organism. With his remarkable classical knowledge and sense of artistic appreciation, he was an instigator of the revival in studies of growth and form even beyond a restatement of the problem in terms of transformation grids. He thought that mathematical methods had been slow to aid morphology because of the hostility, or rather want of understanding of the nature of the problem, implicit in Bergson's (1911) comment "Organic creation . . . the evolutionary phenomenon which properly constitutes life, we cannot in any way subject to a mathematical treatment". The problem was consistently tackled on an inappropriate basis, attempting to describe the form of a given species mathematically, instead of describing the contrasts of form between the two affine species in vector terms.

After Thompson, or rather toward the close of his life, which included 64 years in university chairs, came the development of studies of allometric growth, with which the names of Huxley and Teissier are closely associated (Huxley and Teissier, 1935). Although almost entirely confined to the studies of the joint variation of two characters, it became apparent at an early stage that the generalization of the allometric relationship was going to mark a milestone in the history of growth studies. This step has been taken mainly as a result of the work of French-speaking biologists, amongst whom we may mention Teissier (1938) for one of the earliest applications of properly multivariate methods to invertebrate studies; Cousin (1961) for an extension of the use of factor analyses to studies of inheritance, already investigated using allometric methods by Bocquet (1953); and Jolicoeur (1963d) for placing the multivariate generalization on a mathematical footing.

It should not be thought that Pythagorean ideas disappeared entirely in the third century BC: ideas which meet a human need are tenacious of life, and these were partly modified, partly continued, by Persian and Arabic-speaking philosophers and then by mathematicians such as Leonardo Fibonacci of Pisa (twelfth century) and artists such as Albrecht Durer (1471–1528) (see Durer, 1613) and Leonardo da Vinci (1452–1519) whose passionate concern for the harmony of form lead them to make sketches which stem from Pythagorean preoccupations with form yet anticipate the ideas of D'Arcy Thompson. Medicine afforded a further line of intellectual descent, since Galen of Pergamon (131–201 BC) taught that the incidence and course of disease was dependent on the body-build of the patient, a perennial subject of medical interest now known as somatotypy.

Whereas, up to about 1750, it had seemed proper to select characters for taxonomic study on Aristotelian *a priori* principles, connected with their supposed importance in determining the way of life of the animal or plant,

Adanson (1759) produced a radically new departure, allotting equal *a priori* weighting to all characters. In this he was, as Cain (1959) remarks, much ahead of his time, and we refer the reader to Cain's work for an appreciation of the state of taxonomic theory in Adanson's times. Adanson's earlier progress towards equal weighting came from his studies on molluscs; it came to fruition in his great botanical treatise (Adanson, 1763).

Independently of Adanson, Scopoli (1777) who also studied plants as well as animals was to publish very similar views on equal weighting, although his contribution has received much less notice during the revival of what is now often called Adansonian taxonomy. Scopoli's work was brought to our attention by Mr R. G. Davies of Imperial College, London.

This reinforcement of Adanson's ideas was not, however, sufficient. The claims of Aristotelian logic, the belief that by taking thought one could construct the natural system, without recourse to mundane observation, remained too strong. A whole series of *a priori* weighting systems came to the fore, only to fall as the next fashion came on the scene. This is not the place to discuss the circularity of much *a priori* thinking, which has been cogently argued by Cain (1959), Sneath and Sokal (1973) and Colless (1967) amongst others. Darwin's *a priori* system was, however, based on observations linked to evolutionary criteria. It has seemed to many workers during the last century so satisfactory that three decades ago few would have predicted that it would be the subject of serious dispute; but so it is and the issue is still open. The swing away from the "new systematics" lies at the heart of the "new taxonomy" initiated by the coming of numerical taxonomy; although the two issues of numerical methods and empirical theory are in fact logically distinct.

2
The Mathematical
Framework of Morphometric
Analysis

An essential problem in morphometrics is to measure the degree of similarity of two forms. Although much can be, and has been, written about each stage of the analysis, the following exposition will suffice to explain the essential ideas.

Consider the two forms shown in Fig. 2.1. To help visualize the idea of a degree of similarity, the smaller of the two forms has been drawn inside the larger one. Imagine the smaller form growing so that the difference in volume between the two forms is reduced. The growth pattern may be of any kind, so that as each of the marked points on the left-hand side of the smaller form approaches its homologue on the larger, we permit each marked point to travel at a rate independent of that of any other. It is intuitively obvious that the two forms will become identical when the two sets of marked points coincide, unless, of course, there are features of either organism which are unlike but which have not been allocated a marked point. This would be a failure of homology. We will assume for the moment that the marked points are sufficient to follow the growth of all salient features of the organisms. The volume occluded between the two forms may be regarded provisionally as one measure of the dissimilarity of the forms. If we consider the expansion of the smaller form as being accompanied by a displacement of every marked point, the vector of displacements of the marked points on the smaller form towards the homologous points on the larger form may be used as the basis for a measure of divergence between the forms. Essentially, we have set our analysis in motion in terms of multivariate vectors defining the marked parts

8

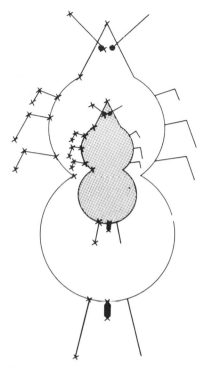

FIG. 2.1. Affine forms of a hypothetical animal: the marked points on the smaller, shaded, form move to their homologues on the larger, unshaded, form as the smaller form "grows".

for the smaller and larger forms—this is the essence of all truly multivariate studies of the form.

Formally, instead of considering the measurement of the volume displaced by the expansion of the smaller form, we can set up a p-dimensional set of Cartesian coordinates in which the axes are the directions of displacement of the marked points. Each of the two forms can then be located within this framework of reference by noting the amount of the displacement of each marked point along its axis of variation. One way of defining distance between the two forms is to apply the p-dimensional equivalent of Pythagoras' equation. At least three epithets are used to describe this metric: Euclidean, Pythagorean and Sokal's metric. We have retained the designation used by the original author for the metric employed in his paper.

The distance as measured in this Euclidean space has been suggested as suitable for use in taxonomic work by Sokal (1962). For the multivariate extension of Pythagoras' theorem to provide a realistic measure of distance, the movement of any one marked point must be independent of the movement of any other marked point, with no correlation between the measured characters. In general, we can expect substantial correlations to exist and these will seriously bias such a measure of distance.

It is preferable to be able to compute a distance which takes into account the correlations between the characters. This, in effect, involves setting the angles between the coordinate axes so that the cosine of the angle between any two axes is equal to the coefficient of correlation between the characters whose displacement they represent.

The distance so calculated, the generalized statistical distance between the two forms, adjusts for any correlations between the measured characters by introducing the (inverse of the) matrix of dispersions of the characters into the calculations. This matrix is also referred to as the "covariance matrix" and, sometimes, as the "variance-covariance matrix".

With the ready availability of statistical packages, multivariate analyses may now be performed on data without any comprehension of their mathematical basis. The effort to understand, even if the understanding is incomplete, is, we feel, rewarding, not only in the sense of providing some control over the calculations, but also of affording a glimpse of the workings of an approach of wide generality. We feel it is in place to sound a note of warning about the use of the computer by research workers who lack insight into the function of the techniques involved. With very little effort it is possible for somebody with a collection of data to have these processed by multivariate statistical programs. Unfortunately, even well-tried programs may contain faults, and unless users have a fair knowledge of what should be happening, they run some chance of being saddled with incorrect results.

The need for multivariate calculations in statistical work provides a good example of how a crucial problem, long pondered over, is suddenly solved, in different ways and independently, in different countries. In England, Fisher developed the idea of the discriminant function, in America, Hotelling generalized the t-test for use in the multivariate case, and in India, Mahalanobis was responsible for the evolution of the generalized distance between two multivariate populations. Although Hotelling's generalizations of the t-test appeared in 1931, key papers from all three authors appeared within a few months of one another in 1936. The D^2 had been in use for a decade in Mahalanobis' anthropological papers, but without the theoretical background necessary for its full exploitation.

We should emphasize that, whereas much of the development of multivariate theory has been concerned with significance testing, such tests play a minor, though assuredly not unimportant, role in morphometrics. For many purposes, such multivariate exploration is concerned with the construction rather than the testing of hypotheses.

The Mahalanobis' generalized distance concept lies at the heart of morphometrics. In our view, it provides the only realistic measure of multivariate distance. It is the fundamental measure for multivariate techniques involving discrimination and allocation, and provides the basis for techniques which can be used to examine distributional assumptions. More recent advances in techniques which are less sensitive to atypical observations also depend on the Mahalanobis' distance as the basic measure of distance.

3
The Biological Background to Multivariate Analysis

For the benefit of readers coming to the problems of analysing multiple measurements for the first time, or with rather limited experience of these topics, we discuss what each of the techniques actually does, expressed in terms of a study of the shapes and sizes of Australian rock crabs. Experience suggests that the problems of growth and form require for their understanding the capacity to conceptualize the mathematical manipulations in concrete terms. We believe that until research workers know what the mathematical techniques are capable of doing for them, they are most unlikely to master them.

The types of biological problems amenable to multivariate morphometrical analysis are frequently of a taxonomic kind. It is often desired to compare the range of variability of two or more species thought to be closely related; the possibilities offered by multivariate methods for providing useful graphical representations of such relationships are unsurpassed. The approach is particularly useful when comparing the relationships between and within genera, and between and within species; for example, Hopper (1978) used multivariate techniques to describe the similarities and differences between populations within species, between species within genera, and between genera of the kangaroo paw *Anigozanthos* and *Macropidia*. The same approach was used to examine the composition of possible hybrid populations, so providing further insight into the evolution of the species. Other work in evolutionary biology may demand that polymorphs of a particular species be contrasted; this is again a situation in which it would be difficult to obtain unequivocal results without recourse to multivariate statistics. The obvious advantage of a multivariate approach to such problems over analyses of isolated variables is that it permits an integrated assessment of variation in the organism in which due regard is given to covariation between variables.

DISCRIMINANT FUNCTIONS, CANONICAL VARIATES AND ALLOCATION

The various concepts to be discussed in this chapter will be illustrated by a discussion of data on two closely related species of the rock crab genus *Leptograpsus* (Campbell and Mahon, 1974). The species were originally considered to be morphs of *L. variegatus*, though ecological and genetic studies by Mahon (1974) established their specific distinctness. These two species, distributed around the coast of Australia, are designated here as the "blue species" and the "orange species". The blue species is in fact *Leptograpsus variegatus* while the orange species is as yet unnamed. Present evidence shows that the blue species extends around the coast of Australia, from the north-west coast of Western Australia to the Queensland coast. The orange species is sympatric with the blue species on the Western Australian and South Australian coasts.

When freshly sampled, the two species are readily identified by colour. However, leptograpsid crabs which have been chemically preserved lose their distinguishing colour. In the first part of the study, samples of the two species were collected from Fremantle, Western Australia, for both males and females. The aim was to determine whether the species could be distinguished on the basis of five carapace and body measurements; the carapace measurements used in the study are illustrated in Fig. 3.1. While differences in the sculpture and shape of the carapace appear to be evident on close examination of typical specimens, it proved impossible to establish non-quantitative characters (other than colour) which would lead to correct identification of the specimens. Separate analyses of the carapace and body measurements showed virtually complete overlap between the species, as shown in Table 1 of Campbell and Mahon (1974). However, a multivariate approach established complete separation between the species, highlighting subtle differences in the shape of the carapace.

The success of this first stage led to the collection of samples around the coast of Australia, to study the nature of the geographic variation in the measured characters. In the third stage, samples were obtained from New Zealand, Chile and Easter Island, to examine further the variation in the blue species. The results of analyses of the data from these various stages established that separation between the blue species and orange species was substantial. Moreover, there was no evidence of geographic variation, with the separation between the species being of a different nature to the

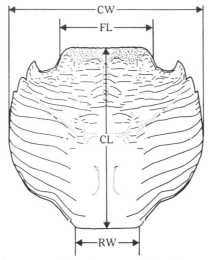

Fig 3.1. Dorsal view of carapace of *Leptograpsus* showing measurements used. *FL*, width of frontal region just anterior to frontal tubercles. *RW*, width of posterior region. *CL*, length along midline. *CW*, maximum width. The body depth was also measured; in females but not in males the abdomen was first displaced. Reprinted from *Aust. J. Zool.* **22**, 417–25 (1974) with permission.

differences between samples from different localities. Given this result, the final stage was to identify museum material currently designated as *L. variegatus* to the blue or orange species.

Linear Discriminant Function

The data from the first stage of the study (at Fremantle, Western Australia) consist of four populations (two species by two sexes), with the main interest being in the differences between the species. As a first approach, it would seem reasonable to examine the nature of the separation between the two species for males and for females separately; we discuss below ways of examining the variation between the four populations simultaneously.

The differences between the two species for each of the sexes could be examined by means of ratios among the five carapace and body measurements. However, as we discuss in chapter 4, ratios can be very unsatisfactory. For one thing, with five characters, there are ten ratios that can be calculated, with no guarantee that any of them will provide good separation. A more satisfactory approach is to form a linear compound of the characters. As a first approach, consider

$$z = x_1 + x_2 + x_3 + x_4 + x_5$$

where z denotes the resulting linear compound (often referred to as a score), x_1 denotes the measurement for the width of the frontal region, x_2 denotes the measurement for the width of the posterior region, and so on. The hope is that larger values of the linear compound will correspond to one of the species and smaller values to the other. However, it is unlikely that all characters will contribute equally to the separation, or indeed that the contributions will all be positive. Moreover, in its current form, the linear compound is likely to be size-oriented. If, as is often the case, shape changes are involved, then some contributions will be negative.

In order to be able to express the different amounts of information supplied by each of the variables to the linear compound, it is necessary to calculate a set of weights to be attached to each of the characters giving the relative contributions of the characters. Thus weighting coefficients b_i will need to be calculated so as to increase the separation of the scores of the two species on the basis of the five measurements used. Optimally, the coefficients can be chosen to maximize the separation between the two species. The linear compound then becomes

$$z = b_1x_1 + b_2x_2 + b_3x_3 + b_4x_4 + b_5x_5$$

The actual form adopted by the calculation is explained in Chapter 6. An important aspect of the method is that the correlations between the characters are taken into account in the calculation, so that if a new character were to be included, it would bring only that information which is not already carried by the earlier ones.

For males at Fremantle, the linear compound which best separates the two species is

$$-0.89 \, x_1 - 0.62 \, x_2 - 0.62 \, x_3 + 1.83 \, x_4 - 1.91 \, x_5$$

while for females, the linear compound is

$$-1.80 \, x_1 - 0.93 \, x_2 + 0.21 \, x_3 + 1.49 \, x_4 - 1.36 \, x_5$$

Each of the linear compounds is a discriminant function (Fisher, 1936). The discriminant function provides a means of describing, in quantitative terms, the maximum difference between the two species in terms of the five variables. It is the best linear function for discriminating between the two species.

Once we have the discriminant coefficients for a particular discriminant function, all we need to do to obtain the score of an individual along this function is to multiply each of the coefficients by the value of the corresponding character for that individual, with due regard for signs, and sum the products. If new individuals become available, their scores can also be calculated and plotted, to see where they lie in relation to the original samples from the two species.

Generalized Statistical Distances

Once the inter-species differences are described in terms of the vector of coefficients in the appropriate discriminant function, it is useful to know how far the populations are, in fact, separated by the function, since this allows a comparison of the separation achieved in different analyses. For example, what is the distance between the sample of males of the blue species and the sample of males of the orange species? And what is the distance between the females of the two species? Very large numbers of measures of similarity have been proposed, and at least 30 are known to us in the literature (cf. Lamont and Grant, 1979). However, there is one which stems from the same approach as the discriminant function and is, in fact, readily computable from it: this is the generalized statistical distance (D^2) of Mahalanobis (1936), which can be obtained by multiplying the vector of coefficients that constitutes the linear discriminant function by the vector of differences between the means for the two groups (see p. 43). As an outcome of its connection with the discriminant function, the generalized distance allows each character to carry only its proper amount of information about the separation of the groups. For males of the two species of crab considered in our example, $D^2 = 36.7$ and $D = 6.1$, while for the females, $D = 5.3$. Should a test for the statistical significance of the difference between the species be required, the D^2 can be related to another multivariate statistic, the Hotelling (1933) T^2 (which is an analogue of Student's t of univariate analysis). This, in turn, can be transformed into the well known F distribution (see, e.g. Mardia, Kent and Bibby, 1979, p.77). In the present case, the values of F are so high that they indicate, unequivocally, that for both males and females, the two species are indeed different with respect to the characters studied.

Canonical Variates and Displays of Interrelationships between more than Two Groups

When several populations are involved in the analysis, the generalized distances can be calculated between all pairs of populations, and an attempt can be made to display these distances in a low-dimensional representation (such as a bivariate scatter diagram). For many examples, it will be possible to represent the distances quite adequately in two or three dimensions. Having done this, we can then attempt to gain some insight into the underlying dimensions of variation that separate the populations, and to place some biological interpretation on the nature of the variation.

For example, in the crab study it is of interest to be able to examine the separation between the species and the sexes simultaneously. The relevant generalized distances are depicted in Fig. 3.2. We can see that for these data

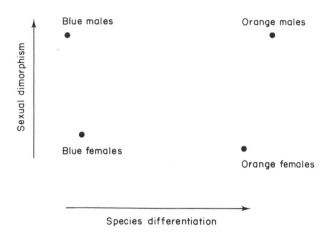

FIG. 3.2. Display of generalized distances between the four populations of rock crabs.

there are two distinct dimensions of variation, corresponding to differences between the species, and to differences between the sexes.

The calculation of generalized distances between all pairs of populations can provide useful insights into the major sources of variation. A more direct approach is to calculate vectors representing underlying dimensions of variation that best separate the populations, and use these as the axes of charts on which the mean positions of the populations may be plotted (Rao, 1952). Such underlying axes of variation are called *canonical variates*. This technique has the advantage that, should the arrangement of the populations require three or more dimensions for its proper expression, the canonical variates can be taken two at a time so that different aspects of the relationships can be examined in detail.

The method of canonical variates chooses new variates which are again linear combinations of the original variables. These new variates are chosen to be uncorrelated within populations and so may be displayed in a rectangular coordinate system. Hence the canonical variate representation provides a useful framework for displaying the interrelationships between the populations, which can be studied once their positions have been plotted. The value of a canonical variate analysis becomes apparent when dealing with large numbers of populations; the problem of unravelling relationships for a large number of generalized distances can quickly become formidable. Moreover, the degree to which the interrelationships can be successfully displayed in two, three (or more) dimensions is apparent from the results of the analysis—specifically, from an examination of the magnitudes of the canonical roots (see chapter 7).

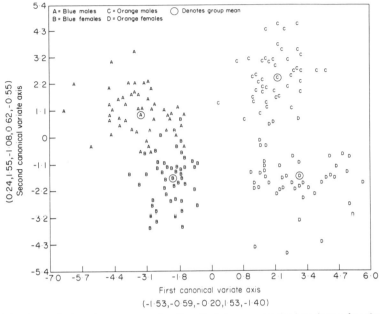

FIG. 3.3. Canonical variate scores for the first two canonical variates for the four samples of rock crabs. The circled symbols represent the canonical variate means. First standardized canonical vector is (−1.53, −0.59, −0.20, 1.53, −1.40). Second standardized canonical vector is (0.24, 1.55, −1.08, 0.62, −0.55). A, blue males; B, blue females; C, orange males; D, orange females; O, denotes group mean.

Figure 3.3 shows a plot of the canonical variate scores in the plane of the first two canonical variates, together with the population means, for males and females of the blue and orange species. There is clear separation of all four populations, with the species delineated along the first axis, and the sexes along the second (cf. the positions of the populations in Fig. 3.2).

Examination of the relative magnitudes of the components of the standardized canonical vectors can be used to indicate the importance of the variables. The standardized components are given by the elements of the canonical vector multiplied by the corresponding within-populations standard deviations. For the crab data, the first standardized canonical vector has a large positive component for the width of the carapace, and large negative components for the width of the front lip and the depth of the body; this indicates that separation of the species is effected by a contrast of the width of the carapace with the width of the front lip and the depth of the body, the blue species having a relatively narrower carapace. Similar inspection indicates that separation of the sexes is effected by a contrast of the length of the carapace with the width of the carapace and the width of the posterior, with males being less elongate.

The second and third stages of the analysis of the crab data again involve a study of the underlying sources of variation between a number of populations. In the second stage, populations throughout Australia were included, while at the third stage, Pacific Ocean populations were also sampled. Canonical variate analysis was again used to examine the nature of the variation, and provide a graphical representation of the distances between the populations—procedures for displaying data in this way are often referred to as ordination techniques.

Allocation of Individuals

Canonical variate analysis examines the degree of separation between populations which can be identified at the outset; for the crab data, populations of the blue species and the orange species can be erected on the basis of colour. For these data, the final stage of the analysis involved identifying *Leptograpsus* specimens held in museum collections throughout Australia, since the colour had been lost on preservation. In other examples, distinguishing taxonomic characteristics may be totally lacking. For example, members of the species complex of *Anopheles gambiae* can only be identified using electrophoretic markers. Morphometric measurements of wing and body characters can be taken for populations of specimens erected using the electrophoretic criteria; the morphometric criteria can then be used in an attempt to identify or allocate new specimens to one or other of the species.

In general, the allocation stage requires the use of procedures to make an assignment of one or more individuals (usually of unknown origin) to one (or more) of a number of reference groups. The allocation procedures can also be applied to the specimens which form the reference groups, to see how many would be correctly identified. Such calculations lead to probabilities of misallocation (also referred to in the literature as probabilities of "misclassification"). It is our view that allocation procedures should be treated as being distinct from descriptive or separatory procedures such as canonical variate analysis and the linear discriminant function (see also Geisser, 1977). Allocation procedures involve the calculation of probabilities of membership of each of the reference populations; these probabilities can be calculated directly from the individual squared Mahalanobis' distances (see, for example, Campbell, 1983).

Homogeneity of Dispersion

The subject of the homogeneity of the dispersion of two or more populations is an important one. Many of the multivariate statistical procedures used in this book have the theoretical requirement of homogeneity in the dispersion matrices in order to be legitimately applicable with respect to the calculation of probabilities and tests of significance. For the crab data, inspection of Fig. 3.3 suggests that the canonical variate scores are approximately uncorrelated within each of the four clusters, and have similar (approximately unit) standard deviations. To all intents and purposes, the associated dispersion matrices (the matrices of variances and covariances within each of the populations) for the four populations can be considered homogeneous.

Heterogeneity of dispersion matrices may have geological or biological significance. In palaeontological morphometric analyses, significant heterogeneity in covariance matrices may be a reflection of sexual dimorphism in the material, changes in breadth of variation deriving from different ontogenetic phases (particularly in arthropods), and sorting of hard parts by geological agencies. Fortunately, many of the multivariate methods used in this book seem to be reasonably robust to moderate deviations from homogeneity.

There is no fully satisfactory method of examining the homogeneity of dispersion matrices. The commonly used likelihood ratio test (see, for example, Kshirsagar, 1972, Section 10.1) is sensitive to departures from multivariate normality. A graphically-based procedure proposed by Campbell (1981) examines the variances and correlations separately and, therefore, requires a little care and forethought if a correct analysis of heterogeneity in dispersions is to be made. As in all statistical investigations, it is highly desirable that a graphical appraisal of the data be made (Healy, 1968). Procedures for examining multivariate normality of the data are discussed in Gnanadesikan (1977, Sections 5.3, 5.4). The interpretation of such analyses can be misleading if atypical observations are present; Campbell (1980) has suggested modifications which are less sensitive to such observations (see also Campbell and Reyment, 1980). Pronounced non-normality in one or more variables will be readily picked up from a probability plot of the data for each variable separately (and, if deemed necessary, from univariate tests of normality). When combined with the procedures discussed by Gnanadesikan, Healy and Campbell, these are adequate for most purposes.

Reyment (1971b) has considered the question of multivariate normality in relation to palaeobiological analyses, including an empirical study of the often negative effects on normality of the commonly-used transformation to the logarithms of the data (see also Humphries et al., 1981).

PRINCIPAL COMPONENTS: VARIATION WITHIN A POPULATION

So far we have been concerned with the contrasts of form between two or more distinct populations. Variation within a single population (a species,

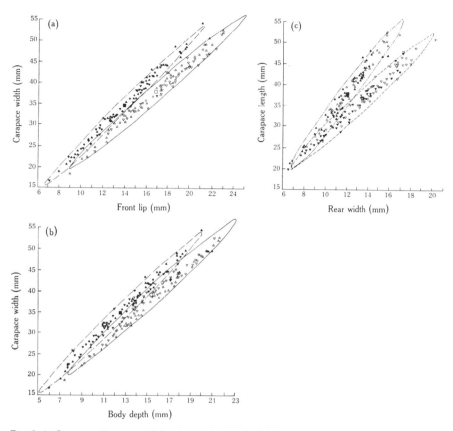

FIG 3.4. Scatter diagrams (bivariate plots) of: (a) front lip width against carapace width; (b) body depth against carapacc width; (c) carapace length against rear width. The dotted lines are the 95% confidence ellipsoids for each species with sexes combined (a and b), or for each sex with species combined (c). Blue males (▲), blue females (▼), orange males (△) and orange females (▽). Reprinted from *Aust. J. Zool.* **22**, 417–25 (1974) with permission.

for example) will also be canalized along certain dimensions of biological interest; Pearce (1959) has stressed this in the context of growth studies. Bivariate plots of pairs of characters for a single population typically form elongated swarms of points, often roughly elliptical in shape if the two characters are correlated. Plots for the crab data are shown in Fig. 3.4. We can once again seek to determine the major sources of underlying variation in such data; an appropriate approach is to examine the axes corresponding to directions of maximum variation of the swarm of points. These axes and the corresponding scores of the points along the axes can be determined by a principal component analysis. The variation of the scores along an axis

TABLE 3.1. First latent vectors and roots for principal component analyses for the correlation matrices for the four samples of rock crabs.

	Blue male	Blue female	Orange male	Orange female
FL	0.448	0.447	0.447	0.447
RW	0.442	0.445	0.445	0.443
CL	0.449	0.449	0.448	0.450
CW	0.449	0.449	0.448	0.450
BD	0.448	0.446	0.448	0.447
latent root	4.95	4.95	4.97	4.93

defines a quantity known as the latent root, while the orientation of a new principal axis relative to the original coordinate axes defines the components of a latent vector (see chapter 9 for details).

The canalized patterns of growth and form which exist within a supposedly homogeneous population can be obtained from a principal component analysis of the dispersion matrix. Since the technique is sensitive to differences in the scales of the characters, which can then dominate the calculations, it is often preferable to employ the correlation matrix. Generally, the latent vector of the correlation matrix corresponding to the largest latent root reflects variation in the size of the organisms. This concept is, however, considered by some to be a rather tenuous one and not unchallengeable (Mosimann, 1970). Table 3.1 presents the first latent vectors (which provide the coefficients yielding the first principal components) for the four samples of rock crabs considered in this chapter. The possibility of using the first latent vectors of the correlation matrices to describe size variation in the five variables is clearly apparent, since the components of each vector are positive, and very similar in magnitude. The second latent vectors contain elements with different signs and can, in consequence, be thought of as being a kind of representation of shape variation, although with some reservations (cf. Humphries et al., 1981). The third, fourth and fifth latent vectors are similarly composed. Each of the vectors in order is connected with a decreasing amount of the variation. In many situations, the latent vector associated with the smallest latent root is virtually representative of that direction which is connected with practically no variability (if the latent root is almost zero); such a vector can be a useful ancillary taxonomic tool. For the crab data, the smallest latent root accounts for less than 0.05 per cent of the variation in each case. The corresponding latent vector contrasts carapace length and carapace width, suggesting that the overall shape of the carapace is virtually invariant.

Other terms in use for latent vectors are eigenvectors and proper vectors (a translation of the German word rendered into French, thence English).

In accordance with a terminology that has grown out of psychometric interpretations, the latent vector analysis of a correlation matrix, which is oriented towards studying relationships between variables, is often referred to as an R-mode analysis (where R derives from a common symbol for the sample correlation matrix). Many workers have, however, been interested in studying the relationships between the individuals constituting a multivariate sample. For example, a museum taxonomist, while sometimes hampered by not having access to living populations of an organism under study, will still wish to examine the similarities and differences between the preserved specimens in the collection. An analysis of the relationship between individuals is usually referred to as a Q-mode analysis (because Q precedes R in the alphabet). The calculations for a Q-mode analysis on a matrix of associations between specimens of a sample involve a latent vector analysis, the same as for an R-mode analysis, though the interpretation of the resulting vectors is quite different (see chapter 9).

Under circumstances in which no knowledge of population structure is available, Q- and R-mode techniques can be used to examine the relationships between the individuals. In order to illustrate the effects of the two kinds of computations on the same set of observations, we have performed various R-mode principal component analyses on the pooled sample of crabs and a Q-mode principal coordinate analysis on the same set of data.

It must be emphasized, however, that we do not in general advocate analysing such data by these techniques. When there is known structure in the data, such as species or even samples of species at different locations, it is important that this structure be recognized at the outset, and that the information be used in subsequent analyses. Techniques which examine variation within an unstructured sample, such as principal component analysis, are less effective in determining subtle structure in the data than techniques which take known grouping into account, such as canonical variate analysis. There is also the statistical objection that formal models underlie the use of these techniques; these models should be kept in mind and not be ignored completely when the procedures are being applied.

A plot of the principal component scores obtained from an analysis of the crab data based on the correlation matrix is shown in Fig. 3.5; separation of species and of sexes is evident along the third and second principal components respectively, though the overlap is considerably greater than that observed in Fig. 3.3. For these data, very similar plots are obtained for the analyses based on the covariance matrix and on the covariance matrix for the log data. That this is so is evident from Table 3.2, which shows that the second and third latent vectors resulting from the various analyses are similar in form. The third latent vectors have a large positive component for

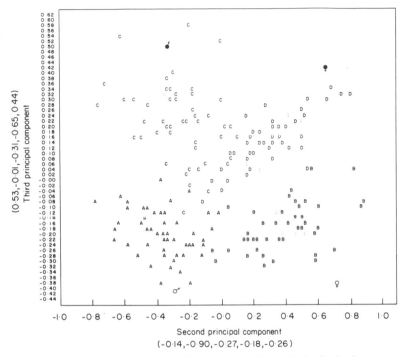

FIG. 3.5. Principal component scores for the second and third principal components for the analysis of the pooled sample of rock crabs based on the correlation matrix. Third latent vector is (0.53, −0.01, −0.31, −0.65, 0.44). Second latent vector is (−0.14, 0.90, −0.27, −0.18, −0.26). Symbols as in Fig. 3.3.

the carapace width and large negative components for the front lip and body depth; this is of the same form as that for the first canonical vector. The second latent vectors all have a large positive component for the rear width of the carapace, and a smaller negative component for the carapace length; the second canonical vector also reflects a contrast between these variables.

The first latent vector for the pooled data points to a strong "size component". Virtually all the discrimination between the species and sexes is in a plane orthogonal to this vector, so that "shape differences" dominate. For these data this conclusion is largely independent of the precise form of data transformation/"association" matrix adopted. In particular, the analysis based on the covariance matrix for the log-transformed data is very similar to that based on the standardized variables (i.e. on the correlation matrix for the untransformed data) and shows that effective discrimination between the species is not evident from a plot of the first two components. Careful examination of all the principal components is necessary to realize the separation between the species that exists.

TABLE 3.2. First three latent vectors for principal component analyses of the pooled sample of crabs.

		Covariance matrix	Covariance matrix-logs	Correlation matrix
I	FL	0.29	0.45	0.45
	RW	0.20	0.39	0.43
	CL	0.60	0.45	0.45
	CW	0.66	0.44	0.45
	BD	0.28	0.50	0.45
% variation		98.3	96.9	95.8
II	FL	0.33	−0.16	−0.14
	RW	0.86	0.91	0.90
	CL	−0.20	−0.21	−0.27
	CW	−0.29	−0.07	−0.18
	BD	0.17	−0.31	−0.26
% variation		0.9	2.0	3.0
III	FL	−0.51	−0.44	−0.53
	RW	0.42	−0.09	0.01
	CL	−0.18	0.37	0.31
	CW	0.49	0.67	0.65
	BD	−0.54	−0.46	−0.44
% variation		0.7	0.8	0.9

A principal component analysis is equivalent to a Q-mode analysis, with the usual Euclidean distance as the metric; the scores from the principal component analysis then define the principal coordinates (see chapter 9, p. 95). It might be expected that other similarity measures of like form would give much the same ordination, and this indeed appears to be the case for these data. A Q-mode principal coordinate analysis with Gower's metric (Gower, 1967) as the measure of similarity again shows separation of the species along the third principal coordinate and separation of the sexes along the second principal coordinate (Fig. 3.6), though again with considerable overlap compared with Fig. 3.3.

A correspondence analysis (Greenacre, 1978, 1980) of the crab data is also revealing. The differentiation along the second coordinate axis for the individuals separates the species (Fig. 3.7), while the sexes are largely separated along the first coordinate. Superimposing the coordinates for the variables reveals that body depth and width of the front lip coincide with the position of the orange species, while width of carapace coincides with the position of the blue species. Recall from the discussion of the results of the

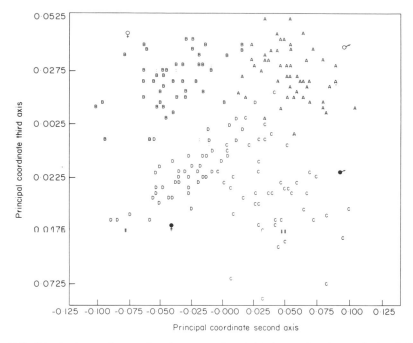

FIG. 3.6. Principal coordinates for the second and third axes for the analysis of the pooled sample of rock crabs based on Gower's metric. Symbols as in Fig. 3.3.

canonical variate analysis that the blue species has the narrower carapace, relative to the width of the front lip. The position of the variables along the first axis shows the posterior width of the carapace to be distinct from the remaining variables, and close to the females. Recall again that females have a relatively narrower posterior width of the carapace. The graphical analysis by canonical variates (Fig. 3.3) arrives at a similar pictorial representation, though the separation of the species is considerably more marked.

FACTOR ANALYSIS: VARIATION ALONG AXES DETERMINED BY AN *A PRIORI* THEORETICAL ANALYSIS

In the techniques for analysis which have been discussed so far, the choice of a suitable method, determined by the experimenter's needs, is the only choice that has to be made apart from technical matters like the number of characters, and the number of organisms on which to assess them. Once the technique has been chosen, all analysts will arrive at the same result, though

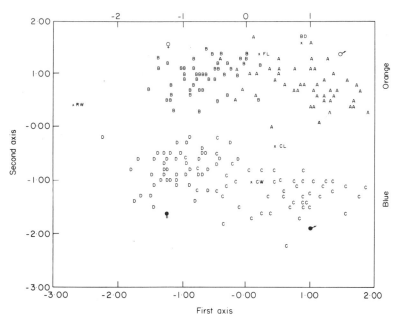

FIG. 3.7. Plot of the first two axes of the ordination of individual crabs and, superimposed, of the ordination of the variables (*FL, RW, CL, CW, BD*) by correspondence analysis. Symbols as in Fig. 3.3.

they may differ in their ability to interpret it. With factor analytical methods as commonly employed (see Chapter 10), however, a new element of choice enters into the calculation, for in factor methods the experimenter often sets out to orient his axes of variation according to his theoretical knowledge, real or assumed, about the nature of variation in his material.

CANONICAL CORRELATION: MORPHOMETRY AND THE ENVIRONMENT

The example considered here could have been made more informative from the biological aspect if it had been possible to couple the morphometrical methods on the crab species to factors of the environment. This is not easy to do in palaeontology (although work on microfossils can be made to include this aspect in some situations), but is relatively easier for work on living organisms. Thus one would have on the one hand a set of morphometric measurements, and on the other, a set of variables such as temperature measures and other ecological factors.

A reasonable method of analysis would then require, first, a measure of the correlation between the two sets and, secondly, the contribution of each variable to the correlation, albeit in approximate terms.

The multivariate statistical method of canonical correlation was developed by Hotelling (1936) to carry out this kind of analysis. Details are given in Chapter 8. Multiple regression is a special case of canonical correlation.

4
Some Simple Forerunners of Multivariate Morphometrics

The earliest attempts at a quantitative assessment of shapes were almost all conducted with the aid of ratios of characters; often there was, underlying this use, the idea that one of the characters could be regarded as indicative of some feature of primary interest whereas the other effectively standardized its variation by providing a measure of absolute size. The use of ratios continues, partly because of a misconceived dogma that one cannot handle more than two characters at once and that ratios are the method of choice for two characters (neither part of this belief being well founded), but mainly because their use constitutes what Blackwelder (1964) has called "an acceptance", that is, a proposition which cannot be shown to be true but which serves to meet a need, and hence generally passes unquestioned because it is thought of as a "standard" technique.

In fact the use of ratios implies certain prior knowledge about the material under examination which, if set out explicitly, would almost certainly be denied by any experienced worker in the field. The weaknesses of ratios include:

(1) The fact that a ratio will not be constant for organisms of the same species unless these are also of the same size, by virtue of the almost universal occurrence of differential growth; of course, the effects of allometry may be small in relation to the differences between species.

(2) As generally used, ratios contain only two characters and thus afford a poor appreciation of what may turn out to be an involved contrast between forms.

(3) To compound two characters into a ratio implies that there is only one contrast of form to be studied, and that that unique contrast is well assessed

29

in terms of two characters of equal weights, but opposite in sign. The assumption that only one contrast of form accounts for all the observable variation in the material is particularly misconceived because almost all properly conducted experiments designed to test this assumption have shown it to be untrue, as the illustrations given later in this book show. The use of ratios of characters has been discussed by Barraclough and Blackith (1962), Christensen (1954) and by Jeffers (1967a). A particularly detailed and thorough investigation has been conducted by Atchley et al. (1976). There are also valuable contributions to the subject by Pimental (1979) and Atchley and Anderson (1978) (but see also Albrecht (1978)). However, Corrucini (1977) dissents from some of the findings of Atchley and his coworkers, and considers that ratios of characters are not quite so unsuitable as those authors consider them to be. Hills (1978) is more cautious, recommending in particular the use of logarithms, which has the result that the difference of the log-characters is utilized. Mosimann and his colleagues (see, for example, Mosimann and James (1979) and references contained therein) also tend to work with the logarithms of the data, although within the framework of a valid mathematical justification for so doing. Nevertheless, one should not lose sight of the usefulness of some quite simple techniques for shape measurement. For example, the seeds of *Menyanthes,* found in rocks of the order of 10^6-years old in Japan, become flattened by the internal pressure in the sediments. They thus become a built-in indicator of the age of the sediments, and Kokawa (1958) has constructed a nomogram to relate the age of the sediment to the length and thickness of the seeds it contains. In general terms, the thickness of the *Menyanthes* seeds decreases exponentially with the square of the age of the sedimentary rock.

Traditionally, morphometric analysis has centred on the study of allometric growth. If much of the recent emphasis in growth studies has moved away from allometry, as expressed in the classical equations, towards multivariate methods, there is still a substantial amount of research devoted to the exploration of differential growth. This work has been well summarized by Gould (1966). Laird *et al.* (1965) have shown how the equation for allometric growth can be related to, and deduced from, the relativity relations of biological time. The work of Jolicoeur (1963b, 1968) and of Jolicoeur and Mosimann (1960) has taken the allometric relationship far along the multivariate road, so that from the usual bivariate allometric relationship we proceed to its generalization in terms of the major axis of the dispersion matrix.

There are some remarkably difficult, even paradoxical, situations in the theory of allometric growth which have been thoughtfully reviewed by Martin (1960). Sprent (1968, 1972) has observed that Jolicoeur's model, using essentially the first principal component as a growth vector, is acceptable to workers who believe that the remaining components define the error space. If one believes that this defines a space relevant to the shape of the organisms, the problems sharpen acutely.

Some of the problems with which biologists, and in particular, palaeontologists, are familiar continue to require attention in the multivariate field as they did in the univariate one. Burnaby (1966a) has remarked that biological taxonomists are often reluctant to employ multivariate methods in cases where the organism continues to grow throughout life, as for pelecypods, gastropods, and foraminifers. He also pointed out that growth is not the sole generator of "nuisance factors" in taxonomy, but that other components of variation in which the investigator may be quite uninterested may occur. Burnaby (1966a) gives details of a procedure to adjust the differences in unwanted components of variation in a canonical variate analysis framework. Gower (1976) and Reyment and Banfield (1976) have investigated this further. Humphries *et al.* (1981) have cast doubt on the generality of Burnaby's procedure, although the arguments they present do not strike us as being conclusive.

Sometimes, the nuisance components interfere with the analysis of only some of the parts of the organisms under investigation. Hopkins (1966) reported that in rats, the liver, heart, spleen, lungs, skin and adrenal weights develop in a pattern consistent with a single allometric pattern, whereas the growth of the kidneys, genitalia and brain developed in a manner which was size-sensitive.

Although the allometry model put forward by Hopkins in the cited paper seems to us to be logical and useful, in actual practice the differences between coefficients found by principal components seems to differ very little from Hopkins' factor-analytic values.

Ideas for the separation of shape and size components of the various contrasts of form with which one may have to deal date back to Penrose (1954). The extent to which it is preferable to eliminate such components of variation rather than to keep them in the calculations, and hence in the representation of the results, and so to study them on the same footing as other components of, perhaps, more obvious concern, is debatable. The present authors tend towards the practice of retaining all discoverable components of variation in the analysis at all stages.

More recently, serious attempts have been made by Mosimann (1970), Bookstein (1978) and Brower and Veinus (1978) to solve the problem posed by the difficulty of giving a fully valid mathematical definition of variation in shape. This subject is taken up later (p. 120).

FROM RATIOS TO DISCRIMINANTS

Our experience suggests that biology students are often "left standing" at the beginning of a course on multivariate techniques, because the very idea of adding and subtracting characters in a discriminant function seems out of touch with their needs. The idea of standardizing the dimensions of an animal by dividing by some measured representative of the size of the animal is easily accepted, but adding and subtracting dimensions seems bizarre.

Middleton (1962) set out a simple relationship, which helps biologists and geologists who find it easy to think in terms of ratios of characters to make the conceptual jump required to manipulate discriminant functions and other linear compounds. We offer this simple example whose logic applies just as much to the measurements of animals and plants as it does to the constituents of rocks. Middleton points out that knowledge of the ratio $y = Na_2O/K_2O$ in the analysis of a sandstone often enables the geologist to say what its tectonic origin (eugeosynclinal, taphrogeosynclinal or exogeosynclinal) may be. Omitting irrelevant constants this equation may be written

$$\log y = (\log Na_2O) - (\log K_2O)$$

in which the potash content is subtracted from the soda content. If we write this equation in the more general form

$$\log y = b_1 (\log Na_2O) + b_2 (\log K_2O)$$

we can see that the relationship remains unchanged only if $b_1 = +1$ and $b_2 = -1$. To make this equation as powerful a discriminator as possible we can change it in one of four ways.

First, we can calculate the values of b_1 and b_2 which do the best job of discriminating between the broad categories of sandstone. Secondly, we can, if we have the necessary data, enlarge the equation by including the magnesia content, the lime content and so on in the hope that each component will help us to distinguish the nature of the sedimentation processes (eugeosynclinal or taphrogeosynclinal) that gave rise to the sandstone. To add more elements in such a linear compound is easy for we then have

$$\log y = b_1 (\log Na_2O) + b_2 (\log K_2O) \\ + b_3 (\log CaO) + b_4 (\log MgO) \ldots$$

whereas to incorporate this extra information in the original ratio would be cumbersome to say the least. Thirdly, we can use our new-found freedom from the restraints of ratios by enquiring whether the logarithms of the characters are the most appropriate transformations to use. We might do just as well, or better, with the original arithmetic values or with some other transformation.

Middleton included the content of eight oxides in his analysis, finding that, when the simple arithmetic values were used, the soda content was most useful for distinguishing eugeosynclinal sandstones from exogeosynclinal ones, whereas the potash content best separated exogeosynclinal sandstones from taphrogeosynclinal ones; this last distinction was substantially improved in the case where logarithmic transformations of the data were used, by including the magnesia content of the rock.

Fourthly, there is no necessity to restrict ourselves to a linear discriminant function, since some or all of the characters may be raised to higher powers. A function of the type

$$y = b_1 x_1 + b_2 x_1^2 + b_3 x_2 + b_4 x_2^2 + \ldots$$

is a quadratic discriminant, and has been exploited for morphometric purposes by Burnaby (1966b). There is also an excellent theoretical paper by Cooper (1965).

Quadratic discriminant functions are sometimes used in medical research (Michaelis, 1972). They can also be useful in some kinds of morphometric work as, for example, the typification of ornamental morphs by means of discriminant scores in cases where the covariance matrices are strongly heterogeneous (cf. Reyment (1982f) who applied quadratic discriminant scores to the interpretation of the evolutionary significance of ostracod morphs).

5
Some Practical Considerations

Once the material for a morphometric analysis is arrayed, the decision has to be taken as to the nature and number of characters to be measured.

Where the organisms are closely related, and all the measurements are quantitative, few problems concerning the nature of the characters are likely to arise. Homologies are generally evident in such cases. It is usually desirable to try to spread the measurements as evenly as possible over the animal or plant, ensuring that at least something is measured on most of the physically available components of the organism. The spread of the characters will, of course, depend on the aims of the study. If the goal is to provide quantitative identification, then a suite of highly correlated characters on one part of the body may provide the best solution. In fact, in multivariate evolutionary studies of fossils, highly integrated sets of characters ("one-organ systems") may provide the only avenue of approach for many kinds of analysis (Lande, 1979).

Quite small numbers of characters may suffice. Satisfactory analyses have been reported with as few as three to six characters, although ten might be regarded as near optimal in a preliminary experiment, where not very much is known about the discriminating power of the various characters. As with many other aspects of morphometrics, it is hard to give advice in the abstract, but certain rules are helpful in particular instances. For example, with animals it is undesirable to measure nothing but the lengths of parts; some breadths should be included, or there is a risk that the separation will result from little but pure size. It may be undesirable to measure parts that are susceptible to shrinkage or expansion in specimens preserved in different ways, for example, dried or pickled insects. If in a preliminary investigation only a small fraction of some potentially larger body of material is used, it is inconvenient to employ characters that may be difficult or impossible to measure on material that one may want to include at a future date.

Where the organisms are less closely related, problems of homology arise. Sometimes, as in the basidiomycetes, homologies may be so difficult to determine that satisfactory characters outside the stable sexual ones may be almost impossible to find (Kendrick and Weresub, 1966): the situation varies greatly from group to group. The character "length of wing" in Diptera (flies) is not homologous with the same nominal measurement in Strepsiptera (stylopids). Key (1967) has produced an operational definition of homology which seems entirely satisfactory, so long as one is prepared to accept a purely phenetic approach to the subject. This definition is "Feature a_1 of organism A is said to be homologous with feature b_1 of organism B if comparison of a_1 and b_1 with each other, rather than with any third feature, is a necessary condition for minimizing the overall difference between A and B." A more detailed discussion of homologies is given by Jardine (1969).

If many or all of the characters are qualitative (dichotomous; presence or absence) so that the analysis takes on some of the characteristics often associated with numerical taxonomy, the requisite number may be increased. Sokal and Sneath (1963) have suggested 100 as the upper limit of the range of numbers of characters over which there is an appreciable improvement in precision as the number of characters is increased.

The topic of "how many characters" is inseparable from the question of redundancy in the character-suite. For qualitative characters, the limit beyond which redundancy makes further accumulation of data less profitable depends on whether new characters contribute new information; whether, in fact, the addition of some new suite of, say, internal anatomic characters would add information that could not have been recovered from the indefinite accumulation of external morphological characters. Drastically new characters, such as would arise if biochemical characters were to be added to morphological ones, are likely to differentiate the material along new axes of variation.

Generally, redundancy is invited when a great number of variables are instituted for parts that are closely bound to each other. For example, there is no particular difficulty in finding redundant variables for bones of the skull of mammals. If a great number of purely morphological variables have been measured, many of them may be expected to be redundant. If, however, it is possible to quantify qualitative data as well and to include dichotomous variables, these will almost always enrich the total suite of variables with characters that contribute additional information.

Palaeontology poses quite a difficult problem at times. Whereas zoologists often have to worry about which of a multitude of characters should be selected for measurement in order to obtain greatest efficiency in an analysis, the palaeontologist is frequently hard put to find a sufficient number of variables so as to make his analysis worthwhile. This is admittedly a rather extreme situation and should be accepted as such, but may be brought home by considering such a genus as *Cytherella,* the species of which ostracod are mostly as smooth as an egg. Some palaeontologists, in their attempts to find a reasonably large number of variables for a study, may become uncritical and manufacture new "variables" by making ratios of some of their measurements.

A distinction between two different analytical situations is required to be made in almost all studies. In the first case, there are well defined and adequately identified groups which we wish to study on the basis of a set of multivariate measurements. In the second situation, the analyst is faced with an undefined sample of multivariate measurements on a sample of individuals and it is desired to examine the material for the possible existence of clusters.

As examples of the first type of study we may take evolutionary analyses involving related species and separated either by time or space or both. Another example could be the study of taxonomic relationships in related species.

The second type of situation is usually less well defined. Often the analyst has very little knowledge of the taxonomic condition of his material and he hopes to expose structure in the data by performing some kind of analysis involving multiple measurements on the presumably heterogeneous sample. This latter approach has tended to be less popular of late on the grounds that it is held to be basically unscientific in that the operator lacks a useful hypothesis for justifying his analysis. However, as an exploratory technique, such an approach can be very useful, particularly in the initial phases of a quantitatively-oriented palaeontological study.

Moreover, science proceeds by continuous feedback, at least in the experimental disciplines with which this book is concerned, and to expect hypotheses to be generated *sui generis* in these disciplines is to misunderstand the nature of experimental activity.

THE EFFECTS OF CORRELATED CHARACTERS

Although there are very many references in the literature to the fact that characters are often correlated, positively or negatively, and to the possibility of using such correlations to determine the degree of "primitiveness" of an organism (Sporne, 1960), there is little information about the effect of varying degrees of correlation on any form of quantitative taxonomy. Indeed, there appear to be few examples in the literature where a taxonomic distance has been computed with and without an appropriate allowance for the degree of correlation between the characters used. Campbell (1976) discusses the effect of character correlation within groups on an ordination of *Wuchereria lewisi* and demes of *W. bancrofti*.

Rohlf (1967) claims that, in a numerical taxonomic analysis of the pupae of 45 species of mosquitoes, the only effect of using characters which were chosen for their high correlation was that the generic clusters appeared to be unduly elongated. It is hard to see the basis for this claim, because the pupal mosquitoes were not also classified on the basis of less correlated attributes, and if they had been, the validity of the comparison would have depended on the non-specificity hypothesis (Sneath and Sokal, 1973). The effective basis of the argument must lie in the fact that the pupae were classified by these characters into a hierarchy not very different from that obtained by classifying the adults of the same species, on the basis of characters that appeared to be less strongly correlated. In any event the elongation remarked upon by Rohlf seems to be confined to one of the five generic clusters, that for *Anopheles*.

In the early stages of numerical taxonomy, there was a marked emphasis on the production of phenograms (dendrograms representing relationships based on phenetic similarity rather than on common descent, as in a conventional phylogenetic diagram). More recently, there has been a movement towards the use of ordinations based on distance measurements or on principal component or canonical variate analyses because of the distortion that may by introduced when the relationships are forced into a dendrogram in two dimensions. There is a need to consider how the correlations between characters influence distance estimates.

Bartlett (1965) gives a theoretical discussion of the problem based on Cochran's analysis of the situation (Cochran, 1962). Taking the simple case of two variables, we begin with the familiar Pythagorean distance formula

$$D_p^2 = d_1^2 + d_2^2$$

where d_1 and d_2 denote the difference between the means in standardized units for the first and second variables respectively, and rearrange this to give

$$D_p^2 = d_1^2 (1 + f^2) \quad \text{where } f = d_2/d_1$$

If the variables are correlated, then the generalized Mahalanobis' distance is the appropriate one, giving

$$D_m^2 = d_1^2 + d_1^2 \frac{(f - r)^2}{1 - r^2}$$

Hence the increase in D_m^2 due to the inclusion of the second variable (relative to the degree of separation, d_1^2, provided by the first variable) is

$$(f - r)^2/(1 - r^2)$$

The effects of correlation will increase the D_m^2 provided this relative incre-
ment is greater than that when there is no correlation, namely f^2. That is,
correlation will be helpful provided that

$$(f - r)^2 > f^2(1 - r^2)$$

In morphometric studies, it is usual to find that $r > 0$, whereas f may be
positive or negative (note that Cochran assumes $f > 0$, and, if necessary,
adjusts r to achieve this).

So assume positive correlation and consider first the case with $f < 0$. In
this situation, correlation will always be helpful; in fact, the higher the
correlation, the better the separation, with shape differences dominating.
When $f > 0$, the situation is more complicated, with the D_m^2 being
augmented if $r > 2f/(1 + f^2)$. For example, if the major axes of the concentration ellipses (the elliptical clouds of points) are collinear ($f = 1$), with size
differences dominating, then positive correlation will never be helpful. It is
worth looking further at some of the numerical consequences of the relationship. If there is a marked disparity between d_1 and d_2 so that $f = 0.1$, any
positive correlation coefficient greater than 0.20 will augment the D_m^2. For
$f = 0.4$, only positive correlations above 0.69 will have this effect.

Table 5.1 provides an empirical examination of the effects of computing
generalized distances, with or without adjusting for the effects of correlation
between the characters.

The example is a morphometric analysis of the shapes of two species of
European wasps, *Vespula germanica* and *V. rufa*. The original data, with the
D (not D^2) values, has already been published (Blackith, 1958), together
with a generalized distance chart which was a three dimensional model. The
same set of D values has now been calculated without adjusting for the
effects of inter-character correlation.

The main influence of ignoring the correlation (i.e. of using Pythagorean
distance) has been to greatly increase the distances between those groups
which were already well separated; that is to say, the sexual dimorphism in
each species has been effectively doubled. The distances from queens to
workers have also been much exaggerated, whereas those between males
and workers are uniformly reduced. The net effect of these changes would
be to give the impression that the workers are very much more like the males
than is in fact the case. The reason for this distortion is that the sexual
dimorphism is essentially a matter of size (from the point of view of the gross
biometry of the insects), whereas the contrast between the workers and the
males is mainly a matter of differential growth patterns leading to insects of
the same size, broadly speaking, but with distinct shapes. Where one is

TABLE 5.1. Empirical examination of effects of correlations on generalized distances

	Queens	*Vespula germanica* Workers (large)	Workers (small)
Vespula germanica			
Males	13.32 (24.87)	6.60 (2.74)	6.57 (5.05)
Queens	0	10.57 (29.37)	12.01 (29.45)
Workers (large)	0	0	3.05 (5.63)
	Males	*Vespula rufa* Queens	Workers
Vespula germanica			
Males	4.43 (4.15)	11.01 (21.24)	7.95 (5.17)
Queens	14.37 (27.49)	5.95 (7.49)	12.39 (29.48)
Workers (large)	8.09 (3.68)	8.12 (19.69)	4.68 (5.62)
Workers (small)	6.80 (2.85)	12.59 (25.30)	2.30 (1.57)
Vespula rufa			
Males	0	11.29 (22.97)	6.89 (2.52)
Queens	0	0	9.96 (24.88)
Workers	0	0	0

Generalized Mahalanobis' distances between groups of wasps and, in parentheses, the distance without taking into account the correlations between characters.

dealing with large size differences, the correlations between the attributes are almost sure to be all positive, but where there is marked allometric growth leading to organisms of the same size, but different shapes, the correlations may contain negative elements. It is noteworthy that the difference between the two species is not very seriously changed by failure to take into account the effects of correlation between the attributes.

In another example, the comparisons were made between the Mahalanobis' generalized distances and the Pythagorean distances for the skull measurements published by Mukherjee *et al.* (1955). Taking contrasts between the Jebel Moya people and several other African groups, in almost all cases the Pythagorean distance is at least twice as great as the generalized distance. If this increase were uniform, the effect would not be serious, since it would act as if there had been a scale change to a chart on which all dissimilarities were multiplied by some such factor as three. However, the augmentation is not uniform, with a tendency for those groups which are least removed from the Jebel Moya people to seem more remote than they really are. This tendency would be evident as a systematic distortion of the chart. As Seal (1964) comments, Sokal's (1961) suggestion of the Pythagorean distance for use in taxonomy may lead to unjustifiable conclusions.

The effect of taking into account the correlations between characters has been considered more in anthropology than in any other branch of biometry. Earlier, Pearson's Coefficient of Racial Likeness (CRL) was popular, though it fails to allow for the correlations between characters. Mulhall, in Talbot and Mulhall (1962), reaches the general conclusion that "allowance for correlation alters the magnitude of the coefficients significantly but not enormously", and again, that "the CRL is an unsatisfactory measure of group divergence, and that neglect of the mutual intercorrelations between the characters can result in false conclusions being drawn regarding racial affinity". He appends a table of distances between various groups of Nigerians computed with and without the allowance for correlations between characters.

SOME COMMENTS ON THE PLACE OF SIGNIFICANCE TESTS IN MORPHOMETRICS

In many of the situations with which an experimental scientist has to deal, tests of significance are secondary to the problems of ascertaining the structure of the experiment in multidimensional space. If a group of organisms is not known *a priori* to consist of definable subgroups, then a principal coordinate analysis seems an appropriate tool for probing its morphometric structure. Once the group is known *a priori* to consist of definable subgroups, an analysis along canonical variates seems to be a reasonable choice for such an investigation. Once the decision to use canonical variates has been taken, tests of multivariate significance lose some of their point, for we know in advance that the subgroups differ. The outcome of the test will be influenced by the sample sizes, by the wise or unwise choice of variables to be measured, and by the way in which variation along particular axes of growth or development happens to be represented in the material to hand, to name only some obvious factors. It is as much a mark of immaturity in the experimenter to leave multivariate problems at the level of the significance test, without proper estimation and interpretation, as it is commonly agreed to be in other areas of statistics. Of course, appropriate tests of significance do play an important role, to prevent overinterpretation of the data.

The point at issue is not primarily a statistical one: an entomologist investigating the form of insects in a bisexual species would rarely be well advised to test the significance of the sexual dimorphism, for a glance at the genitalia will settle the question of sex in most instances. A palaeontologist concerned with the question of whether or not some sexual dimorphism exists in fossil brachiopods might well take a quite opposite view, if by dissecting living forms he could gain an insight into the nature of the fossil record.

In many biological situations, further inferences may be required. It may be of interest to ascertain whether the groups follow some trend (e.g. with latitude, time, altitude, depth). It can also be of interest to test whether means are collinear, etc. In practical experimental situations it may be desirable to assess associated probabilities, for example in connexion with determining polynomial trends in relation to specific treatments of an ecological nature.

In applications involving multivariate techniques, there is at the one extreme a tendency to accept all latent roots which are statistically significant as being of biological importance; at the other extreme, arbitrary rules of thumb are applied, irrespective of the sample size and the significance of the latent root. Assessment of appropriate probabilities is necessary to take account of sampling variability. Some unpublished empirical work on the size and distribution of the sample latent roots from a canonical variate analysis is instructive here. For within-groups degrees of freedom of around 200, there is an upward bias of about 10% in the average sample value, with an approximately symmetric distribution of values. However, for degrees of freedom around 20, there is an upward bias of more than 100% in the median sample value; none of the sample values was below the corresponding population value.

For degrees of freedom of the order of 200, the application of significance tests will tend to indicate that latent roots from a canonical variate analysis which are of little practical value are nevertheless statistically significant at conventional significance levels. Here rules of thumb can provide a useful complement. For smaller sample sizes, tests of significance become important.

There has been steady development in the general area of discrimination of tests and procedures which give added insight into the nature of the group separation. The texts by Kshirsagar (1972) and Mardia, Kent and Bibby (1979) contain most of the relevant advances in the area.

6
Discriminant Function Analysis and Allocation

This chapter considers the allocation of specimens to one of two populations and the separation of the populations. For some biological applications, interest centres around the problem of allocation of specimens to one of the underlying universes. Although the linear discriminant function proposed by Fisher (1936) was designed to maximize the separation between two populations, it provides a first step on which to base the allocation, and is often used for this purpose. While we will suggest that it is preferable to base an allocation procedure directly on the associated Mahalanobis' distances, the widespread adoption of the discriminant function makes it worthwhile discussing this latter approach first.

The ideas underlying the method of discriminant functions, as usually practiced by biologists, may be discussed in terms of two universes. Reference samples are assumed to be available from each of the two universes and a linear discriminant function is constructed on the basis of these samples. The researcher then often wishes to assign a new individual, on which the same measurements are available, to one of the universes. The assumption that the individual actually comes from either of the universes may make poor sense biologically and may be even worse in the context of palaeontology and geology, and should be examined in the light of the available data. It is more realistic to assume that a newly acquired specimen may derive from either of the universes, it may be morphometrically close to either of them, or it may come from another universe. The biological utility of such an approach is obviously greater than that of assigning the material to one or other of two universes.

Unfortunately, a solution cannot be pursued solely in the context of allocation via the linear discriminant function. The discriminant function is satisfactory for deciding which of the two universes is the more appropriate, but not for examining whether either universe is appropriate. What the discriminant function does is to offer a useful procedure for a graphical display of populations with respect to the degree of morphometric likeness between samples, drawn from various geographical and, or, stratigraphical locations.

A generalized concept of what is done by a discriminant analysis under ideal conditions is given in Fig. 6.1. This geometric interpretation of the procedure applies to two universes and two variables. The discriminant function—the line M in Fig. 6.1—is chosen to maximize the separation between the universes, relative to the variation within each universe. The discriminant function can be constructed geometrically as follows. Consider the line through the points at which the ellipsoids of scatter S_1 and S_2 cut each other, denoted by L, and a perpendicular, M, to it, conveniently drawn to pass through the origin of the coordinate system, O. The major axes of the ellipses are parallel and of equal length (i.e. the ellipses are equally inflated). The projection of the ellipses onto the perpendicular gives the two univariate distributions S_1' and S_2' shown in the diagram. The transformed values of the observational vectors—the discriminant scores—may be used in a univariate display. The two distributions of the example in Fig. 6.1 display a slight degree of overlap.

The linear discriminant function between two samples may be defined as

$$y = (\bar{\mathbf{x}}_1 - \bar{\mathbf{x}}_2)^{\mathsf{T}} \mathbf{S}^{-1} \mathbf{x}$$

where $\bar{\mathbf{x}}_1$ and $\bar{\mathbf{x}}_2$ are the mean vectors for the respective samples, $\mathbf{S}^{-1}$ is the inverse of the pooled sample dispersion matrix, and $\mathbf{x}$ is a vector of variables (for three dimensions it would be $(x_1, x_2, x_3)^{\mathsf{T}}$). The value y is the discriminant score corresponding to the observation vector $\mathbf{x}$.

The coefficients of the discriminant function are defined as

$$\mathbf{a} = \mathbf{S}^{-1} (\bar{\mathbf{x}}_1 - \bar{\mathbf{x}}_2)$$

where $\mathbf{a}$ is the vector of discriminant coefficients. If the variances of the variables are almost equal, the discriminant coefficients give an approximate idea of the relative importance of each variable to the discriminant function; otherwise multiplying each coefficient by the standard deviation of the corresponding variable can be used to give some idea.

The linear discriminant function is connected with the Mahalanobis' generalized distance by the relationship

$$D^2 = (\bar{\mathbf{x}}_1 - \bar{\mathbf{x}}_2)^{\mathsf{T}} \mathbf{S}^{-1} (\bar{\mathbf{x}}_1 - \bar{\mathbf{x}}_2)$$
$$= \mathbf{d}^{\mathsf{T}} \mathbf{a}$$

where the vector $\mathbf{d}$ is the difference between the two sample mean vectors.

The discriminant root—the ratio of the between-populations to the

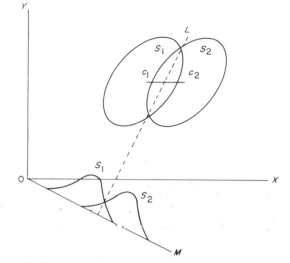

FIG. 6.1. Schematic diagram indicating part of the concept underlying discriminant functions (adapted from a diagram in Cooley and Lohnes, 1962).

within-populations sums of squares for the discriminant scores for the individuals from the two samples—is given by

$$(n_1 + n_2 - 2)^{-1} n_1 n_2 (n_1 + n_2)^{-1} D^2$$

For those interested in a detailed exposé of discriminant functions, books to study are those of Rao (1952), Kshirsagar (1972), Cacoullos (1973) and Morrison (1976). Discriminant analysis with presence or absence data has been illustrated by Cochran and Hopkins (1961), and there is a book on the subject by Goldstein and Dillon (1978).

DISCRIMINATION AND SEVERAL GROUPS

The logical extension of the case of two populations is to a concept of linear discriminant functions for more than two populations. Thus, one could construct a so-called discriminant function for each population (see, e.g. Rao, 1952, Section 8b.9). Equivalently, the squared Mahalanobis' distance associated with the individual specimen could be calculated for each of the populations, with the calculation of the associated probabilities being based directly on these values. In some studies, there may be good *a priori* reasons for weighting the associated probabilities so that the likelihood of a certain specimen falling into one or more of the populations is greater than for its landing in others.

A second possible line of attack which is sometimes used is by means of the method of canonical variates. This analysis consists of a transformation of the original observation space, which results in a reduction of the within-population ellipsoids of scatter to spheres (see Chapter 7). The canonical variate approach produces a set of discriminant functions which are the latent vectors corresponding to the latent roots or canonical roots. The coefficients of the variables (characters) in the discriminators are the corresponding elements of the latent vectors. The individual scores and the population means are plotted, and scores are calculated and plotted for new specimens. While the canonical variates are often used to identify individual specimens, they should more logically be employed to provide an ordination of the populations (see Chapter 3).

As Kendall and Stuart (1966, p.314) aptly stress, the discriminant function identifies and does not classify; the classification problem is logically quite distinct from the identification concept. It is unfortunate that many speak of "classifying" when they work with discriminatory problems.

A MORE GENERAL APPROACH TO ALLOCATION AND SEPARATION

We shall now attempt a synopsis of the ideas expressed up to this point for the analysis of two populations, and extend the discussion to more general aspects of allocation. The underlying statistical theory assumes that there are two universes, usually taken to be multivariate Gaussian (i.e. multivariate normal). Reference or training samples are usually available from each of the universes.

If the aim is to provide maximum separation of the populations, a formal model can be postulated, with an estimate of the discriminant vector being formed from the samples. It should be emphasized that this sample discriminant vector is chosen to maximize the separation between the populations. Moreover, the assumption is made that each of the sample vectors of observations comes from one or other of the two universes. In some cases, there may be doubt about the "closeness" of some of the specimens in a sample to the underlying universe; robust procedures could be introduced to accommodate these values and provide estimates little affected by the doubtful specimens.

If there is some doubt about the composition of the samples themselves (it might, for example, be suspected that two closely-related species make up the sample), then the homogeneity of the sample should be examined statistically. One approach is via probability plots for each variable separately and for suitable summary statistics. In particular, probability plots of D^2 values have been found to be useful (Gnanadesikan, 1977), particularly if robust procedures are incorporated into the calculation of the means and covariances (Campbell, 1980; Campbell and Reyment, 1980). In essence, examining the statistical homogeneity of the samples acknowledges that while such an assumption is a necessary part of the underlying theory, the biological reality is that the sample data do not always accord with the ideal situation, and so such assumptions should be examined at the outset.

For allocation (or identification) of specimens, two questions arise naturally. The first is: "given that a new specimen actually comes from one of the two universes, which of the universes is the more likely?" The relevant approach involves the calculation of population membership probabilities, and these involve the relative magnitudes of the multivariate Gaussian densities for the vector of observations for the new specimen. In biological applications, the assumption that a new specimen comes from one or other of the universes may be a rather strong one, and certainly should be examined. The question then to be asked is: "just how likely is it that the specimen belongs to either universe?" The relevant approach here involves the calculation of typicality indices or probabilities, and these depend on the magnitudes of the individual Mahalanobis' distances for the specimen relative to each of the two populations (see, e.g. Aitchison, Habbema and Kay, 1977).

While the population membership probabilities can be calculated from the results of a discriminant analysis, the typicality indices cannot be computed simply (see, e.g., Campbell, 1980, p.533). Since both population membership and typicality probabilities can be obtained directly from the Mahalanobis' distances, and these can be computed directly from the means and covariance matrix (or matrices), it is often more convenient to go directly to the distances. Moreover, unequal covariance matrices can be accommodated readily, while the discriminant function approach involves a quadratic discriminant function.

The allocation approach outlined above, involving the concepts of population membership probabilities and typicality probabilities, extends directly to the case where there are more than two groups and to unequal covariance matrices.

HOMOGENEITY OF DISPERSION MATRICES

Here, as in so many other connections in multivariate work involving two or more populations, the theory formally presupposes that the universes are multivariate normal (see above) and, for the linear discriminant function, that the dispersion matrices are statistically homogeneous. (Of course, the linear discriminant function can also be derived in a distribution-free manner, as was done by Fisher (1936) in his pioneering work.) In many cases this requirement is met, particularly for biological data, but deviations from the ideal state are sufficiently common to warrant attention. Providing the heterogeneity is not grotesquely manifested, little harm is likely to be done in performing the calculations in the usual way. In most situations of a doubtful or marginal nature, a few exploratory empirical studies of the data will often be sufficient to make the experimenter realize that he should make use of the procedures available for analyses involving heterogeneous dispersion matrices.

CASE HISTORIES INVOLVING DISCRIMINANT FUNCTIONS

Distinctions Between Morphometrically Similar Beetles

Some species of animals are almost indistinguishable in external appearance (sibling species). Identifications, to be reliable, need to delve more deeply into the morphological properties of the organism than is done in the conventional key. Lyubischev (1959) has studied the situation in the chrysomelid beetles of the genus *Halticus*. Here, the beetles may differ only in respect of the male genitalia, but the dissection and study of these is a lengthy task, and may be quite impracticable if, for example, museum regulations forbid the dismembering of specimens.

For two species of *Halticus* from the European part of the USSR, Lyubischev examined 21 characters and found that none of them gave frequency distributions for the two species *(H. oleracea* and *H. carduorum)* that did not overlap seriously. However, the four best of these characters, when combined into a discriminant function, avoided all overlap between the two species in the sample on hand, and reduced the potential overlap in very large samples to about one specimen in every 33 examined.

With computer facilities available, one might be inclined to include all the characters in the preliminary discriminant function, to save all the eliminatory trials that Lyubischev carried out by hand. If, for example, a more concise function were needed, the less useful characters (for example, those that carry the lowest standardized coefficients in the discriminant vector, or those identified by all-possible-subset calculations) could be dropped, until the frequency of misidentification reaches an acceptable level.

A Discriminant Function to Distinguish Asiatic Wild Asses

Nine measurements of the skulls of two samples of wild asses, Khurs from the Little Rann of Cutch, and Khiangs from Ladakh and Western Tibet, were converted into a discriminant function by Groves (1963). Nine Khurs and twelve Khiangs were available, together with much smaller numbers of several other groups of Asiatic wild ass. This discriminant, despite the small samples and the distortion of some of the skulls because the animals had lived on unsuitable diets in captivity, clearly distinguished the two groups. Three of the attributes had positive signs in the discriminant and six had negative ones, so that the differences involve changes of shape in addition to possible differences in size. A certain amount of sexual dimorphism was noted in the Khiangs but not in the Khurs. On the other hand, the Khurs showed most distortion of the skull due to life in captivity, with only a small overlap of scores between the wild and the captive specimens. Once the framework of reference had been set by calculating the discriminant, the other isolated skulls could be fitted into the picture. Some care is needed here, since the typicality of a new specimen should also be examined to ensure that the new material does not differ from the earlier material in the multivariate space orthogonal to that defined by the discriminant function, a condition which the discriminant function of course does not take cognizance of. With this proviso, a single specimen can often be made to serve a useful purpose in a multivariate analysis if it is fitted into a framework constructed with more plentiful material of the same general form.

Changes of Shape in Whitefly Growing on Different Host-plants

Whiteflies of the genus *Bemisia* are pests of cassava, cotton and tobacco plants in West Africa. It has been strongly suspected that the Cassava, Cotton, and Tobacco Whitefly are all members of the same species, but that they grow up to look different according to the host-plant. To clarify this issue, Mound (1963) measured seven characters of the fourth-stage larva drawn from populations on cassava and tobacco plants, all larvae being originally descendants of a single parthenogenetically reproducing female. These seven characters (all in the same units) were: length and breadth of pupal case (the fourth instar larva is commonly referred to as a pupa), length and breadth of vasiform orifice, length and breadth of the lingula tip, and length of the caudal furrow. The discriminant vector which optimally separates the two populations on this basis is: $(0.23, -0.38, 0.35, 2.69, 2.08, -3.99, -0.73)$. The corresponding variance ratio for 562 degrees of freedom was found to be 14.23, so that there can be little doubt of the reality of the host-correlated variation.

The nature of the discriminant function shows that the form from tobacco is longer and narrower than that from cassava, and its vasiform orifice is much broader and slightly longer. The lingula tip is narrower and longer in the form from tobacco, and the caudal furrow shorter. The fact that three of the seven coefficients in the discriminant function are negative, whereas four are positive makes it clear that a general change of shape is involved. If all, or nearly all, the coefficients were of the same sign, one would suspect that the difference between the two forms was mainly one of size, and hence probably due to differences in the nutritional status of the two populations. In fact, it is the growth pattern, rather than the amount of growth, that is changed by the host-plant.

The significance of analyses of this type for evolutionary studies involving changes in the phenotype is great and it is to be hoped that more work of this kind will appear in the future.

In a similar way, Fraisse and Arnoux (1954) found that the shape of the cocoon of the silkworm varies systematically according to the diet of the developing larvae, as shown by discriminant analysis.

The Assessment of Ear-formation in Barley Breeding Experiments

In breeding plants for better yield, Mather and Philip (in Mather, 1949) note that just as a single character can be resolved into subcharacters, so can characters be compounded into supercharacters such as yield, ear conformation etc., which, taken together, form an estimator of the total merit of a variety. They in fact compounded the length of the ears (neglecting the awns); the maximum breadth; and the combined length of the central six internodes. Their objective was not so much the discrimination of one plant variety from another, but to maximize variation between plants, relative to that within plants, so as to produce a measure of ear conformation whose genetical variation is at a maximum compared with at least one important kind of non-heritable variation. This vector turned out to be proportional to $(1, -5.3, 5.8)$ when the two varieties, Spratt and Goldthorpe, were contrasted. These varieties differ in the genetical architecture of the ears, but no simple Mendelian differences could be detected between them, variation being continuous in the F_2 generation.

COMPARISONS OF LINEAR AND QUADRATIC DISCRIMINANT FUNCTIONS

The input spectrum from ten vocalized monosyllabic words (Bit, Bet, Bat, Bot, But, Bert, Beet, Boot, Book and Bought) was assessed in terms of the output from 35 band-pass filters by Smith and Klem (1961). A machine was programmed to recognize each sound by means of a discriminant function, with variables made up of these outputs. The machine calculated the likelihood that each input sound belonged to a particular one of the ten words and allocated the sound to the most probable word. Using a linear discriminant function, the machine was able to allocate 87% of its input correctly. With a quadratic function, it improved its performance to 94%, but it scored the most difficult word (But) correctly only 71% of all the times that this word was vocalized, irrespective of whether a linear or quadratic function was used. The use of the quadratic discriminant function hardly seems justified for a 7% improvement in performance, especially as this improvement applies to recognition of words that are only rarely wrongly allocated by the machine. Every case must be judged on its merits, and other authors, such as Welch and Wimpress (1961), have chosen linear discriminants for vowel recognition.

Where the discriminant function is to be used in systematic work, however, the quadratic form may often be superior to the linear function (Burnaby, 1966b; Cooper, 1963, 1965; Lachenbruch, 1968; Lerman, 1965). When the individual Mahalanobis' distances are used directly, the calculations are essentially the same for both equal and unequal covariance matrices.

RATIOS OF GENERALIZED DISTANCES IN MORPHOMETRIC STUDIES

Dempster (1964) introduced the concept of what he terms "covariance structure". The idea is based on the comparison of ratios of generalized statistical distances and distance-like quantities. Reyment (1969f) has applied the method to fossil material as a means of tracking down the possible sources of the variability in fossil associations. Variation in a species of living, agglutinating foraminifers was used as a guide and the results obtained for that material were interpreted as being of possible genetic origin. Two species of Devonian brachiopods were also analysed in this paper. One of these was concluded not to vary significantly from sampling locality to sampling locality, while the second species could be shown to reflect sample differences of putative genetic origin.

PARTITION OF MULTIVARIATE VARIATION AND SHAPE AND SIZE CHANGES IN AN OSTRACOD SPECIES

Reyment (1966a, p.118) used Dempster's (1963) method of principal variable analysis, a form of stepwise multivariate analysis of variance, for studying the variation in the carapace of the Paleocene ostracod *Iorubaella ologuni,* in order to attempt the identification of possible environmental influences on the morphology of the shell. The method is based on an analysis of variance in steps, related to each of the latent roots. The analysis disclosed that significant differences in mean vectors have resulted from pure size differences as well as shape differences.

THE USE OF ANGLES BETWEEN VECTORS

For certain purposes, it is helpful to be able to pick out some kind of biological contrast, common to a number of species, and compare the contrast in one species with that in another. For instance, sexual dimorphism is often similar in several related species, the males often being consistently larger or smaller than the females (Reyment, 1969d). Or one may be interested in the change of habit of several species of plants growing either in their normal habitat or on top of a windswept, overgrazed, mountain, where the habit of growth may be quite different, leading to stunted, prostrate forms, at first sight difficult to reconcile with the normal appearance of the plant.

One can, of course, proceed to analyse all available measurements made on all the material, so as to be in a position to construct a canonical variate plot or other model in order to display the relationships between the entities studied. However, such an elaborate procedure may not meet the experimenter's needs in cases where he is concerned only with one kind of variation. Such a case arose when it became desirable to examine the effects of crowding on the shapes of adult locusts (Blackith, 1962). A peculiar feature of the locusts is that although several species are known which respond to crowding by changes of shape, physiology and behaviour leading ultimately to swarm formation, other species do not respond in this way. Moreover, some genera include both swarming and non-swarming species, whereas the swarming species are to be found in different subfamilies, so that they are not of necessity closely related.

By calculating the discriminant functions between the populations of any given species which had been reared in crowded conditions of crowded parents, or had been relatively isolated for two generations, it was possible to obtain a vector describing the resulting change of shape ("phase" change) for each species. For two vectors, $\mathbf{a}$ and $\mathbf{b}$, the angle θ between the vectors is given by

$$\cos \theta = \mathbf{a}^{\mathrm{T}}\mathbf{b}/(\mathbf{a}^{\mathrm{T}}\mathbf{a}\,.\,\mathbf{b}^{\mathrm{T}}\mathbf{b})^{1/2}$$

In general, for two discriminant vectors which represent unrelated patterns of growth, the angle will approach 90°, whereas when the two vectors are substantially parallel, the angle will approach zero. The angle thus measures the extent to which two taxonomic comparisons are alike.

When the discriminant vectors had been computed for the two phases of

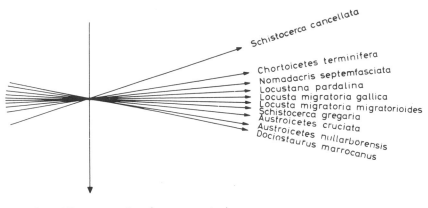

Anacridium aegyptium (non-swarming)

FIG. 6.2. Directions of vectors describing the response to crowding in ten species of swarming locust and one species of non-swarming grasshopper, based on three morphometric characters. Redrawn from Blackith (1962).

each of ten species of swarming locust, they were arranged as shown in Fig. 6.2 with the angles between them corresponding to those calculated. It can be seen that the entire group of vectors forms a fan whose arc subtends an angle of less than 20°, which confirms the remarkable homogeneity of this phenomenon. Furthermore, when the same vector was computed for a non-swarming species *(Anacridium aegyptium)*, the new vector was practically at right angles to the fan-shaped bundle of the "phase" vectors. This large "tree-locust" often forms aggregations in bushes in arid scrubland, but is not known to form true swarms, although a related species *(A. melanorhodon)* does so. However, *A. aegyptium* does differ in shape when reared in crowds or in isolation. Even more striking was the vector representing the effects of crowding in an insect that was not a locust at all, but a cotton stainer (a kind of bug), whose vector was found to project from the plane of the paper (not shown in the diagram).

A word of warning should perhaps be sounded in this connection. The angles between latent vectors or other discriminators will naturally be subject to quite considerable variability, which will be most pronounced where the sample sizes are small. In order to be reasonably reliable, angles calculated between vectors should therefore be a technique reserved for larger samples, except for work of a more exploratory kind. Campbell and Tomenson (1983) have developed likelihood-based procedures for making detailed comparisons of discriminant analyses for several sets of data.

7
Canonical Variates

One of the more interesting morphometric applications of multivariate statistics concerns the method usually known as canonical variate analysis. The technique is used to examine the interrelationships between a number of populations simultaneously, with the end in view of representing the interrelationships graphically in only a few dimensions (ideally only two or three). The axes of variation are chosen to maximize the separation between the populations, relative to the variation within each of the populations. Successive axes are chosen so that the resulting plot of individual specimens in the new coordinate system should exhibit a lack of correlation within each population.

The geometrical representation of the approach follows essentially the same steps as that for the discriminant function. For example, imagine that there are three universes in Fig. 6.1 rather than the two. In principle, the individual observations can again be projected onto lines representing all possible directions of vectors in the X_1–X_2 plane. The first canonical vector will be given by that direction for which the separation between the means of the projected points for each universe is greatest, relative to the scatter of the projected points within each universe. Further discussion is given in Campbell and Atchley (1981, p.271).

More formally, the first canonical variate is that linear combination of the characters which maximizes the ratio of the between-populations sum of squares to the within-populations sum of squares for a one-way analysis of variance of the resulting canonical variate scores. The ratio of the between-populations to within-populations sums of squares is referred to as the canonical root. Subsequent canonical variates satisfy the same criterion, subject to being uncorrelated both within populations and between populations with the previous ones.

For v characters and g populations, the canonical variate scores are given by

$$y_{km} = \mathbf{c}^{\mathrm{T}}\mathbf{x}_{km}$$

where $\mathbf{x}_{km}$ denotes the mth of n_k observations for the kth group.

The first canonical vector $\mathbf{c}$ is chosen to maximize the ratio

$$f = \mathbf{c}^{\mathrm{T}}\mathbf{B}\,\mathbf{c} / \mathbf{c}^{\mathrm{T}}\mathbf{W}\,\mathbf{c}$$

where $\mathbf{B}$ is the between-populations (or among-populations) sums of squares and products matrix, and $\mathbf{W}$ is the within-populations sums of squares and products matrix.

The canonical vectors $\mathbf{c}$ and canonical roots f satisfy

$$(\mathbf{B}-f\,\mathbf{W})\,\mathbf{c} = \mathbf{0}$$

The vectors are scaled so that

$$\mathbf{c}^{\mathrm{T}}\mathbf{W}\,\mathbf{c} = n_w$$

where n_w denotes the within-populations degrees of freedom $(= \Sigma_{k=1}^{g}(n_k-1))$. There are $h = \min(g-1, v)$ non-zero canonical roots.

There is another useful geometrical interpretation of the calculations involved in a canonical variate analysis, this time as a two-stage principal component analysis (or more generally as a two-stage rotation).

The first stage involves rotation of the original characters to their principal components, with the latent vectors being those of the pooled within-populations covariance or correlation matrix. (If the correlation matrix is used, then the population means must be standardized by the corresponding within-populations standard deviations.) The principal components are then scaled by their standard deviations—in this case, the square roots of the latent roots—to have unit standard deviation.

This rotation and scaling to give orthonormal variables defines the first stage. In essence, the transformation allows the relative distances between the populations to be examined in the context of the familiar Pythagorean distance, rather than the generalized distance which is necessary in the original coordinate system.

The rotated and scaled axes become the reference axes for the second stage of the analysis. The variation between the population means is considered in this coordinate system. Since the first-stage analysis can be considered as describing the nature of the within-populations variation, the second stage of the analysis is in essence describing the variation between the populations relative to the variation within populations.

The second-stage analysis again consists of a principal component analysis, this time of the population means for the new orthonormal variables. The resulting latent roots give the usual canonical roots, while the latent vectors given the canonical vectors for the orthonormal variables. The canonical vectors for the original characters are found by first reversing the scaling and then the rotation.

The second-stage principal component analysis involves the calculation of a matrix describing the variation of the population means about the grand mean. Usually the population means are weighted by the corresponding numbers in each population; this results in the maximum likelihood estimates of the canonical roots and vectors. If each vector of means is given the same weight, irrespective of the sample size, then the solution proposed by Rao (1952, Section 9c) results; here the overall generalized Mahalanobis' distance is maximized for the chosen number of dimensions.

Once the calculations have been carried out, the next stage is to display the results graphically and determine the characters which contribute to the separation between the populations.

Plots of the population means for the first few canonical variates display the major differences and similarities between the populations. Common representations of the canonical variate means are to plot the mean scores on the first canonical variate or axis against those on the second and third (and higher) axes, and perhaps plot the scores on the second axis against those on the third. A useful representation is to plot the population means for canonical variate II against those for canonical variate I, and represent the third axis by a line above or below the point representing the population in the I–II plane. (The fourth axis could be represented by a horizontal line to the left or right.)

A less appealing representation is the use of a perspective diagram, with canonical variates I and II defining the horizontal plane, and the third axis being represented by the height above the plane. This representation tends to give too much visual emphasis to the third axis, the least important of the three.

To give some idea of the actual degree of separation, it is often worthwhile plotting the canonical variate scores for the individual specimens. Some authors choose to superimpose approximate 95% confidence contours for each population, though this may be unnecessary in many applications.

The components of the standardized canonical vectors can often be used to indicate the characters which contribute most to the separation along each axis. The standardized canonical vectors are the vectors for the characters standardized to unit standard deviation within populations, and are found by multiplying the components of the canonical vectors by the corresponding pooled within-populations standard deviations. Those characters with the smallest absolute values of the standardized coefficients generally contribute little to the discrimination.

Some care is needed with this approach, since high within-populations correlations between characters can sometimes give a misleading interpretation. This will occur when a subset of highly correlated characters results in a latent vector with a very small latent root, with the between-populations variation in the direction of the latent vector also being small. The effect of this can be to produce components of the canonical vector for the subset of characters which are similar in magnitude but of opposite sign. If one or more of the subset of characters is eliminated, or if the principal component corresponding to the latent vector with small latent value is eliminated (see Campbell and Reyment, 1978, for technical details), there will often be little change in the canonical roots, but a marked change in the components of the canonical vectors corresponding to the subset of characters.

Lack of stability in the components will only occur when the between-populations variation in the direction of a latent vector(s) with small latent root(s) is small. When there is a considerable between-populations variation in such directions, the resulting standardized canonical vectors will in general give a good indication of the important characters effecting the discrimination.

As an example, consider the analysis of divergence in wing shape between females of two species of blowfly, *Calliphora albifrontalis* and *C. stygia*. The measurements chosen describe the length of the wing from position h along various veins (S_c, R_1, R_{4+5}, M_1) (see "Insects of Australia", Fig. 34.7C, p.665), together with overall length and width, and the width at vein (M_{3+4}). The venation pattern is such that vein M_1 finishes near the tip of the wing. If there were subtle differences in the shape of the wing tip from species to species, then a contrast of this character with the overall length would provide useful discrimination. A first examination of the canonical vectors would seem to suggest that this is in fact the case. The standardized canonical vector has components 1.13 (for S_c), -3.06 (R_1), -1.82 (R_{4+5}), -2.77 (M_1), 3.80 (L), 4.70 (W), -2.15 (W at M_{3+4}), perhaps suggesting in part a contrast between L and M_1. The canonical root is 1.43. However, a closer examination shows that the latent vector corresponding to the smallest latent root of the correlation matrix derived from the pooled within-populations covariance matrix has the form (0.02, -0.00, 0.04, -0.71, 0.70, -0.02, -0.02), with a latent root of 0.001, the latter accounting for only 0.02% of the variation within populations. An analysis with the contribution of this latent vector removed gives a canonical root of 1.39 and standardized components for M_1 and L of 0.47 and 0.65. Note that the sum of the two standardized components is roughly the same for both canonical vectors. None of the other standardized components is altered substantially (e.g. that for W changes from 4.70 to 4.93). Excluding M_1 from the analysis gives a canonical root of 1.41, with the standardized component for L being 1.24.

By way of contrast, consider the analysis of divergence between populations of three species of grasshopper reported by Campbell and Dearn (1980). The first canonical variate shows complete separation between the three species, with populations of each species grouped tightly along the first axis. The first canonical root is 8.19. Much of the variation in this direction can be summarized in terms of three head characters—eye width, eye depth, and width of head across the genae (see Campbell and Dearn, 1980, Fig. 2 for details); these characters give a first canonical root of 6.84 (see Table 4 of that paper). An analysis based on the last two (of the three) principal components of the pooled correlation matrix gives a first canonical root of 6.77, while an analysis based on the third principal component gives a canonical root of 6.08. For these data, there is substantial between-populations variation in the direction of the latent vector with the smallest within-populations variation, and the magnitude of the components of the canonical vector(s) can be used to indicate the nature of the vector.

It is also important to find out how many of the canonical variates correspond to biologically or geologically meaningful sources of variation, that is, how many such variates so order the data that the analyst can learn something of consequence from the way in which the data are ordered.

The magnitudes of the canonical roots can be used to give some idea of the likely biological importance of the corresponding canonical variates. For overall sample sizes of the order of 150 to 200 specimens, our experience suggests that canonical roots less than about 0.5–0.75 are rarely associated with separation of any biological importance. Canonical variates with roots less than about 0.5 usually exhibit considerable overlap between the scores for the different populations. In general, unless there is some particular interest in one or more of these canonical variates, they can be ignored in the interpretation.

Sometimes, the first canonical variate (corresponding to the largest canonical root) takes up such a large fraction of the total amount of variation in the material that merely plotting the positions of the various entities along this first axis gives good insight as to the nature of the separation along the axis. Eyles and Blackith (1965) performed a canonical variate analysis of ten measurements made on several species of lygaeid bugs and their hybrids. The first canonical variate represented nearly all the variation in size, displaying the hybrids as intermediate in size between the parent species.

A clear distinction should be drawn between the statistical and the biological importance of an axis of variation. Statistically, the importance is measured by the size of the canonical root which generates the vector; roots which are not significant can normally be disregarded. From the biologist's standpoint, such an assessment of "importance" may be as misleading for what it does not do as for what it does. For example, the first canonical vector

may represent some quantity which is numerically large but of little consequence to the biologist. In morphometric analyses, "size variation" may
occupy one of the canonical variates corresponding to the largest canonical
roots, but size variation may not be an object of study in itself.

CASE HISTORIES INVOLVING THE USE OF CANONICAL VARIATES

The Nature of Variation in a "Difficult" Grasshopper Genus

It is a commonplace that, amongst any group of animals, some genera or
other subdivisions are particularly difficult to classify. The reasons vary;
among parasitic animals the great importance of behaviour patterns
integrated with those of the host ensures that species are differentiated, all
too readily from the taxonomist's point of view, by behavioural traits which
may leave little record in the anatomy. Even in non-parasitic groups,
however, "difficult" genera are to be found. One such is the pyrgomorphid
genus *Chrotogonus*, with a vast range from Egypt to Soviet Central Asia and
down to Ceylon. As Blackith and Kevan (1967) comment, the genus shows
a morphological plasticity, even within individual species, which is bewilder-
ing. These authors measured seven morphological characters (length and
width of the pronotum; length and width of the hind femur; width of the
mesosternal interspace; elytron length; and head width) on 1093 insects of
both sexes. This mass of information related to 34 different predetermined
taxa according to the species, subspecies, sex, and degree of alary
polymorphism, since *Chrotogonus* species may include individuals with
well-developed wings and elytra or with these organs rudimentary.

A canonical variate analysis of the data showed that there were effectively
only two basic dimensions of variation, the first accounting for 94.6% of the
total variation and the second for 4.1%. The two vectors (-0.147, $+0.297$,
$+0.359$, -0.832, $+0.108$, -0.001, $+1.000$) and (-0.220, $+0.018$, $+0.006$,
$+1.000$, -0.921, $+0.350$, -0.826) make an angle of some 40° in three-
dimensional Euclidean space.

The interpretation of the chart constructed from these vectors, using them
as ordinate and abscissa respectively and plotting the positions of each group
along the axes, is unusual. Each canonical vector subsumes two biologically
distinct patterns of variation.

In Fig. 7.1 the subspecific distinction between *C. senegalensis brevipennis*
and *C. s. abyssinicus* occupies the same (second) variate as the alary
polymorphism contrasting the long-winged and short-winged forms. In Fig.
7.2 the first variate subsumes both sexual dimorphism and some of the
differences between species, since *C. homalodemus* always has higher scores
along this variate than the various elements of *C. hemipterus*. Many other
examples can be seen in the original paper.

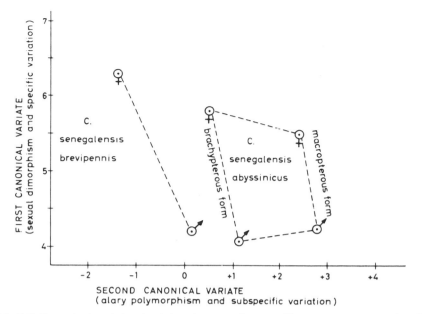

FIG. 7.1 Canonical variate chart for the grasshopper *Chrotogonus senegalensis* showing that subspecific variation is symbatic with alary polymorphism. Adapted from Blackith and Kevan (1967).

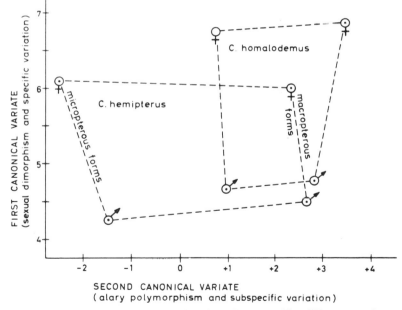

FIG. 7.2. Canonical variate chart showing that specific differences between *Chrotogonus hemipterus* and *C. homalodemus* are symbatic with sexual dimorphism. Based on seven morphometric characters. Adapted from Blackith and Kevan (1967).

We thus have a disquieting situation for the taxonomist; sexual variation is, at least on the basis of these seven characters, indistinguishable from the specific contrasts of form corresponding to the division between species with slender ovipositors and species with stouter ovipositors. Moreover, alary polymorphism is inextricably bound up with many of the differences between subspecies. Patterns of growth which are biologically quite distinct in their causation but phenotypically identical in their expressions have been called "symbatic" by Blackith and Albrecht (1959) who found another example in the polymorphism of the red locust. It seems as if the genus *Chrotogonus* is incapable of exhibiting more than these two patterns of contrasting forms, so that all the ecological and environmental variation has to be expressed along these two variates. As a consequence, the range of variation of any one species or subspecies along them is enormous; within one species individuals can be found whose shape and size correspond with any of the other species (sex for sex) within the genus. Little wonder that the subtle appreciation of shape which is normally the basis of a skilled taxonomist's work should have led earlier workers further into difficulties. It is, in fact, virtually impossible to be sure of an identification of any individual *Chrotogonus* unless the provenance is known, since the forms which are most alike often have different geographical origins.

Reyment (1969c) revealed that a somewhat similar situation arises in the ammonites of the Jurassic genus *Promicroceras* where five putative species can be splayed out on principal coordinates charts without any apparent grouping into species, at least on the basis of the seven measured characters.

It is possible that the taxonomic situation for these ammonites could have been complicated by unresolved sexual dimorphism in one of the species *(P. planicostum)*.

The Simultaneous Analysis of many Different Groups of Grasshoppers

When large numbers of groups of organisms are to be analysed, problems can arise that are rare when few groups are involved. Blackith and Blackith (1969) analysed 196 groups belonging to the Morabinae, a subfamily of the eumastacid grasshoppers, by means of a canonical variate analysis based on ten measurements; 1450 insects were measured altogether. Each axis of variation reflects a biologically meaningful contrast of form; however, some of these contrasts will not affect some of the groups. For instance, sexual dimorphism is a relevant contrast of form for all except the single parthenogenetic species of the subfamily. The fifth of the canonical variates contrasts the two divisions of the genus *Keyacris,* for instance, and is thus irrelevant to members of other genera. All these other groups will, however, have scores along the fifth axis and it becomes necessary to pick out the meaningful contrasts of form from the apparently meaningless ones.

In such circumstances, it may be more effective to analyse subsets of the groups (perhaps using the pooled covariance matrix for all groups if they are reasonably homogeneous).

A Psychometric Example Using Dichotomous Variables

Maxwell (1961) has pointed out that one can employ canonical variate analysis even when the data are dichotomous. Indeed, as Claringbold (1958) has shown, most of the standard techniques of multivariate analysis can be worked successfully with data of the presence-absence type. In Maxwell's example, 224 schizophrenics, 279 manic depressives, and 117 patients with various anxiety states were scored for the presence or absence of the following four symptoms: (a) anxiety, (b) suspicion, (c) schizophrenic type of thought disorder, (d) delusions of guilt.

When these scores were converted into a matrix, and the canonical variates extracted, a graph was constructed on which the positions of the three groups of patients could be plotted (Fig. 7.3). The manic depressives

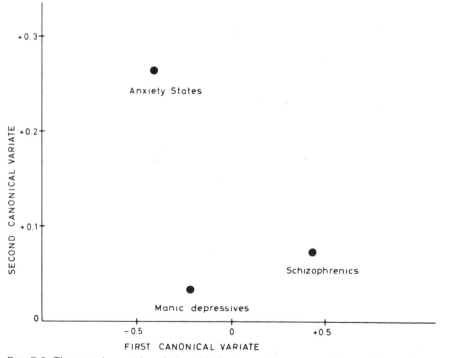

FIG. 7.3. The psychometric relationship between three sets of mentally disturbed patients, assessed along two canonical variates. Based on four dichotomous characters. Redrawn from Maxwell (1961). Note the non-standard scaling of the canonical vectors.

and patients with anxiety states form one pole of the first canonical variate, accounting for 83% of the total variation, and the schizophrenics form the other pole. The manic depressives are separated from the patients with anxiety states by the second of the canonical variates, accounting for 17% of the total variation. It should be noted that since there are only three groups under comparison, only two variates can be extracted from the data. The weights of the characters in these two variates show that the distinction between schizophrenics and the rest depends mainly on the first three of the characters, whereas that between manic depressives and anxious patients depends mainly on the first and last characters.

Naturally, this kind of experiment can be performed with continuously distributed data, instead of dichotomized data; an example is given by Rao and Slater (1949).

The Primate Shoulder in Locomotion

Nine measurements of the scapulae of a number of Anthropoidea and Prosimii were compounded into nine canonical variates by Ashton *et al.*, 1965; Oxnard, 1967, 1968. The first three of these variates were found to have a readily interpretable biological significance. The separation afforded by the first two of the canonical variates is shown in Fig. 7.4.

Although differentiation of the material at the highest (subordinal) level was poor, good discrimination was obtained within the Hominoidea, separating the genus *Homo* from the four genera of apes in the Pongidae. These apes are all brachiators. Again, within the Cercopithecoidea, the semibrachiators of the Colobinae are distinct from the quadrupedal Cercopithecinae. The basic method of locomotion adopted by a species influences, as might be expected, the shape of its scapula. Among the Prosimii, a similar contrast of the forms of the scapulae is shown by the separation of the quadrupedal Galaginae from two genera of Lorisinae which are hangers. This distinction is pervasive, and turns up again in the lemurs so that *Lemur*, a quadrupedal genus, is clearly separated from the genus *Propithecus*, which is a hanger. Most of these distinctions are made by the first variate, which reflects primarily those anatomical changes which have occurred as a response to the amount of use of the forelimbs for suspension. The second variate, however, seems mainly to differentiate ground-dwellers from arboreal dwellers, a distinction which involves changes of the scapula apart from those involved in the use of the forelimb in hanging, for there are important aspects of locomotion other than the suspensory ones. The almost exclusively arboreal forms *Pongo, Hylobates, Brachyteles,* and *Rhinopithecus* cluster at one extreme of this ordination, whereas the more terrestrial members cluster at the opposite extreme.

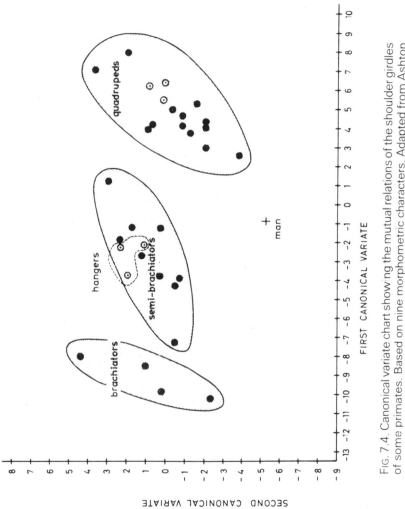

FIG. 7.4. Canonical variate chart showing the mutual relations of the shoulder girdles of some primates. Based on nine morphometric characters. Adapted from Ashton *et al.* (1965). Open circles, prosimiids; closed circles, anthropoids.

The third of the canonical variates does not distinguish between the subhuman primate genera, but it does clearly separate man from them. *Homo* is thus unique in being the only genus to stand out from the plane formed by the first two canonical variates into a third dimension of variation. In some respects, man has a shoulder girdle like that of the quadrupeds, in other respects it is like the brachiators, but it is a mosaic of these trends and not, as are the semibrachiators, intermediate between the two extremes in respect of separate features. This analysis is an excellent illustration of the use of multivariate analyses in sorting out what has been termed mosaic evolutionary trends, where advances (or retrogressions) have proceeded simultaneously along several quite distinct axes of variation. The success of the technique is all the more striking because considerable efforts had previously been made to recover the information outlined above from a study of the individual measurements and their ratios, without conspicuous success. It may be noted that the canonical variates are often referred to in these papers as discriminant functions.

This usage is meant to convey the idea that linear functions are being used for discrimination; however, it is perhaps likely to mislead the uninitiated into thinking that each is a discriminant function between a pair of groups.

Functional relationships of the kind revealed by these studies can be traced in the evolution of the shoulder girdle in marsupials, edentates, rodents and carnivores (Oxnard, 1968) and in the teeth of hominoids (Ashton *et al.,* 1957). Oxnard, in fact, goes on to suggest that these influences on the form of the shoulder girdles may have become established so early in the evolution of the mammalia as to restrict their development to a limited number of pathways.

It seems that the multivariate analysis of mammalian skeletal structures is able to throw light on the evolution of the group in a way which is quite distinctive and which has escaped the more classical approaches to the subject. The multiplicity of forms in the shoulder girdle that have evolved can be reduced, in fact, to the results of the interplay of only a very few, essentially two, uncorrelated patterns of development. These do not appear to differ appreciably even in such widely separated groups as the marsupials, the edentates, and the primates. The adaptation to a gliding habit gives a third pattern of variation. Some classical workers had suggested that these functionally adaptive features of the skeletal joints were superimposed on the basic pattern of development of the animals: the facts seem to be exactly the contrary; the functional adaptations, as reflected in the interplay of the two patterns of development, measurable along the first two canonical variates illustrated by Oxnard (1968), and repeated here for a small part of his total findings, *are* the basic patterns of development whose degree of

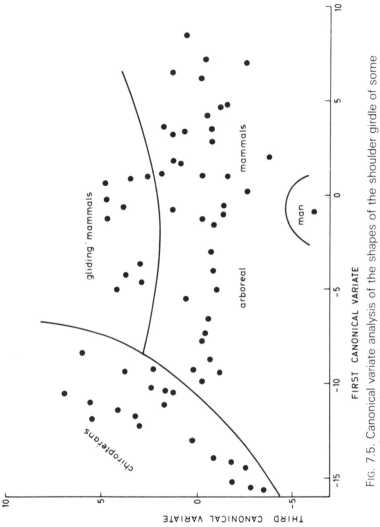

FIG. 7.5. Canonical variate analysis of the shapes of the shoulder girdle of some mammals associated with trees, based on nine morphometric measurements. Scale units are one standard deviation along each axis. Redrawn and adapted from Oxnard (1968).

interlocking is controlled by the evolutionary process. That the evolution of form depends essentially on the degree of interlocking of a few distinctive patterns of growth has already been suggested for insects by Blackith (1965). Thus multivariate studies of growth and adaptation can make a contribution to functional anatomical studies which more classically oriented methods have not seemed capable of making.

Oxnard has shown (Fig. 7.5) that the adaptation to actual gliding is distinct from the adaptations that go with living arboreally. Moreover, some of the bats which have a butterfly-like flight are separated along the third axis from those which have either powerful long-distance flight or swift straight flight. This axis also separates man from the other primates.

Within the arboreal carnivores, almost all the useful separation was along the first axis: this axis seems to separate those forms in which the forelimbs are designed to withstand pulling from those whose forelimbs are designed to take thrusts, this is the essential distinction between the terrestrial and arboreal ways of life.

As often happens when very extensive bodies of data are analysed, some of the later canonical variates, in Oxnard's work up to the ninth, afforded useful contrasts of form with evident biological interpretations.

Subspeciation in South American Anteaters; a Confused Taxonomy

The development of the taxonomy of the South American anteaters of the genus *Tamandua* seems to have been unusually confused, authorities disagreeing flatly about the number and nature of the subspecies of *T. tetradactyla*. Appreciating that much of this confusion arose from the failure to realize that individuals of different sizes will often have distinct morphometric ratios of attributes, Reeve (1941) quite rightly stressed the importance of considering the taxonomy of the genus in the light of what was at that time known about allometric growth patterns, and making a suitable allowance for the size of the animal.

Later, Seal (1964) reanalysed the data collected by Reeve, involving the measurements of the basal length of the skull excluding the premaxilla; the occipitonasal length; and the greatest length of the nasals. Seal found that the two canonical vectors in which the three attributes could be linked to describe variation in the group consisted of (108.918, −33.823, −35.497) and (−46.608, 59.952, 22.821). Seal notes that the picture which emerged from the canonical variate analysis was substantially different from the conclusions drawn by Reeve, who felt that the subspecies *chapadensis* and *mexicana* were the only ones with a claim to distinct entities on the basis of the skull measurements. However, Seal's analysis showed that *chiriquensis* and *instabilis* were distinct from one another and from the first two sub-species. Moreover, Reeve's conclusion that *instabilis* seemed to be only a small edition of *mexicana* and *chiriquensis* is not upheld by the later analysis, but it is only fair to mention that Reeve was here speaking solely in terms of skull measurements, and considered that on the basis of tail-length, and other body characters, "this group clearly deserves to be placed in a separate subspecies".

The almost total discrepancy between the interpretation of Seal's analysis and that of Reeve based on the same data is disconcerting, and it is worth while trying to find out how such a situation could arise, because, although quantitative multivariate analyses often supplement and clarify more conventional ones, they rarely run flatly counter to them. We have here a situation where Reeve's analysis also disagreed with those of earlier workers and Reeve is probably correct in considering that it is the earlier workers who were at fault in forgetting that a morphometric ratio can be misleading when taken on animals of different sizes, because of allometric growth; see, for example, Christensen (1954). On the other hand, examination of the nature of Seal's canonical variates shows that neither is representative of "size", since the vector $(1,1,1)$ evidently makes a large angle with each of the canonical variates.

There seem to be two variates, neither strongly size-sensitive, which are capable of ordering the putative sub-species of *T. tetradactyla*; allometric studies such as those of Reeve are particularly at risk when more than one underlying pattern of development has to be assessed.

A further analysis of Reeve's data shows some interesting features. The summary statistics were taken from Seal (1964, p.134) (we suspect that the $(1, 2)$ element of the covariance matrix for *chapadensis* from Minas Geraes should be 0.0008347, otherwise the resulting correlation is > 1). This correction has in itself quite a marked effect on the analysis. For example, the first two canonical roots for the data in Seal are 2.91 and 0.62 while those for the corrected data are 2.56 and 0.91 (see Table 7.1). Note also the change in the canonical vectors (compare "usual" column with "error-usual" column). An analysis of the corrected data shows inherent instability in the canonical vectors. The smallest latent root of the pooled correlation matrix is 0.032, with corresponding latent vector $(0.65, -0.75, 0.12)^T$. The principal component corresponding to this latent vector/root combination (referred to below as the smallest principal component) contributes only 8% to the between-populations variation. When the contribution of this smallest principal component is eliminated, the canonical roots are little changed (2.56 vs 2.35, 0.91 vs 0.88). However, there is a marked change in the

TABLE 7.1. Summary of canonical vectors for various analyses of the data published by Reeve (1941). Canonical vectors are standardized unless indicated otherwise.

	usual	$k_3 = \infty$	$k_3 = \infty$ − unstd	error − usual	error −$k_3 = \infty$
CVI					
$b\,l$	1.992	0.964	53.6	2.435	0.959
$o-n\,l$	−0.612	0.643	36.7	−1.322	0.644
$l\,n$	−1.083	−1.302	−44.8	−1.112	−1.300
f	2.564	2.350	2.350	2.913	2.339
CVII					
$b\,l$	−0.592	−0.017	− 0.93	−3.047	−0.016
$o-n\,l$	0.932	0.126	7.17	4.226	0.125
$l\,n$	0.661	0.909	31.25	−1.179	0.909
f	0.907	0.875	0.875	0.616	0.075

components of the canonical vectors for $b\,l$ and $o-n\,l$; these two characters dominate the third latent vector given above. The differences in the results for the data in Seal and for the corrected data also virtually disappear when the third principal component is eliminated (compare columns headed "$k_3 = \infty$" and "error − $k_3 = \infty$" in Table 7.1). The comments in the previous paragraph on size-shape variation can also be examined further. The first latent vector of the pooled correlation matrix is (0.59, 0.60, 0.55), which differs little from the normalized unit vector (with all components equal to 0.58). The first principal component accounts for 31% of the overall between-populations variation. When this component is eliminated from the analysis, the canonical roots are 2.37 and 0.05, indicating that most of the variation along the first canonical variate is due to differences in shape.

Anthropometry of some Nigerians

A very extensive collection of cranial measurements of various subdivisions of the people of Eastern Nigeria was made by Tablot, and analysed after his death by Mulhall in Talbot and Mulhall, 1962, who used generalized distances to make an excellent arrangement of the material based on eight characters. Subsequently, Reyment and Ramdén (1970) have reanalysed the data, using canonical variates to construct an ordination. In view of the relationship between the canonical vectors and the principal coordinates of the matrix of generalized distances, this analysis should closely resemble that of Mulhall. Mulhall had originally examined the homogeneity of the basic dispersion matrices for each group and found an encouraging level of concordance.

In the event, all Mulhall's interpretations were confirmed. The ordination required at least three dimensions for its proper representation, but the first two canonical variates, accounting for 68% of the total variation, gave a chart which seemed to include all the essential features of the relationships between the groups. As is almost always the case with anthropometric data, which deals with material differing at much less than subspecific level even by the generous standards employed in hominoid taxonomy for the erection of categories, there is little apparent "structure" to the analysis. The first canonical variate contrasted facial height against head length and bigonial breadth. The second canonical variate contrasted head length and facial height against minimum frontal breadth and bizygomatic breadth. The third canonical variate contrasted head length against all the remaining characters and appears to have many of the features of a "size" vector. It singled out the Etche and Abakaliki peoples as being, in this sense, small-headed.

The Form of some Brachiopods from the Devonian

Figure 7.6 shows the canonical variate analysis based on three morphometric measurements for four samples of the brachiopod *Martinia inflata* (Schnur) from the Devonian of Bergisch-Gladbach, Germany (Jux and Strauch, 1966). Large numbers of these brachiopods were measured, 212 from the Wipperfürtherstrasse exposure at Flora, 214 from the Spitze exposure, 8 from Schlade, and 103 from the quarry at Flora. The variables measured were length, height and breadth of the shell.

The first canonical variate, accounting for about 84% of the total variation, contrasts the first character against the remaining pair, whereas the second canonical variate takes up most of the remaining 16% of the variation and contrasts the second character against the first and third.

This variation seems to be of two distinct kinds, that along the first canonical variate, which separates the first, second and fourth localities clearly, and that along the second variate which distinguishes the third, Schlade, sample from all the others. The three grouped samples come from localities fairly close to each other, whereas the small sample from Schlade is not from the same vicinity. A note of caution is needed with the interpretation in view of the small sample size for Schlade in relation to the rest of the material.

A further point of some palaeoecological significance is worth mentioning. The three more or less linearly located points, Wipperfürtherstrasse, Flora quarry (located close to each other) and Spitze, seem to reflect some kind of homogeneous development of the brachiopods at these three localities. We wish to thank Professor Jux for making this material available.

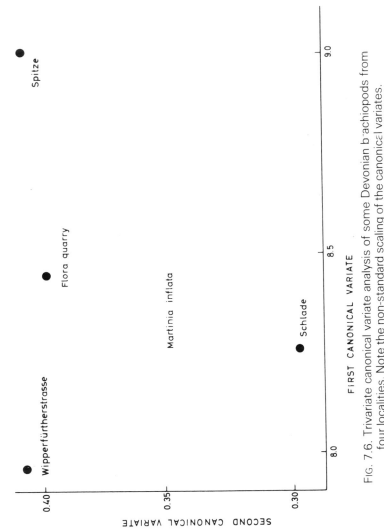

FIG. 7.6. Trivariate canonical variate analysis of some Devonian brachiopods from four localities. Note the non-standard scaling of the canonical variates.

Phenotypic Flexibility of Shape in the Desert Locust

In a very full investigation of the form of the Desert Locust, as assessed by the three characters, head width, length of hind femur, and elytron length, Stower *et al.* (1960) acquired samples of locusts from a variety of East African localities. These localities included some that were particularly hot whilst the locusts were breeding and some that were cooler. In some places the locusts were breeding at high densities, in which case the adults have a characteristic body-shape, colour, and other characters which place them in the phase *"gregaria"*. At other places, the locusts were relatively isolated during their development, leading to adults of the phase *"solitaria"*. In some instances, it was possible to obtain samples from a parental population and from its progeny, which enabled the authors to examine the remarkable phenotypic plasticity of locusts, leading to changes of body-shape when the progeny are reared under conditions different from those in which the parental population had been reared.

By computing the generalized distances between each sample, and constructing an ordination showing the relationships, many interesting facts were established. Basically, there are two fundamentally distinct dimensions of variation, the one dominated by sexual dimorphism, the other dominated by "phase" variation reflecting the density at which the locusts were reared. Both these dimensions of variation, however, can reflect changes of shape associated with factors other than sexual dimorphism and phase variation. The "phase" axis, which is almost orthogonal to the "sex" axis, also carries the effects of temperature differences during the period of development of the young locusts, in such a way that those reared under hot conditions resemble *"solitaria"* locusts, whatever their rearing density. The effects of rearing density and rearing temperature are thus "symbatic" in the sense of Blackith and Roberts (1958).

It seems likely that the use of only three characters rather limits the discriminatory power of the technique. The authors suggest the inclusion of a further character, the length of the pronotum, a suggestion which unfortunately has not been adopted.

Even three characters enabled Stower *et al.* (1960) to determine the nature of populations of the phase *"congregans"* (i.e. parental population isolated, progeny crowded) and those of the phase *"dissocians"* (i.e. parents crowded, progeny isolated) from the fact that *congregans* populations have larger adults and *dissocians* populations have smaller adults, thus shifting the populations along the axis of variation representing sexual dimorphism because this axis also, to a large extent, reflects the size differences. A somewhat similar analysis using canonical variates (Symmons, 1969) shows that the first variate (of three examined) can be used to order locusts according to their "phase" status.

Later, Davies and Jones (unpublished) showed that a third dimension of variation reflected differences between the Desert Locust *(Schistocerca gregaria)* and other species of the same genus such as *S. americana* and another subspecies of *S. gregaria, flaviventris.* In this, Davies and Jones agreed with Albrecht and Blackith (1957) who found that the contrast of form between the genera *Schistocerca* and *Nomadacris* lay at right angles to the "phase" and "sexual dimorphism" axes within each genus. That these axes of variation can be found almost unchanged in different species and genera is of obvious genetic interest. We are most grateful for permission to use this additional information.

The Ecology of Insects Living on Broom Plants

Eight attributes of some mirid bugs living on broom plants were measured by Waloff (1966) for populations in South East England, California, USA and British Columbia, Canada. The first three canonical variates were found to be of interest in the biological sense. Waloff has drawn on her chart the underlying dimensions of sexual dimorphism and specific differences (between the measured two species of the same genus *Orthotylus*) as reasonably inferred from the fact that all the male insects are located by these two canonical variates at the top left-hand of the chart, with the females down in the bottom right-hand portion, whereas all *O. virescens* fall into the top right-hand portion and all *O. concolor* in the bottom left-hand.

The method of canonical variates is not always thought of as a clustering technique, because it operates on measurements made on groups of organisms (groups which are known prior to the analysis). However, in this example it affords an effective method of clustering the groups into biologically meaningful entities (see Chapter 3). The four entities here are determined by sex and species. The axes drawn inside the chart are purely inferential and do not require computation; they are in fact no more than a guide to the reader to tell him what to search for in the interpretation.

Within each of the four main clusters there is a visible structure. In each one of the four, the sample from England has the highest score on the first canonical axis, whereas the samples from Canada have the lowest scores on the same axis, with the remaining groups intermediate. Geographical variation is thus symbatic with the first axis, and so far as geographically determined changes of form are concerned there is much in common between them and the sexual dimorphism.

This interpretation, at least in its main features as outlined here, is quite straightforward and presents few problems. Reference to the original publication will show that there is more to be extracted from the data than has been discussed above, but the paper does show that the introduction of

Orthotylus to British Columbia has changed the shape of this British bug in both species.

As was intimated in Chapter 3, canonical variate analysis may be thought of as a method of ordination and as such would qualify as a clustering method for groups, as in the present example. It can, however, serve as a means of clustering specimens.

The Identification of Sex and Race from Skeletal Measurements

The sexing of skeletal human material has not been without its difficulties: when even the most experienced workers have tested themselves "blind" against material of known sex, the scorings were generally in the range 75–85% correct. According to Giles and Elliot (1963), the introduction of a nine-variable discriminant function only raises this rate to some 82–89%, but it does enable this determination to be made by less experienced workers. These authors suggest that the skulls for which wrong identifications were made, either visually or as a result of the application of the discriminant function, included a fair proportion of material from individuals whose hormonal balance was tipped away from that appropriate to their "chromosomal" sex and whose skeletal conformation reflects this history.

As soon as judgements, built up on the basis of abundant contemporary material, have to be transferred to fossils, or subfossils, difficulties may arise. The application of discriminant functions may not resolve all these difficulties, but it does sometimes correct doubtful subjective judgements. Giles and Elliot quoted an example of a palaeo-Indian skeleton, previously judged subjectively to be female, which seemed to be distinctively male when measured and assessed on the multivariate basis. Moreover, Irish skeletal material from a monastery burial ground (sixth to sixteenth centuries) gave a lower proportion of females than had been assessed visually.

An extension of this approach to the medico-legal topic of race identification is also reported by Giles and Elliott (1962).

Craniometric and other anthropological studies have the honour of being the first field in which the generalized distance was applied (Mahalanobis, 1928) and have since attracted the attention of many workers interested in multivariate methods (Martin, 1936; Pons, 1955; Rao, 1961; Giles and Bleibtrau, 1961; Campbell, 1962; Defrise-Gussenhoven, 1957, to name only a few exponents).

Anthropometry of the Skulls of the Jebel Moya People of the Sudan

The Jebel Moya people of the Sudan were a negroid people who moved into the southern Sudan during the first millenium BC and whose affinities have

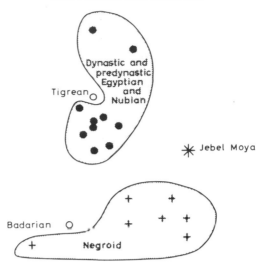

FIG. 7.7. Generalized distance chart showing the affinities of the Jebel Moya people of the Sudan, based on seven cranial characters. Adapted from Mukherjee *et al.* (1955).

long been a source of speculation. One approach was made by Mukherjee *et al.* (1955), who compared as many of the skulls as could be recovered more or less intact (from a large collection with an unfortunate history), with skulls drawn for candidate populations from which, at some historical or prehistorical era, the Jebel Moya people might have arisen. The method of comparison chosen was to compute the generalized distances between all the various groups of skulls, using seven morphometric characters. We have adapted the diagram presented in the publication of Mukherjee *et al.* to show more clearly the fact that the populations chosen for comparison with the Jebel Moya skulls form distinct clusters according to their racial affinities, the two main groups being the dynastic and predynastic Egyptian and Nubian peoples, whose skulls were derived from tombs of at least approximately known data, and the peoples of negroid stock who include a population of Nilotic negroes from Egypt (Fig. 7.7).

The Jebel Moya people stand out as distinct from all the candidate populations chosen for comparison, although presenting some negroid features. It is interesting that many of the skeletal features, and also some of the cultural traits, found in these ancient graves could be found in the local population of the area when the excavations were performed between 1911 and 1914. Two other peoples, the Tigréans and the Badarians, seem difficult to place on the chart. The Tigré-speaking people of Eritrea have some local traditions of connections with the classical peoples of Egypt, which seems to be supported by the morphometric findings of Mukherjee *et al.* who do not mention these traditions. Although much mixed ethnically and culturally nowadays, as an East African coastal population is bound to be, the Tigré seem to form a distinct entity. The Badari population comprises the earliest known collection of skulls of any substantial numbers from Egypt. Obviously they represent predynastic peoples, but little seems to be known of their origins. On the multivariate evidence they seem to be negroid, but on classical archaeological grounds several of the populations which cluster with the Egyptian/Nubian populations have been considered negroid and the criteria perhaps leave something to be desired.

This investigation stands as a warning not to take material for granted: out of the remains of 3137 people excavated in the Sudan, only 15 complete sets of seven measurements could be obtained for the work reported here. Indeed, much of the multivariate effort in this work went into the correction of the earlier anthropologists' attempts to sex the bones.

Discriminant Functions and Time-series of Fossils

The unique contribution of palaeontology to morphometrics is the possibility it offers of studying morphological changes in a species over a period of time. Early examples are those of Reyment and Naidin (1962) who analysed chronologically separated occurrences of the belemnite *Actinocamax verus* (Miller) from the Senonian of the Russian platform and Reyment (1963a) who analysed chronological variation in the Paleocene ostracod species *Trachyleberis teiskotensis* by means of a discriminant function approach. This method of analysis has led to a breakthrough in the quantification of biostratigraphy and it will now be briefly reviewed.

DISCRIMINATION AND BIOSTRATIGRAPHY

Despite its importance in much geological work, the methodology of biostratigraphy has hardly changed in any essential detail over the last 150 years. There is even a tendency for the most conservative element among biostratigraphers to discount any attempts at quantification as being alien to the concepts of their subject.

Reyment (1980a) has demonstrated how biological variation in series in time of fossil species can be used in the same manner as the electric logs of borehole analysis for some stratigraphical studies (so-called biologs). The level-by-level points needed for plotting the biolog may be most expediently obtained from the mean canonical variate scores for the first canonical

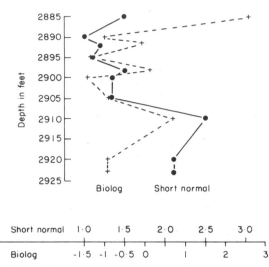

FIG. 7.8. Biolog formed from the growth-reduced principal coordinates of ten samples of *Afrobolivina afra*. The right hand curve is the short normal electrical resistivity log (cf. Lynch, 1964). After Reyment (1978b).

variate computed for a set of morphometric characters measured on a species of microfossils for all samples of the borehole.

The success of biologging is naturally greatly dependent on the correct recognition of the nature of the causes of the observed variability and it is therefore necessary to appraise the role of ecophenotypic variation in relation to genetic (evolutionary) effects (Eldredge and Gould, 1972; Campbell and Reyment, 1978; Lande, 1979; Reyment, 1982a, 1982b). An example of the biostratigraphical application of the methods reviewed in this chapter is illustrated by Fig. 7.8 which shows the biolog computed for the Late Cretaceous (Maastrichtian) foraminiferal species *Afrobolivina afra* plotted alongside the short-normal electrical resistivity curve for the same sampling levels in the Gbekebo borehole of western Nigeria (details of the method are given in Reyment (1980b)). It will be seen that the morphological variation is closely associated with the variation displayed by the electrical log (which here mirrors environmental fluctuations) and is thus ecophenotypic in origin and, consequently of significance for palaeoecological reconstructions, in addition to providing a means of expressing stratigraphical information.

8
Canonical Correlations

In the course of morphometric studies, and also of a wide range of ecological investigations, we may wish to decide whether one set of variables, taken as a whole, varies with another set of variables, and, if the answer is positive, to discover the nature of this joint variation.

In ecological investigations, there may be special interest attached to the changes of the elements of the fauna, such as the collembolan fauna of a peat bog, as the bog is drained, a process which in itself will entrain a large number of other changes such as the depth of the anaerobic layers, of root systems, etc. In strictly morphometric work, there are occasions when we need a linear compound of the measurements of an organism which are, as far as possible, uncorrelated with environmentally induced changes of shape, which in turn will be assessed in terms of another linear compound of measurements. To watch the mutual fluctuations of two linear compounds in this way is to take, in effect, a synoptic view of the processes at work; not every worker approves of the synoptic approach to biological problems, probably because in the past this approach has been associated with very generalized statements which, in practice, have proved essentially non-operational in that the hypotheses generated by synoptic approaches have been untestable in any sufficiently critical quantitative terms. Canonical variates, and particularly canonical correlations, do afford a means of testing hypotheses; if we set up the hypothesis that, say, the collembola of a bog vary together with the nematodes of the bog, as the latter is drained, we can estimate these two groups of animals in the bog at different stages of draining, and determine whether there is or is not an association between the two sets of estimates. The price paid for this general approach is a certain loss of detail, but we can agree that the synoptic view is inadequate for special purposes without denying that such a view is in other respects rewarding.

77

Potential uses in geology have been suggested by Lee (1969) and in meteorology by Glahn (1968).

Cooley and Lohnes (1971) have given the subject of canonical correlation close attention and recommended the use of certain developments of the original theory, couched in terms of redundancy analysis.

FRAMEWORK OF CANONICAL CORRELATION

The method of canonical correlation, which is in effect a generalization of the concept of multiple regression, was developed by Hotelling (1935, 1936) for analysing the interrelationships between two sets of measurements. Canonical correlation analysis may be described as a way of finding the maximum correlation between linear functions of the two sets of variables. The two sets of measurements are transformed simultaneously by forming a linear combination of each set of measurements so that the correlation between the transformed variables is a maximum. The coefficients defining the linear combinations give the components of the canonical vectors, while the resulting correlation is the canonical correlation coefficient. Successive linear combinations are chosen so that the resulting transformed variables again have greatest correlation between the two sets, subject to the new variables being uncorrelated with previously obtained transformed variables within each set.

It is convenient to discuss the technique of canonical correlation in terms of a morphometric study in ecology. One set of variables (say, p in number) may be thought of as being composed of measurements on some species. The other set of variables (say, q in number) is a set of ecological measurements. Thus each observation vector $(\mathbf{x}_1^T, \mathbf{x}_2^T)$ is made up of p morphometrical elements and q ecological elements.

The pair of canonical vectors, $\mathbf{a}$, $\mathbf{b}$ is found by forming the linear combinations $\mathbf{a}^T\mathbf{x}_1$, $\mathbf{b}^T\mathbf{x}_2$ and maximizing the correlation $r_c(\mathbf{a}^T\mathbf{x}_1, \mathbf{b}^T\mathbf{x}_2)$, subject to the constraints that the resulting canonical variables have unit standard deviations.

With $\mathbf{S}_{11}$ and $\mathbf{S}_{22}$ denoting the sample dispersion matrices corresponding to the p variables and the q variables respectively, and $\mathbf{S}_{12}$ denoting the matrix of covariances between the two sets, the required canonical vectors and correlation coefficient satisfy the matrix equations

$$(\mathbf{S}_{12}\mathbf{S}_{22}^{-1}\mathbf{S}_{12}^T - r_c^2\mathbf{S}_{11})\mathbf{a} = \mathbf{0}$$

and

$$(\mathbf{S}_{12}^T\mathbf{S}_{11}^{-1}\mathbf{S}_{12} - r_c^2\mathbf{S}_{22})\mathbf{b} = \mathbf{0}$$

For the case in which $q = 1$ and $p > 1$, the canonical correlation structure reduces to one of multiple regression.

The correlation matrix could be employed to find the (same) canonical correlations as have just been obtained using the dispersion matrix. The evaluation of the linear combination for producing transformed observations should then be made in terms of the standardized and mean-centred variables; the resulting vectors are those that would be obtained by multiplying the components of the canonical vectors for the original variables by the corresponding standard deviations.

CASE HISTORIES INVOLVING THE USE OF CANONICAL CORRELATIONS

Influence of the Superincumbent Water on the Pore Water of a Sediment

The chemical composition of interstitial water in the pores of a sediment, at least in the top layers of sediment, will be influenced by a variety of factors including the amount of exchange with the superincumbent water and its composition. Reyment (1968) measured the pH, Eh, free oxygen, carbonate and water contents of the sediment, and the pH, Eh (reduction-oxidation potential), and free oxygen of the superincumbent water.

The linear compound for the sediment which was most closely correlated with the equivalent compound for the superincumbent water was 0.12; -0.51; 0.53; 0.35; -0.56 and that for the superincumbent water was 0.12; -0.65; 0.75. Evidently pH plays only a minor part in this canonical correlation, which borders on significance for its upper value (0.50).

Nepionic Variation in Upper Cretaceous Orbitoides

During the study of growth in the Cretaceous larger foraminiferal genus *Orbitoides* the question arose as to whether the dimensions of the principal chambers, the deuteroconch and the protoconch, depended on the number of chambers in the megalospheric test. The material, mainly from the French Upper Cretaceous, was measured by Professor J. E. van Hinte of the Nieuwe Vrije Universiteit, Amsterdam.

Two sets of variables were measured. The first comprises: outer maximum diameter of the initial four chambers, measured across the principal auxiliary chambers; inner maximum diameter, measured across the principal auxiliary chambers; outer maximum diameter, measured across the protoconch and deuteroconch; inner maximum diameter, measured across the protoconch and deuteroconch.

The second set of variables comprises: budding number of youngest pseudonepionic closing chambers; budding number of youngest periembryonic closing chambers; number of periembryonic chambers, excluding the epiauxiliary and closing chambers.

The calculations were made on a pooled sample of 314 specimens. A highly significant first canonical correlation coefficient of 0.953 was found, corresponding to the two vectors (0.48, 0.29, 0.76, 0.31) and (0.92, 0.02, 0.38). The linear dimensions of the early part of the test, especially that of the outer maximum diameter across the protoconch and deuteroconch, depend fairly strongly on the number of chambers, especially the number of youngest pseudonepionic closing chambers.

The pooled samples do not tell the whole story, however. When nine of the samples, from successively younger horizons, were analysed separately, the first three were found to be dominated by the inner diameter across the protoconch and deuteroconch, but in the last six samples, the outer diameter of the initial four chambers predominates, either alone, or with characters other than the inner diameter across the protoconch and deuteroconch.

Applications to Ammonites and Ostracods

Reyment (1975a) applied canonical correlation analysis to an evolutionary problem involving an elusive trans-generic change in which there is a directional shift in ribbing frequencies in passing from the Late Turonian (Cretaceous) ammonite genus *Subprionocyclus* to *Reesidites*. Reyment (1972, 1976) has employed a variant of canonical correlation analysis (cf. Cooley and Lohnes, 1971) for palaeoecological and ecological studies in the marine environment. The same method of analysis was used for studying ecological relationships in hemicytherinid and trachyleberinid ostracods from the Niger Delta (Reyment, 1975b). The species involved dwell in the interstitial environment. The canonical correlation analysis showed that the distribution of species is correlated with depth and content of phosphates in the interstitial environment in an inverse relationship between the latter two variables.

CANONICAL CORRELATIONS AND THE ENVIRONMENT

The last examples indicated how canonical correlation can be a useful way of computing the strength of the interrelationship between the occurrence of species and environmental factors. The method also offers a way of analysing the interaction between variation in morphological characters and environmental factors. This useful approach to ecological work has not been made use of to any great extent, although Gittins (1979) has produced an excellent monographic type of work in which canonical correlations are applied to ecological problems. One of the reasons for the apparent lack of interest is that it is often difficult to interpret the results. The output from canonical correlation studies is less straightforward than for most other multivariate methods, as it is not always clear what the various associations between linear combinations of "predictors" and "responses" really signify.

Most morphometrical examples known to us come from the sphere of palaeoecology. For example, Ivert (1980) studied variation over time in the Maastrichtian (Cretaceous) foraminiferal species *Gabonita elongata* in relation to the geochemical composition of host sediments. The morphometrical measurements on the test of the foraminifer were weighted against the concentrations of 13 chemical elements determined at each stratigraphical level in a borehole. The analysis by canonical correlations indicated that *G. elongata* developed larger tests in an arenaceous environment, whereas carbonaceous, argillaceous sediments seem to have caused the development of significantly smaller tests. Especially the microspheric generation shows a greater reduction of growth in the last chamber in carbon-rich, argillaceous sediments.

9
R- and Q-mode Methods of Analysis: Principal Components and Principal Coordinates

The methods of analysis treated in this chapter are of the kind commonly known as R-mode and Q-mode analysis, depending on whether relationships between variables or relationships between the specimens of a sample are of interest. Jöreskog *et al.* (1976, Chapter 4) provide a detailed account of R-mode methods.

PRINCIPAL COMPONENT ANALYSIS

Principal component analysis is suitable for the analysis of the structure of multivariate observations, and in particular for investigating the dependence structure occurring in a suite of observations when no *a priori* patterns of interrelationship can be suggested or are suspected.

Principal component analysis is used as a technique to provide a low-dimensional representation of the data. Linear combinations of the original variables are formed so that most of the variation between the individuals is contained in the first few linear combinations. The resulting linear combinations are usually referred to as the principal component scores.

The first linear combination is chosen to maximize the variance of the resulting scores. The coefficients defining the linear combination give the components of the latent vector while the variance of the scores gives the latent root. Successive linear combinations are chosen with the property that they maximize the variation of the resulting scores, subject to the condition that they are uncorrelated with previous linear combinations.

Geometrically, a principal component analysis can be considered as a rotation of the axes of the original coordinate system to new orthogonal axes, called principal axes, with the new axes coinciding with directions of maximum variation of the original observations. The components of the latent vectors are given by the cosines of the angles between the corresponding principal axes and the axes of the original coordinate system, while the variances of the projected points give the latent roots.

Algebraically, the first principal component y_1 of the vector of observations, $\mathbf{x}$, is defined as that p-variate linear combination

$$y_1 = a_{11}x_1 + a_{12}x_2 + \ldots + a_{1p}x_p = \mathbf{a}_1^T\mathbf{x}$$

such that the variance of the scores

$$V(y_1) = \mathbf{a}_1^T\mathbf{S}\mathbf{a}_1 = e_1$$

is a maximum, where $\mathbf{S}$ is the sample dispersion matrix. The solution is given by the matrix equation

$$(\mathbf{S} - e_1\mathbf{I})\mathbf{a}_1 = \mathbf{0} \tag{9.1}$$

where $\mathbf{I}$ denotes the identity matrix.

The vector $\mathbf{a}_1$ is the first latent vector of $\mathbf{S}$, and e_1 its first latent root. The foregoing presentation can be easily extended to any of the subsequent latent roots and vectors of $\mathbf{S}$, with the additional requirement of orthogonality of the latent vectors (or lack of correlation of the principal component scores) being added.

If $\mathbf{A} = (\mathbf{a}_1, \ldots, \mathbf{a}_p)$ denotes the matrix of latent vectors and $\mathbf{E} = \text{diag}(e_1, \ldots, e_p)$ denotes the diagonal matrix of latent roots, the latent vector equation in (9.1) generalizes to

$$\mathbf{SA} = \mathbf{AE}$$

from which follows that

$$\mathbf{S} = \mathbf{AEAT}$$

with

$$\mathbf{A}\mathbf{A}^T = \mathbf{A}^T\mathbf{A} = \mathbf{I}$$

The sign (in relation to those of the remaining components) and magnitude of a vectorial element indicate the direction and importance of the contribution of a particular variable to a particular component.

In morphometric applications of principal component analysis, it is common practice to take the logarithms of the observations and to perform the calculations on these transformed vectors. Gould (1967) has pointed out that, providing growth effects are the major source of variation in the material, the first latent vector of the dispersion matrix of logarithmically transformed observations of morphometric variables is intuitively acceptable as a "growth vector". In this case, the ellipsoidal cloud of points (if the observations are multivariate normally distributed, the cloud will be ellipsoidal) is spread around a strongly elongated major axis (namely the direction of the first principal component).

Scale Dependence

Changing the unit of measurement in a variable corresponds to multiplying all the observed values for the variable by a scale factor. The difficulty attaching to this for principal component analysis is that in general the latent roots and vectors of the covariance matrix for the scaled variables cannot be directly related to those of the covariance matrix before scaling. A principal component analysis of the variables on the original scale is therefore not recommended when it cannot be assumed that the variances are the same for all variables (they will be roughly the same if all variables are measured in the same units and the measurements are of the same order of magnitude). A common way out of the problem is to use the correlation matrix instead of the covariance matrix. While this is satisfactory for exploratory work, it does have the slight drawback that the sampling theory available for the covariance matrix can only in a few cases be transferred to the correlation matrix. Hence for more theoretical work, in general, developments such as tests of significance, confidence intervals of latent roots and the like are not available for the correlation matrix.

Geometrical Interpretation of Latent Roots and Latent Vectors

Consider a swarm of data points in two dimensions. The data points can be enclosed by equal-density contours; it is well known that these contours form ellipses for bivariate, normally distributed variables. If the variables are uncorrelated, the confidence contours will be circular; if the variables are perfectly correlated, the ellipse collapses into a straight line. For more than two variables, the data points will lie within a p-dimensional hyperellipsoid.

The major axis of the hyperellipsoid corresponds to the direction of greatest variation of the projected values of the original data points onto the axis. Subsequent axes satisfy the same criterion, subject to their being orthogonal with previous axes. By maximizing the variation of the projected values, the technique in effect minimizes the perpendicular deviations of the original data points from the major axis, and hence may be considered as defining a line of best fit. The first two principal axes may be considered as defining a plane of best fit, and so on.

As noted above, the angles between the original axes and a new principal axis are related to the components of the corresponding latent vector, while the variation of the projected points is related to the latent root.

Factor analysis versus principal component analysis

The fundamental differences between principal component and factor analysis depend on the ways in which factors are defined and on the assumptions concerning the nature of the residuals. In principal component analysis, factors are determined so as to account for maximum variance of all the observed variables, whereas in the factor analysis model, the factors account maximally for the intercorrelations of variables. Thus, principal component analysis can be said to be variance-oriented, whereas factor analysis is correlation-oriented.

The residual terms are considered to be small in principal component analysis; this need not be so in the factor-analytical model. In both methods, the residuals are assumed to be uncorrelated with the factors. However, in principal component analysis, there is no assumption concerning correlations among residuals, whereas in the factor analysis model, it is assumed that the residuals are uncorrelated among themselves.

The smallest principal component

The latent vector attached to the smallest latent root defines a linear combination of variables which is relatively invariant in the sample. This relatively invariant vector can often provide valuable information on the occurrence of growth invariant relationships in the data (Reyment, 1980a), providing the latent root is close to zero.

Basic structure of a matrix and the Eckart-Young theorem

Fundamental to R- and Q-mode analysis is the Eckart-Young (1936) theorem which describes the basic structure of an arbitrary rectangular matrix via its singular value decomposition. This gives the original data matrix directly in terms of the latent vectors (square roots of), the latent roots and (scaled) principal component scores. A summary of the concepts involved is given by Jöreskog et al. (1976, pp.47–50).

If **X** is an $n \times p$ data matrix, the analysis of the $p \times p$ minor product matrix, $\mathbf{X}^\mathsf{T}\mathbf{X}$, is referred to as R-mode analysis and that of the $n \times n$ major product matrix, $\mathbf{X}\,\mathbf{X}^\mathsf{T}$, as Q-mode analysis. There are certain basic structural relationships between these product moment matrices, namely that both matrices have the same non-zero latent roots, while the latent vectors of one are related to the scores of the other, provided that the latent vectors are scaled appropriately. The practical consequences of these two properties are that for the purposes of computation, one may start with the smaller of the two matrices, compute its latent roots and vectors and then find the latent vectors of the larger matrix by a simple calculation (see, for example, the worked example on p.49 in Jöreskog *et al.*, 1976). However, the singular value decomposition provides the latent vectors, roots and scores directly.

The method of correspondence analysis is an application of the Eckart-Young theorem, with a more complicated scaling of axes (Fisher, 1940; Benzécri, 1973; Gower, 1966; Greenacre, 1978, 1981); the R- and Q-mode relationships can be represented on the same diagram. A useful reference on the subject is the paper by Hill (1974)).

A fuller, practically-oriented account of correspondence analysis is given by Lefèbvre (1980). A useful consequence of the dual nature of the plots arising from this form of analysis is the juxtaposition of outliers from the data set together with those characters most responsible for the eccentric position of the point in question; this renders identification of the features of the structure of the plot easier.

Closely connected with the concepts of correspondence analysis and principal coordinates is that of multidimensional scaling; for information on this topic we refer the reader to a comprehensive collection of standard papers (Davies and Coxon, 1982).

An objective of many biological investigations is to classify a sample of objects on the basis of several compositional properties. In order to achieve this goal, it is necessary to investigate the interrelationships between the objects of the sample, this being the design of what is known as Q-mode analysis. In R-mode analysis we are interested in portraying the relationships between the variables. One is then the reverse of the other, as it were.

An important detail of Q-mode analysis lies with the definition of inter-object similarity, and with the subsequent method of ordination. Measures of similarity are discussed in Sneath and Sokal (1973, pp.114–168). Methods of Q-mode ordination include Imbrie's Q-mode factor analysis which defines similarity with respect to proportions of the constituents, Gower's method of principal coordinates, which can use a variety of measures of distances and similarity indices, Benzécri's method of correspondence analysis and the biplot of Gabriel (1971).

CASE HISTORIES OF THE USE OF PRINCIPAL COMPONENTS

The First Principal Component as a Measure of Size

The interpretation of the first principal component as a size vector is an old concept, as old as the history of morphometric analysis. In fact, the size-interpretation of the "first factor" suggested by Teissier (1938) for *Maia squinada* is a principal components approach. Over the last decade, however, that which seemed to be one of the cornerstones of multivariate morphometric analysis has been seriously challenged, particularly with respect to analyses involving allometry. This recent work is of sufficient interest to warrant a chapter on its own (Chapter 11).

Two forms of thrips, so alike that serious doubts about the independent identity of the forms had been expressed, were studied by Ward (1968), who measured three characters: the lengths of the seta at the pronotal posterior angle of the tenth tergite, and of seta No. 3 on the ninth tergite. A principle component analysis of the dispersion matrix gave a first principal component which succeeded in ordering the thrips, apparently by size rather than shape, despite the unusual choice of characters. The distribution of scores for individual thrips along this first principal component was clearly bimodal. Misidentification by the vector (1, 1, 2) was less than by any other method tried. We note, however, that the possible influence of scale-effects was not considered in this study.

That the first principal component often has the character of a size component has also been demonstrated by Jolicoeur and Mosimann (1960), who provided a shape analysis of the painted turtle. These authors emphasized that all coefficients must be of the same sign, whereas those of the other components should generally be of mixed signs for this interpretation to be logical. Rao (1964) attempted a mathematical justification for this reasoning.

Reyment and Sandberg (1963) measured four characters of fossil ammonites on a logarithmic scale (shell diameter; umbilical diameter; height of last whorl, and breadth of last whorl). One Cretaceous and two Triassic species were measured. In each instance, the first principal component of the dispersion matrix accounted for over 90% of the total variation; all four components of the corresponding eigenvector were positive. From the taxonomist's point of view, the interest in this case centres on the remaining components which can be loosely identified as "shape" components. The vector corresponding to the second component was found to be consistently of the general form (0.2, −0.7, 0.3, 0.4), though with minor variations from this representation, implying that the umbilical diameter decreases relative to the height of the last whorl in ammonites standardized as to size.

Reyment (1966b) measured six characters on the tests of Paleocene foraminifers and seven characters on a recent species of foraminifers (Reyment, 1969c) and again found the first principal component of the dispersion matrix to be defined by positive coefficients. The vectors defining the subsequent components are "shape vectors" which are largely influenced by the proportions of gamonts and schizonts (sexual and asexual forms) in the population.

Principal component analysis is also useful in the study of the Foraminifera for morphometrical problems other than polymorphism. For example, Reyment (1969b) was able to demonstrate for the agglutinating living species, *Textilina mexicana*, that variations in the dimensions of the test are not correlated with the diameter of the proloculus (see also Reyment (1982e) in connexion with the study of evolution in *Afrobolivina afra*).

Fries and Matérn (1966) have also used the first principal component as a measure of size in forest trees; Amtmann (1965) extracted a size (first) principal component and shape (second) one in an investigation of the skulls of squirrels. Many other illustrations of this general representation occur in the literature. In some cases the first principal component takes up almost all the variation; Murdie (1969) measured eight characters on 80 aphids and found that the first principal component accounted for no less than 94.24% of all the variation in form, a reflection of the high correlations between the characters. We may also note that Jeffers (1967b) has published a series of case histories using principal component analysis in which this point is taken up. There is no *a priori* reason for the hypothesis that the first principal component has a "size" property; the attitude is largely based on its widespread occurrence in morphometric studies.

Variation in Scale Insects

The scale insects have proved difficult to identify because shape differences form an important part of the distinctions between species, yet the use of ratios and similar devices ("antennal formulae") has become discredited because the ratios changed in magnitude when larger individuals of a given species were compared with smaller ones.

Eighteen characters were measured by Blair *et al.* (1964) on 27 female individuals, all of which were taken from a single plant (because the shape of a scale insect can vary according to the plant from which it is taken). Of the 152 correlation coefficients between these 18 characters, no fewer than 35 were negative and only 52 were of such a magnitude that, when tested for significance as individual correlations, i.e. neglecting the fact that the correlations are not independent of one another, they were "significantly greater than zero". Few of the appendages grow isometrically with the rest of the body, none has a correlation of more than 0.47 with the length, nor more than 0.62 with the breadth. In general, the growth of the parts of the legs were most closely interlinked, that of the segments of the antennae the least so, to the extent that the first, fourth and fifth antennal segments each had eight negative correlations with other characters, which accounts for more than two-thirds of the negative correlations.

The first latent vector was easily identifiable as representing size variation, since it included only three negative coefficients and 15 positive ones. The second vector showed that the legs, in this pattern of growth, elongate disproportionately as compared with the rest of the body, since the parts of the legs have positive coefficients whereas the remainder mainly have negative ones. The third vector similarly shows that there exists a pattern of growth in which the distal antennal segments elongate disproportionately with the rest of the body. The fourth and fifth vectors seem to involve patterns of growth in which the development of the first antennal segment plays an important part, and the sixth vector represents "attenuation", that is, the elongation and (relative) narrowing of the body form, a type of variation which is known to occur when scale insects are transferred from one host-plant species to another. This last vector is thus a "signpost" to a source of variation which can be exploited by the insect in its adaptation to a new environment.

Cheetham (1968) published a thoughtful study of the taxonomic relationships within the Neogene cheilostomate bryozoan genus *Metrarabdotos* by a battery of quantitative methods, including principal components and clustering techniques. The resulting five phenetic groups (based on 23 characters) were related to a time-stratigraphic framework. Taxonomic interpretations were based on inferred phylogenetic relationships within and among groups. He concluded that the morphological overlap among groups has derived from convergent and parallel trends in size, positions, orientation and differentiation of avicularia and in denticulation of the secondary orifice in the American and Eurafrican stocks, considered to have been isolated through most of their history. The principal component analysis of 888 zooecia was enlightening and the first three components could be related to zooecial size, zooecial shape, and avicularian length relative to oral dimensions, respectively. The phenetic comparisons were made using a numerical code for both qualitative and quantitative characters, and analysing these by the methods of Sokal and Sneath (1963). Slightly more than half of the zooecial characters could be adequately expressed in two-state code, while the others needed codes running to as many as five states.

The indications yielded by the principal component study were used to weed out redundant variables from the numerical taxonomic treatment; for example, the means of the first three principal components were used in the stead of the original variables.

The Forms of the Common Salamander: Effects of Coding for Sex

Eighteen characters were measured on 527 salamanders of either sex. The correlation matrix of these characters was subjected to a principal component analysis on two occasions; on the first occasion the sex of the salamanders was coded +1 for males and −1 for females, making in effect a 19th character, whereas on the second occasion the coding for sex was dropped.

Inspection of the correlation matrix showed that most of the correlations between the characters were within the range 0.7 to 1.0 except for character 18 (the breadth of the tail) whose correlations with the remaining characters lay in the range 0.39–0.49, and character 17 (height of tail) in the range 0.60–0.78. We should, therefore, expect the first principal component to be essentially a "size" vector; in fact, the weights of the various characters are remarkably uniform, (0.26; 0.25; 0.24; 0.24; 0.26; 0.25; 0.23; 0.24; 0.23; 0.24; 0.21; 0.25; 0.25; 0.26; 0.23; 0.24; 0.20; 0.13). These weights are not sensibly affected by the inclusion of a coding for sex.

However, the sex coding, when included in the analysis, dominates the second principal component, to such an extent that the corresponding vector has a single large element, with the remaining elements close to zero; the weights for the other characters lie between −0.18 and +0.15. When the sex coding is removed from the analysis, the vector defining the second principal component substantially maintains its quality but it is now the 18th character (breadth of tail) which predominates with a weight of +0.91. The rest of the characters again have small weights. Since individuals with a code of ·−1 generally have broader tails than have individuals with a code of +1, it seems likely that tail-breadth is taking over the function of a sexual discriminator along the second component. The sex code itself seems to serve little useful purpose, and is virtually redundant. We may note that the first character, total length of the animals, is also redundant, since it is the sum of the second and third measurements. The nature of the second vector is to be expected, since characters which have low correlations with the remaining ones will dominate the first few vectors after the first.

While a plot of the second and third principal components shows an almost complete separation between the males and females, this separation is due to the inclusion of the coding for sex, and thus simply reflects the information fed into the analysis. If the coding is omitted, there is little sign of sexual dimorphism in the various plots of pairs of the first three principal components.

Another interesting feature of the salamander data worth a quick mention is the fact that the plots of the principal component scores for both sexes give neatly elliptical data, a reflection of the fact that the observations show good agreement with a multivariate normal distribution.

We wish to express our thanks to Dr. Josef Eiselt of Naturhistorisches Museum, Vienna, for making this suite of observations available to us.

The Growth of Fruit-trees

Two orchards of fruit trees were measured by Pearce and Holland (1960) in respect of four characters; the weight of the mature tree above ground; the basal trunk girth of the mature tree; the total shoot growth in the first four years of age; and the basal trunk girth at four years of age. The logarithms of these quantities were used in the analyses. In each case, the latent vectors corresponding to the two largest roots of the correlation matrix were:

$$a_{11} = (0.489, 0.521, 0.507, 0.482)^T; \ a_{12} = (0.485, 0.551, 0.497, 0.463)$$

and

$$a_{21} = (0.564, 0.416, -0.439, -0.562)^T; a_{22} = (0.550, 0.409, -0.481, -0.547)$$

Pearce and Holland made the wise comment that one way in which one can try to make sense out of the principal components is to see if they are at least consistently defined from one analysis to another. This consistency test is satisfied by the first two components; the corresponding vectors essentially reflect "vigour" (size variation) and make an angle of only 2° 9′ with one another. The same is true for the vectors corresponding to the second principal components; they make an angle of 2° 42′ with each other. The second principal components are reflections of the "establishment" factor in tree-growth, where a tree fails to live up to its early promise, or else exceeds it.

Pearce and Holland (1961) and Holland (1968a) followed this analysis of tree growth with a further principal component analysis of the five chemical measures made on the leaves of the trees; three of the components seemed to be of some consequence, and were reasonably consistent from year to year. Holland (1968b) has also analysed chemical data derived from groundnut and sugarcane plants by means of principal components.

The Use of Principal Component Analysis for Clustering Data

Apart from the explicit use of clustering techniques, principal components can often serve as a clustering technique of great generality. Temple (1968) discovered the presence of two forms in the Silurian brachiopod *Toxorthis* by this means. It is a moot point whether these forms should be interpreted as morphs within a species or species of the same genus living in the same habitat, or even, conceivably, as an example of sexual dimorphism.

Temple took the view that the first principal component essentially reflected general growth, and plotted his material on charts constructed from the second and third principal components derived from a dispersion matrix (in terms of the relations between five measurements of the pedicle valve and seven of the brachial valve). The pedicle and brachial valve were analysed separately. In the case of the pedicle valve the analysis split the data into two distinct groups, corresponding to the two forms as judged visually, but there was a very slight overlap. As Fig. 9.1 shows, there is complete separation between the two groups based almost entirely on the discrimination afforded by the second principal component, which is probably close to the orientation of the optimum discriminant function. A further example of the use of the second principal component to distinguish two putative morphs within the species *Pionodema retusa* is also reported by Temple.

Because the initial visual discrimination of the two morphs was difficult and subjective, it is not surprising that the occasional individual appears to have been wrongly placed in the initial assessment, upon which the principal component analysis in no way depends.

Sociological Components Differentiating British Towns

In an exceptionally well interpreted study of the sociological differences between 157 towns in Britain, which we offer here as an example of generalized or "abstract" morphometrics, Moser and Scott (1961) assessed 57 different characters of the towns. They were limited, naturally, to characters for which adequately comparable statistics had been published, and chose seven characters from the field of population size and structure, eight from that of population change, 15 reflecting the properties of the households and housing, ten economic characters, four social class indicators, five characters concerned with voting behaviour, six concerned with health statistics, and two with educational statistics.

These 57 characters were then submitted to a principal component analysis based on the correlation matrix. There are several features of this analysis which render it worthy of special attention. First, there is no component representing the size of the towns, and indeed only one-fifth of the variance of "size" as a character is taken up by the first four components extracted. This finding should serve as a warning not to treat the first principal components of analyses as "size" components, unless the structure of the experiment clearly indicates such treatment. Secondly, the first four components only accounted for some 60% of the total variation, an unusually low figure, the first component only taking up 30% of the total amount. Thirdly, an interesting attempt at reification was made by splitting the towns into northern and southern locations, and repeating the analysis for each

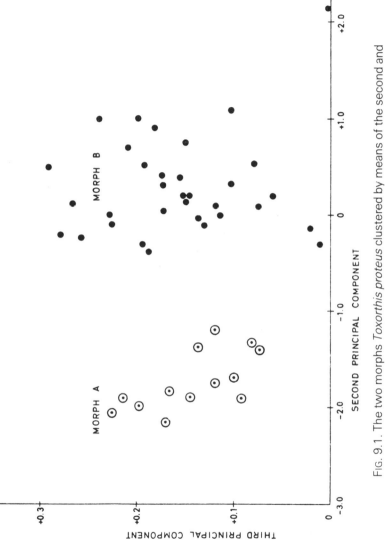

FIG. 9.1. The two morphs *Toxorthis proteus* clustered by means of the second and third principal components of the matrix of the seven morphometric characters of the brachial valve. Redrawn from Temple (1968).

section. Similar results were obtained, which encourages confidence in the general veracity of the analysis. Fourthly, the analysis was repeated using the logarithms of the original data, and also with the rank order of the data. Again, closely comparable results emerged from these three analyses, giving one a feeling of confidence that the interpretation is unlikely to be seriously perturbed by changes of scale or minor differences of method.

When four components were used to classify the 157 towns they fell into clusters with recognizable qualities, such as seaside resorts, spas, commercial centres, railway centres, ports, textile centres, residential suburbs of various kinds, and the depressed towns of the north-eastern seaboard and the Welsh mining towns.

A Sociological Classification of the Regions of Italy

Casetti (1964) used principal component analysis to score various regions of Italy on a sociological basis, after an attempt to eliminate "size differences" by dividing through by the working populations in each area. Out of a total of 25 components, derived from the 25 attributes of the region, the first four were selected for interpretation. These components take up 79% of the total variation in question. They were interpreted as reflecting (i) standards of living; (ii) a component of which one extreme was thought to represent districts with a high tourist attraction and mountains, the other pole representing areas of industrial agriculture; (iii) a component whose high positive scores imply a predominance of intensive peasant agriculture, the low scores implying that of industrial agriculture; and (iv) a component differentiating between regions of intense economic activity and those in a depressed condition. Casetti also used this technique to split up the North American region into climatic areas.

Components of Literary Styles in Latin Elegiac Verse

A great deal of mediaeval and more recent literary work consisted of attempts to write Latin elegiac verse in a style approximating as closely as possible to the verse of the Augustan age in Rome. This style, however, consisted of at least three components which were revealed by a principal component analysis (Blackith, 1963) of four attributes of the verse; the elision frequency (syllables written but not pronounced for reasons of euphony or the demands of the metres); the mean syllable number per word; the variance of the distribution of the syllable number; and the entropy of mixing of words of different syllable number. In this investigation the entity whose "structure" is being examined is style: we may look on it as "abstract morphometrics".

Since poetic composition is at least partly under conscious control, some attributes of the elegiac verse such as elision frequency were closely controlled by the authors, following the dictates of fashion since antiquity, and were almost uncorrelated with the other attributes. However, the third latent vector, in which the mean syllable number plays a predominant part, enables the analyst to determine the date at which a poem was written, in a way which is apparently independent of any attempts by the poet to write pastiche or even to forge texts. Anxiety on the part of the poet to write in an "acquired" style will influence the first two components of variation, but not the third, which appears to be hardly at all under conscious control.

Just how far this third component is influenced by subconscious stress is shown by comparing Ovid's elegiac verse written before and after his banishment by the Emperor Augustus to the Black Sea coast. There is a highly significant shift along this third latent vector, a shift which occurs suddenly at the moment of banishment, and not slowly, as one might expect if isolation and ageing were to be held responsible for the change.

The distinguished 20th century Latin elegiast, Henrici Paoli, has written verse which has scores along the first two axes which compare well with those of classical authors. Paoli's scores along the third axis are, however, dramatically different from any classical author, although they differ in a direction presaged by the shift along this axis of "Silver Age" poets as compared with those of the "Golden Age".

THE METHOD OF PRINCIPAL COORDINATE ANALYSIS

Gower (1966) introduced the term principal coordinate analysis for a Q-mode principal component-type analysis of a matrix of associations or similarities. The same type of technique has been used by psychometricians for some time (cf. Torgerson, 1958, Chapter 11), though without the refinements introduced by Gower. One of the strengths of principal coordinates is that it permits one to mix quantitative, qualitative and dichotomous data in the one connexion, using a suitable similarity measure such as that suggested by Gower (1967), to provide an ordination of the individuals. It is a useful approach for ascertaining whether a multivariate sample is homogeneous in structure.

Principal coordinate analysis is a technique for providing a geometrical representation of the distance or association between individuals. The procedure involves the calculation of the latent roots and vectors of a matrix representing the association between the individuals, with the matrix adjusted so that all rows and columns sum to zero. The latent vectors are scaled so that the sum of squares of the components of the vector is equal to the corresponding latent root. The scaled latent vectors then give a set of coordinates providing an exact representation of the distances between the individuals, in one less dimension than the number of individuals.

When much of the variation is contained in the first two or three dimensions (as indicated by the relative magnitudes of the latent roots), the coordinates given by the corresponding scaled vectors may be plotted, with the distances evident in the graph preserving as nearly as possible the overall distances between individuals.

A heuristic description of the technique is as a decomposition of an $n \times n$ matrix of associations between n individuals (based on a number of attributes or variables) into its latent roots and vectors, followed by a principal component representation in which the latent vectors are considered as n input variables. By a suitable centring of the association matrix, the calculations can be collapsed into a single latent vector analysis.

Moreover, the distance between two individuals can be written in terms of the components of the latent vectors of the association matrix, and this reduces to a representation in terms of the entries in the association matrix corresponding to the two individuals. This representation of the distance in terms of the entries in the association matrix leads directly to an appropriate form for that matrix given the distances between the individuals.

Consider the $n \times n$ *symmetric matrix* $\mathbf{M}$, and let the latent vectors be $\mathbf{a}_1, \ldots,$ $\mathbf{a_n}$ with corresponding latent roots $e_1, \ldots, e_n$. Denote the latent vectors scaled so that the sums of squares of their elements equal the corresponding latent roots as $\mathbf{b}_1, \ldots, \mathbf{b_n}$. Hence the matrix $\mathbf{M}$ can be written as

$$\mathbf{M} = \sum_{i=1}^{n} e_i \mathbf{a}_i \mathbf{a}_i^{\mathsf{T}} = \sum_{i=1}^{n} \mathbf{b}_i \mathbf{b}_i^{\mathsf{T}} \tag{9.2}$$

with $\mathbf{b}_i^{\mathsf{T}} \mathbf{b}_i = e_i$. It follows that

$$m_{ii} = \sum_{r=1}^{n} b_{ir}^2 \text{ and } m_{ij} = \sum_{r=1}^{n} b_{ir} b_{jr}$$

The elements of the ith row of the matrix $\mathbf{B} = (\mathbf{b}_1, \ldots, \mathbf{b}_n)$ can be taken as the coordinates of a point, P_I, in n-space. The distance, Δ_{ij}, between P_i and P_j is then given by

$$\Delta_{ij}^2 = \sum_{r=1}^{n} (b_{ir}^2 + b_{jr}^2 - 2b_{ir} b_{jr})$$

$$= m_{ii} + m_{jj} - 2m_{ij} \tag{9.3}$$

This last relationship provides the means of determining the association matrix $\mathbf{M}$ given the distances d_{ij} between the individuals (see below).

The question still remains as to whether the points can be represented adequately in a small number of dimensions. A principal component analysis provides an obvious way to do this. Hence the next stage of the analysis would be to treat the column vectors $\mathbf{b}_1, , , , ,\mathbf{b}_n$ as n variables and to calculate their dispersion matrix. The column vectors as defined have non-zero means, and these enter the calculations for the dispersion matrix.

If the vectors $\mathbf{b}_i$ were such that the means of the components of each vector equal zero, then the principal component solution would be given directly by the decomposition equivalent to that in (9.2), with the points P_i being the principal component scores. This can be achieved if the matrix $\mathbf{M}$ is replaced by the centred matrix $\mathbf{M}_c$ with typical element

$$m_{c.ij} = m_{ij} - m_{i.} - m_{.j} + m_{..} \qquad (9.4)$$

where the . denotes averaging over the subscript.

The latent vectors of $\mathbf{M}_c$ corresponding to the non-zero latent roots now sum to zero. To see this, note that the rows of $\mathbf{M}_c$ all sum to zero, so that the vector $\mathbf{1}$ is the latent vector of $\mathbf{M}_c$ corresponding to the zero root. Since latent vectors are by definition orthogonal, it follows that $\mathbf{1}^T\mathbf{b}_{c.i} = 0$.

Hence, the two latent vector analyses can be collapsed into the one analysis, the latter having three stages. The first stage involves forming the association matrix $\mathbf{M}$. The second stage involves the calculation of the centred matrix $\mathbf{M}_c$ as in (9.4). The third stage consists of a latent vector analysis of the resulting matrix, with each vector scaled so that the sum of squares of its components is equal to the corresponding latent root.

The ith row of the resulting matrix $\mathbf{B}_c$ of latent vectors can be taken as the coordinates of the point P_i. The distance between the points P_i and P_j gives the best approximation to the overall distance given by (9.3) in the chosen number of dimensions. As in the standard application of principal component analysis (as a latent vector analysis of a correlation or dispersion matrix), the relative magnitudes of the latent roots will tend to indicate the number of dimensions needed for an adequate representation of the distances.

The question of the appropriate choice of $\mathbf{M}$ given the distances between the individuals follows from the relationships in (9.3). For if $\mathbf{M}$ is to represent a distance matrix with zero diagonal elements, then

$$\Delta_{ij}^2 = -2m_{ij}$$

Given that the squared distance between two points is d_{ij}^2, the association matrix corresponding to these distances is given from this simplified form of (9.3) as

$$m_{ij} = -\tfrac{1}{2}d_{ij}^2$$

and

$$m_{ii} = 0 \qquad (9.5)$$

Note that for the decomposition as in (9.2) of the matrix $\mathbf{M}$ defined in (9.5), at least one latent root must be negative, leading to an imaginary set of coordinates. However, the transformation from $\mathbf{M}$ to $\mathbf{M}_c$ overcomes this problem and produces a real set of coordinates.

When k multivariate universes are to be considered, the generalized distance can be found for every pair of universes to give a $k \times k$ symmetric matrix. A principal coordinate analysis of the matrix of D^2-values gives the coordinates of the universes referred to principal axes. The principal coordinate analysis of D^2-values is equivalent to a canonical variate analysis with an unweighted between-groups sums of squares and products matrix, giving the procedure suggested by Rao (1952).

CASE HISTORIES OF PRINCIPAL COORDINATE ANALYSIS

Study of variation in a species of recent marine ostracods

The marine ostracod, *Buntonia olokundudui*, displays a type of meristic variation which is rather common among the cytherid ostracods. A principal coordinate analysis of the association matrix using the Gower metric based on various counts of marginal ornamental spines and the length of the carapace for 503 individuals disclosed the existence of several well-defined groups which could be related to the variation genetics of this species (Reyment and Van Valen, 1969).

There is a considerable concentration of the "variation" in the association matrix to the first latent root. This is a healthy sign and is a good indicator that the analytical method is going to produce some kind of intelligible result.

Analysis of a cyclic sedimentary sequence

Baer (1969) made a study of the palaeoecology of a thin, well-exposed, section in the well-known Eocene Green River Formation of Utah. He found the 40 m section to be divisible into five shale-carbonate sedimentary cycles, which could be further subdivided into categories related to the sedimentary environments deltaic, transitional and lacustrine. The deltas proved to comprise organic limestones, siltstones and shales and low-grade oil shale units. The so-called transitional beds are finely laminated and there are calcite and aragonite paired laminae with occasional laminae which are rich in organic substance. The lagoonal sediments comprise lacustrine carbonates, shales and siltstones.

Baer made a great number of analyses of many kinds. Among his results there is an interesting table of determinations of the elements Si, Al, Fe, Mg, Ca, Na, K, Mn, Rb, Sr, U, and Th. The information is given in the form of mean values and, it was thought, might provide interesting material for a principal coordinate analysis.

Clearly, the various phases of the cyclicity are not exposed by the plot of the first two coordinate axes (Fig. 9.2), which seems to suggest that the

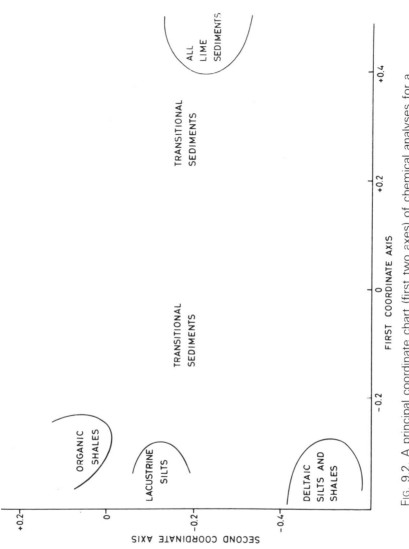

FIG. 9.2. A principal coordinate chart (first two axes) of chemical analyses for a cyclic-sedimentational sequence in the Lower Green River formation of Utah.

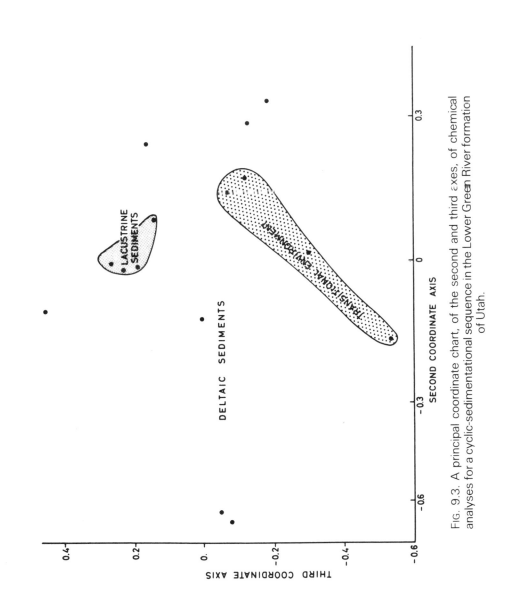

FIG. 9.3. A principal coordinate chart, of the second and third axes, of chemical analyses for a cyclic-sedimentational sequence in the Lower Green River formation of Utah.

chemical properties of the subdivisions at comparable levels do not differ strongly from each other. However, all the limestones cluster neatly to the right of the diagram and the argillaceous sediments rich in organic substance tend to group together, although organic content was not one of the things analysed by Baer in this connection. The lacustrine silts are also seen to group closely; the deltaic silt and shale form an isolated cluster well away from the other groups. It may also be of sedimentological significance that the rocks of the transitional beds lie spread over the centre of the diagram.

It is important to examine the amount of variation explained by the first few principal coordinates, to determine whether an adequate representation of the distances is being obtained. For the geochemical data considered here, the percentage residual, after the first two latent roots have been subtracted, is a low 38%. We dislike categoric statements about the reliability of analytical results in morphometrics, but one would be tempted to place considerable faith in the groupings yielded by this principal coordinate study on what really is difficult material.

A plot of the second and third coordinates given in Fig. 9.3 groups the analyses in accordance with the environmental categories. This is quite a different picture from that shown in the previous figure (Fig. 9.2), in which the grouping was more in agreement with the nature of the sedimentary category than with the environment in which the sediment was formed. The second and third coordinates reflect chemical differences arising from the particular conditions of the three sedimentary environments.

Taxonomic relationships in the coccids

Boratynski and Davies (1971) have investigated the taxonomic value of male coccids by a battery of numerical methods, including numerous Q- and R-mode analyses among which the method of principal coordinates played a prominent rôle. A total of 101 characters of various kinds were employed. These authors found generally good agreement between the numerical methods, not only with one another, but with conventional taxonomic ideas built up through investigations of the females and a more recent, comprehensive investigation of the males. Boratynski and Davies arrived at the tentative view that principal coordinate analysis using various measures of association is perhaps better suited to the treatment of coded, multistate data than other available methods.

10
Factor Analysis

It is very hard to discuss factor analyses (for many different variants are referred to as factor analysis, even though there is only one true model) without generating more heat than light. They are the most controversial of the multivariate methods. Part of the difficulty stems from the widely different terminology of factor analysts as opposed to practitioners of other forms of multivariate analyses. Partly, confusion arises because principal component analysis grew up in much the same context as factor analysis and many factor-analysts seem to consider it as a special case of "their" techniques. For instance, when Ehrenberg (1962) pointed out that, after 50 years of the practise of factor analysis, it was still uncertain whether anything of value had emerged, he was answered by other contributors to the same symposium in terms of benefits accruing from the use of multivariate analysis in general, and it remains unclear to this day whether any considerable service is performed by the use of factor analysis (in any strict sense of the term) in morphometrics that is not better performed by one or other of the multivariate techniques mentioned earlier in this book. Individual psychologists would claim that factor analysis had made possible an orderly simplification of the vagaries of the workings of the human mind: this view we can gladly accept in principle but, even here, there is a disquieting disagreement between schools of psychologists as to the nature of the structure of the mind, which suggests that the interpretative methods are something less than objective. Could it be that factor analysis has maintained its position for so long because, to a considerable extent, it allows the experimenter to impose his preconceived ideas on the raw data?

There seem to be two quite serious difficulties with factor-analytical methods in the strict sense: one is that they fail to include criteria for assessing the agreement of two or more sets of results, the other that they are somewhat less amenable than principal components to the calculation of scores for individual organisms along the relevant vectors; whereas one can easily, and advantageously, compute scores along components, one can only estimate, by methods of dubious utility, the corresponding factor scores. The first problem has been diminished, though not eliminated, by the maximum likelihood methods associated with Lawley (1940, 1958, 1960), Lawley and Maxwell (1963, 1971) and Jöreskog (1963, 1967), and the second remains to a certain extent (cf. Jöreskog et al., 1976, pp.142–146). Readers who wish to pursue the topic further may peruse the symposium comprising of the following references: Lindley (1962, 1964), Warburton (1962, 1964), Jeffers (1962, 1964), Ehrenberg (1962, 1963, 1964). To this rather discouraging body of opinion may be added the paper by Rasch (1962), who summarizes at least one of the fixed case types of factor analysis, and illustrates the summary with an analysis of 13 body measurements of cattle into three factors, as well as those of Cattell (1965a, b) and some thoughtful, and probably apposite, comments on the concept of simple structure by Sokal et al. (1961).

One of the main ways in which the term factor analysis has been used to distinguish its role in applied work from that of principal component analysis is that the factors are rotated to determinable positions in which they are not necessarily, nor even generally, orthogonal. It seems that the rotation of factors following a factor or principal component analysis is perceived to be an essential part of the procedure. There is of course no reason why a principal component analysis should not be followed by a rotation of the axes, although the reason for wanting to do so may be difficult to grasp (cf. Le Maitre, 1982). One way of determining these new positions is to adopt "simple structure", described by Thurstone (1947) as one of the turning points in the solution of the multiple factor problem. Other solutions have been described with unusual clarity by Gould (1967, 1981) and there are also references in Chapter 6 of Jöreskog et al. (1976).

Where the factors are not rotated, there is some evidence that despite the apparently distinct mathematical models, factor analysis and principal component analysis give similar outcomes; indeed, Gower (1966), who specifies some of the conditions under which this statement is approximately correct, considers that the meaningful results obtained when factor analysis is used under apparently unsuitable circumstances may well stem from the extent to which this technique can simulate a principal component analysis. We may note that Dagnelie (1965b) has presented some interesting comments on the use of the technique.

Seal (1964) has distinguished factor analyses from other multivariate techniques in two respects: that the p original characters are analysable into m ($m < p$) factors, usually taken to be orthogonal, with uncorrelated residuals; and that these m factors may be subsequently rotated to conform to a new set of factors, imposed by the experimenter, as a consequence of his theories about the natural processes underlying the measurements he has taken. In the earliest papers by Spearman (1904), the unique underlying factor (the g-factor of intelligence) was the only one considered, but, as special factors of various kinds were introduced, the distinction between imposing a factor model and allowing the data to decide how many factors were at work became blurred. Hotelling (1957) has also considered the relations between factor analysis and other multivariate methods.

Moran (1975) has warned against confusing these two avenues of approach to multivariate data analysis. Factor analysis is, as he emphasizes, not a descriptive technique but an attempt to identify underlying causal factors, and involves the assumptions that certain quantities are known, otherwise identifiability does not occur. He continues "As far as I can see, this information is practically never available in practical applications. Thus the procedure is a very dubious one."

Controversy continues over the practice of rotation of principal component vectors, where the rotation may be orthogonal or non-orthogonal (oblique) and a variety of stopping rules may be employed. These topics and others are discussed by Darton (1980).

Gould (1981) has written an anecdotal and highly readable account of the early intellectual development of factor analysis as a means of quantifying the "vectors of the mind".

Jöreskog et al. (1976), realizing the ambiguity existing in the minds of most users of "factor analysis" in the natural sciences, grouped all of the techniques for R-mode and Q-mode analyses under a single heading "Geological Factor Analysis". Their reasons for so doing have been misunderstood by some (cf. Le Maitre, 1982), despite the fact that the formal factor model was discussed in detail and shown to differ from what is thought of as factor analysis in the natural sciences, namely, principal component analysis, with or without rotation of the resultant axes described by the latent vectors. In the first edition of this book, a rather rigid point of view was adopted in that a clear distinction was called for between the true factor-analytical model, as outlined by Jöreskog et al. (1976, Chapter 3), and various uses made of principal component analysis in a factor-analytical accoutrement. Given the evolution in the application of this type of analysis over the last twelve years, we believe that it is no longer desirable, nor possible, to demand such a sharp distinction in practical work and we are therefore prepared to accept as "factor analysis" s. l. applications of principal components type involving rotation of axes, thus adhering to the broad view taken by Jöreskog et al. (1976).

Accordingly, it is quite immaterial for the concept of "true factor analysis" whether one is claiming to be performing a Q-mode, or R-mode factor analysis (for terms, see p. 86). What one is really considering is whether a

Q-mode or R-mode component analysis is relevant in a particular situation, or has any relevance at all.

The main features of R-mode and Q-mode concepts were taken up in Chapter 9. The reason for breaking factor analysis out of that connexion is that the applications of this technique usually require particular distributional assumptions about the data and a certain *a priori* knowledge of what is to be expected from the analysis. The subject of factor analysis may be conveniently classified into two main spheres, to wit:

1. What should be more precisely referred to as principal component analysis;
2. "true" factor analysis, in which a formal model is assumed and explicit assumptions about the underlying statistical universe are made.

The subject of factor analysis in the broad sense is clearly a complicated one, the ramifications of which lie well outside the scope of the present book. The interested reader is referred to Chapter 4 of Jöreskog *et al.* (1976) for an account of the differences in the fixed and random models.

Attempts to use factor analysis have convinced us that the results rarely differ from those yielded by principal component or principal coordinate analysis in any manner that seriously influences a biologist or a geologist seeking to interpet a mass of data. A somewhat similar conclusion was reached by Crovello (1966c). Insofar as some differences appear, their induction and interpretation involve such a high degree of subjective judgement that we prefer not to give details which might be interpreted as recommendations. We now briefly present two examples of geological factor analysis. Further examples are given in Jöreskog *et al.* (1976).

Middleton (1964) investigated the interrelationships of ten chemical constituents in scapolites, beginning with a principal component analysis and continuing with a "factor analysis", by rotation of the principal components to positions determined by the quartimax procedure (Harman, 1960). The elements fall into two bands roughly at right angles, considered to represent the consequences of substitution in one or two series of solid solutions. The rotational procedure does not seem to have done more than formalize conclusions apparent from inspection of the principal component analysis.

Some genera of fusulinid foraminifers were studied by a "principal component" factor analysis, which involved the calculation of nine factors and the rotation of seven of these according to the varimax criterion (Pitcher, 1966). Only four of these factors were considered to be significant, and all four contributed to the separation of the various genera in successive two-dimensional charts (Fig. 10.1). Since the study of the characters which contributed most to the separation of the genera and species was conducted by means of factor scores, one might think that it would have been easier to interpret the results had a simple principal component analysis been carried

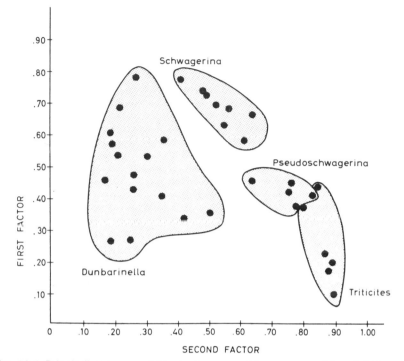

FIG. 10.1. Principal component "factor" scores of four genera of fusulinid foraminifera showing clustering into generic groupings along the first two factors. Redrawn from Pitcher (1966).

out. Pitcher makes the useful general point that when a specimen remains within a particular grouping as the data are plotted on different axes, the credibility of the proposition that the individual belongs to that group is enhanced: conversely, if replotting the data along new axes of variation shuffles the individual into another group, its classification must remain dubious. This concept has been formalized by the techniques concerned with producing minimum spanning trees.

Sokal *et al.* (1980) have discussed aspects of the reification of factor analyses with particular reference to physiological variables measured on human beings.

The process of reification involves reading back on to the multivariate plots the appropriate interpretation. For example, should most of the points on the left of a chart represent male organisms, and those on the right females, then that axis of variation represents sexual dimorphism. Similarly, should objects high up on axis I represent material from Scandinavia, whereas those low down on this axis represent Mediterranean material, we should suspect axis II of reflecting a north-south cline of geographical variation.

11
Patterns of Growth: Genesis and Genetics

In this chapter, we review the representation of complex patterns of growth by a small number of vectors, and the possible relations between such vectors and the genetical properties of the organisms in which they occur.

We now turn to the more nebulous problem of discovering "structure", not in the relations between different groups, but within any one group of undifferentiated material. This step is often treated as a major logical one, and a persistent criticism of discrimination techniques has been that they do not provide for the initial classification into the groups whose mutual relationships may afterwards be studied. For a variety of reasons, this criticism lacks validity and one of these reasons is that the distinction between discrete groups and homogeneous groups represents the extreme forms of a continuum, the intermediate members of which we have to deal with in real life.

Pearce (1959) has raised the substantial point that variation, though random in certain senses, is not necessarily uniform in the multidimensional space, but is in effect canalized along certain patterns of growth whose existence can be demonstrated objectively in multivariate studies.

In other words, we should have, in any supposedly homogeneous sample of organisms, incipient variation which bears a relationship to the contrasts of form between closely affine discrete groups. Looked at in this light, there is a continuum between those situations in which we can easily distinguish the groupings (appropriate for analyses by, say, canonical variates) to those situations where we cannot, or at least have not, distinguished them (appropriate for analyses by principal components or principal coordinates). To a considerable extent, the difference between discrete groups and the continuous, homogeneous, case is subjective rather than objective, for as the contrasts of form between the groups become less and less marked, more intensive and perhaps more sophisticated techniques will be required to detect the heterogeneity of the continuum. Indeed, a homogeneous group has been cynically defined as one in whose heterogeneity the experimenter has not yet become interested.

This is not to say that discontinuous polymorphisms do not exist; they do, and widely. The continuum referred to above relates to our appreciation of their existence, which is another matter. One worker may have noticed that a particular species is polymorphic in respect of some characteristics: another may not have done so. Ideally, the first will choose an analytical technique which takes the structure into account, such as canonical variate analysis, while the second will select one which makes no assumption about structure, such as principal component analysis. Only an excessive preoccupation with purity would condemn either analysis. In fact, one of the cheering features of multivariate analysis is that one would have to be distinctly unlucky to display the material on two or more axes without capturing even a hint of the variation inherent in it. Obviously, no-one would want to choose less effective techniques, but if such a choice is made unwittingly, the consequences may often represent a useful salvage job.

Because this dual approach to multivariate problems is always possible, Blackith (1960) considered the vectors describing variations within a group as "signposts" to possible variation between that group and other groups. Depending on one's views as to the relationship between polymorphism and speciation, one can even describe them as signposts to evolution in the species, though this imagery is less well founded than that of the vectors within a group merging with those between groups created out of the original one.

Great care, however, is needed in the interpretation of vectors describing variation within groups, since objective tests of the nature of this variation are less easy to perform. Moreover, there is considerable disagreement as to the relative importance of various potential sources of trouble. To some authors, departures from multivariate normality may distort the tests. To others, the lack of scale-invariance in principal component analyses is a serious handicap (Gower, 1967), although for many this has never seemed to be important in practice (Stuart, 1964).

If the sexual dimorphism embraced axes other than that representing size, the two analyses would probably differ still more: but, as all but one species in a thousand is bisexual, it would be sensible to suspect hidden sexual dimorphism in an animal that did not overtly show it.

One of the sources of strength in multivariate morphometric procedures is that the vectors which the procedures uncover often correspond to features of the organism which have long been appreciated in an intuitive way; for instance, the concepts of "vigour" and "precocity" in the growth of apple-trees, were found by Moore (1968) to be adequately represented by principal components; so was "taper" in tree-growth (Holland, 1968a; Fries and Matérn, 1966). Holland has emphasized that even where the vectors themselves have no obvious biological interpretation the space which, as axes, they define does.

Similarly, in plant and animal breeding experiments, a quantitative assessment of the body conformation at which the breeders, rightly or wrongly, are aiming is invaluable; Rouvier and Ricard (1965) have shown how component analysis can help to provide this assessment.

GENETICAL ASPECTS OF MORPHOMETRICS

Prior to the first edition of this book, there were relatively few well-conducted multivariate morphometric investigations with a genetic basis, considering the importance for plant and animal breeders of the conformation of the organism. Cock (1966) and Rouvier and Ricard (1965) have summarized much of what has been done in this field, and Cousin (1961) has long insisted on the need for studies of the inheritance of shape and size to be placed on a sound footing. Lefèbvre (1966) has given a review of the studies (multivariate and otherwise) devoted to the growth and conformation of cattle.

Occasionally a direct relationship has been suspected between the action of a single gene and the promotion or suppression of growth along one or other of the principal components of the dispersion matrix of a number of measured variables. For instance, Kraus and Choi (1958) extracted principal components which reduced 12 measured characters of the human foetus to four patterns of growth. Because they were able to obtain teratological specimens in which the failure of a gene, by mutation, entrained the suppression of the pattern of growth constituting its phenotypic expression, they were able to speculate that each of the observed patterns of growth of the long bones of the human foetus were controlled by a single gene. A consideration of matters of this kind is doubly useful; it illuminates genetics by enabling the experimenter to handle large numbers of "unit" characters in meaningful groups, and it illuminates morphometrics and in particular the subdivision known generally as numerical taxonomy, by throwing light on two fundamental hypotheses. These hypotheses are, first, the nexus hypothesis, which assumes that every character is likely to be affected by more than one gene (or, conversely, that most genes affect more than one character, by virtue of pleiotropism) and secondly, the non-specificity hypothesis, that no distinct classes of genes affect one class of characters exclusively. Sokal and Sneath (1963) who are principally responsible for the enunciation of these hypotheses, note that one relies on the non-specificity hypothesis to justify taking characters for a morphometric study from, say, the external morphology of animals without needing to examine the internal anatomy. However, there is now adequate evidence that the non-specificity hypothesis is only partly valid (Michener and Sokal, 1966) so that the experimenter's choice of characters should be spread as widely as possible.

Actual genetically oriented multivariate experiments are few. Eickwort (1969) has commented that multivariate methods are particularly well suited to the task of following the results of genetical experiments on form, and has used ten characters to examine the differential variability of male and female wasps, males being haploid in these insects. The results agree with those of Blackith (1958) in showing that males are more variable than females. White and Andrew (1959) found that in an Australian eumastacid grasshopper which is, in some populations, polymorphic for pericentric inversions in two autosomes, the "standard" sequences serve to augment the weight of the adults, whereas the "Blundell" and "Tidbinbilla" sequences served to decrease the weight. Somewhat surprisingly, these effects were consistent in four separate populations. As White and Andrew comment, this finding implies an amazing evolutionary stability for the size-determining gene complexes over a period probably exceeding a million years.

White (1973) suggested that one of the reasons why major chromosomal rearrangements become established in animal populations, even when clearly deleterious in the heterozygotic condition, is that they serve a useful purpose in protecting co-adapted blocks of genes from disruption. Among these co-adapted blocks one might expect to find those controlling growth and form.

One of the most striking features of a morphometric approach is its ability to uncover patterns of growth, involving all or at any rate some large part of the body, which must have persisted unchanged over such long periods of time that many other evolutionary processes have gone on in the animals concerned without disturbing these stable patterns. A good example is that of phase variation in locusts (see p. 51) where the change of adult form when the immature stages are reared in crowds is almost identical in not very closely related genera, even though each genus may contain species which do not show this phenomenon (Blackith, 1962). It is of course open to question that phase variation has arisen many times in the course of the evolution of the Acrididae, the perpetual choice between a monophyletic and a polyphyletic concept is never wholly to be forgotten, but at least the capacity to undergo such a change may have arisen but once; it is known to be associated with the sex chromosomes. Parallel, or homologous, variation, must often present this dilemma (Blackith and Roberts, 1958) and is of great concern to palaeontologists (Reyment and Van Valen, 1969).

Very consistently, the number of biologically significant axes of variation disclosed by multivariate studies is less, usually much less, than the number of characters employed in the analysis.

There is a certain amount of evidence that natural selection does not operate directly on individual characters but on the degree of interlocking between one dimension of variation and another, the suite of characters undergoing change as a unit in a fashion somewhat analogous to the evolution of "supergenes". This view, of course, is widely held by conventional taxonomists, though it is usually expressed in other terms. Even characters which vary substantially independently within a homogeneous population form part of a more complicated contrast of form as reflected in the discriminant (or canonical variate) linking two distinct groups of organisms. Characters giving rise to this pattern of vectors within and between homogeneous groups of organisms include the weight, number of antennal segments, and elytron length of acridid grasshoppers (Blackith, 1960) and the elytron length of a pyrgomorphid grasshopper (Blackith and Kevan, 1967). Sokal (1962), Grafius (1965) and Schreider (1960) have all, in their different ways and in widely separated fields of study, reached the conclusion that natural selection operates on characters in stable groups which may be discerned by a multivariate analysis. These stable groups are not necessarily the correlated responses to selection that arise in many genetical experiments (see, for example, Scossiroli, 1959) but must often include them. Bailey (1956) has extracted principal components corresponding to the genetical and environmental components of morphogenesis in mice but this distinction may not always be easy to make: Hashiguchi and Morishima (1969) discuss the estimation of such principal components. The subject is of consequence not only because of its interest to animal breeders but also because of the distinctions we draw in this book concerning the possibility of distinguishing "genetic" and "environmental" principal components in palaeoecology.

Even such "obviously" adaptive traits as robust limb-bones in heavier terrestrial animals may, as Jolicoeur (1963d) has found, be associated with a general, probably genetically determined, process throughout the body rather than by adaptive growth as modified by environmental influences. Jolicoeur found that once the principal component reflecting size differences had been removed from the analysis, the remaining "shape" components had markedly unequal variances, and he suggests that those with the lowest variances represent highly canalized patterns of development, which are well buffered against environmental influences.

There is very little evidence for some traditional ideas, such as that changes in body form are mediated through stresses felt in early life before growth has ceased. In the case of Jolicoeur's martens, there seemed some support for an hypothesis of Scott (1957) according to whom robustness depended on the degree of use of the musculature and on the action of sex hormones. Similarly, when Blackith and Kevan (1967) analysed growth in a pyrgomorphid grasshopper, they found that the inverse relationship between the width of the mesosternal interspace and the elytron length formed part of a genetically determined pattern of growth and was not directly influenced by differential muscular stresses in long- or short-winged forms.

That body conformation as a whole tends to be inherited is known to any animal breeder and can be followed quantitatively, as revealed by Lefèbvre's (1966) multivariate studies on Norman cattle. He made 14 measurements on the progeny of 37 bulls. The shapes of the progeny clustered strongly, on a chart whose axes were the first two canonical variates extracted, according to the sire, but a maternal effect of the same general magnitude as that of the sires was noted.

Lefèbvre suggests that the groups of progeny also tend to form clusters representing "typical" conformations equivalent to such concepts as "stocky," "lean" etc. in humans. It may be that different breeders work towards ideal forms and since it is notorious that breeder's concepts of ideal form have had little to do with the economic value of the product and a great deal to do with fashion, it seems plausible that different breeders will be working towards rather different ideal forms. The work of the Centre National de Recherches Zootechniques in France indicates an economically important field for morphometrics in helping to rationalize breeding patterns for stock (see, for example, Rouvier and Ricard, 1965). Multivariate discriminants for the selection of animals and plants were among the first developments of morphometric techniques and reached an advanced stage early in the history of the subject (Smith, 1936; Hazel, 1943; Rao, 1953). Still further progress has been made by Rouvier (1969).

Multivariate methods of analysis have also been able to clarify the situation in the polecats *Mustela putorius* and *M. eversmanni*, which are sympatric, and between which some hybridization has been alleged. Rempe (1965) showed that this suggestion could not be supported from a generalized discriminant analysis of the skulls.

Hybridization between the voles *Apodemus sylvaticus* and *A. tauricus* has been studied by Amtmann (1965) using 17 morphometric characters, of which five accounted for 88% of the discriminating power in the males and 50% of it in the females. Hybrid individuals were few in the zones of contact, which was quite narrow in relation to the mobility of the animals. In such work as this the careful morphometrical studies could perhaps have been enhanced by cytogenetic ones, since it is known that the genus *Apodemus* shows chromosomal rearrangements.

A full study of the nature of hybridization and the selective forces operating at tension zones where two species meet is of the greatest value in understanding the evolution of the species involved: the morphometric aspects of the work are then useful but where a cytogenetic analysis is possible the combination of the two approaches is greatly to be desired. This statement also holds good for experimental crosses, and Lecher's (1967) cytogenetic analysis has added to the value of the morphometric studies of Bocquet and co-workers (Bocquet and Solignac, 1969). For want of the necessary cytological skills, or because the material to hand does not show suitable chromosomal differences, the morphometric approach may be the only one available. Rising (1968) has investigated the body form of two chickadee species in a tension zone in Kansas, on which eight morphometric measurements were taken. It seems likely that interspecific mating, occasionally followed by back-crossing, was encountered, and field observations suggested that the hybrids were at some selective disadvantage during reproduction.

A somewhat similar situation was found in the Louisiana grackles, where putative subspecies showed a stepwise cline in colour but a continuous cline in the body conformation as assessed by discriminant functions based on four morphometric characters (Yang and Selander, 1968).

Some classical correlation studies on the inheritance of human body form can often be expressed to advantage in multivariate form, as de Groot and Li (1966) have shown.

One problem that faces the investigator who is interested in rates of evolution is the measurement of the amount of morphological change involved, and the techniques by which these changes should be assessed. These problems have been discussed by Lerman (1965) and Marcus (1969) both of whom have recommended the use of the generalized distances for measuring the amount of morphological change. Indeed, the analysis of continuous variation when this is supposed to follow Mendelian inheritance has been fully treated by Weber (1959, 1960a, b).

It should never be forgotten that the shape of an organism is the outward and visible manifestation of properties which must very frequently be associated with other inward and often invisible ones; for instance, Clark and Spuhler (1959) found that the fertility of human beings was different in the various body builds, and similar differences must await discovery in many polymorphic populations.

However, there is a deeper problem concerning the extent to which morphological change reflects genetic change; the relationship is far from proportionality, as any taxonomist knows who has to deal with groups that are similar in form but belong to distinct higher taxa, or groups that are radically distinct in form but belong to affine taxa. Some authors using multivariate methods, for instance Grewal (1962) and Brieger *et al.* (1963), have virtually taken a proportional relationship for granted, but Goodman (1969) has begun the long and complex task of considering the factors involved. He makes the point that comparably divergent forms will represent greater genetic divergence if derived from an ancestral population of low variability than from one of high variability. This is an arguable point, since the divergent forms might be derived from the most divergent members, genetically speaking, of the parent population, so that the variation which is, as it were, discounted, is variation inherent in the ancestral population whose translation to the progeny one might well wish to see retained in the measure of overall genetic variance.

Oxnard (1969) reviewed the subject of shape and function and the necessity of the morphometric approach for an adequate appreciation of the subject. The relationship between evolutionary changes, assessed phenotypically, and the underlying genetic differences has been discussed for human populations by Cavalli-Sforza (1966).

Williams *et al.* (1970) analysed four morphometric characters of the spermatozoa from nine strains of mice using both canonical and discriminant analyses. They found that adopting the multivariate approach improved the discrimination between the spermatozoa from the various strains, sometimes markedly so. Nevertheless, an adequate representation of the relationships between the nine strains required all four dimensions, so that this example fails to illustrate the reduction of the effective dimensionality of a set of comparisons which has often been claimed as a major objective of multivariate studies, an objective which is, moreover, usually attained. However, it became clear that a full analysis of these relationships required a multivariate interpretation precisely because of the hypermultivariate nature of the underlying biological situation; for instance, earlier attempts at univariate analysis had shown no significant differences for some of the measurements even though to the experienced eye the spermatozoa were patently distinct. The multivariate approach showed that such univariate tests were maloriented.

Some dominance relationships were discovered when sperm from various hybrids between the strains of mice were measured and analysed. Williams *et al.* suggest that multivariate studies could help in investigations of the epigenetics of spermatozoa.

COMBINED MORPHOMETRIC AND GENETIC ANALYSES

Because of the specialized nature of genetic, and particularly cytogenetic research, there has been an unfortunate historical tendency to look at related species either by cytogenetic means or by morphological ones. There can be little doubt that in at least some instances, problems could have been resolved more readily by a combination of the two approaches. We are particularly pleased to see how rapidly the concerted assault on problems of speciation is developing, almost all the examples given here date from after when the first edition of this book was published. The importance we attach to this development stems partly from the fact that a very high proportion of the events leading to the formation of new species are now known to be accompanied by visible rearrangements of the karyotype (chromosomal rearrangement of various kinds). It used to be thought that such rearrangements were so unlikely to become established that they could be of only minor interest: we now suspect that they are not only fixed fairly frequently, with or without heterosis, but play a critical rôle in the formation of some new species.

Robinson and Hoffman (1975) looked at ground-squirrel speciation and found that their generalized distance analysis gave readily interpretable results, whereas when the Euclidean distance was used as a clustering metric, the results were not so readily interpretable. When they compared the karyotypes of the squirrels with the forms as appreciated by a discriminant function based on 12 morphometric characters, a close association was noted.

There have been two useful studies of the morphometrics of bats in relation to their karyotypes. Davis and Baker (1974) showed that morphometric divergence accompanied chromosomal divergence in bats of the genus *Macrotus,* and Baker *et al.* (1972) did the same for races of *Uroderma.* Atchley (1972) drew attention to the desirability of studying the genomes in hybrids in parallel with the morphometrics, where the degree of genetic similarity can be reliably assessed from the homologies and directionality of chromosomal changes whose relevance to speciation mechanisms then becomes clearer.

In some lizards, the genetic divergence as measured electrophoretically is not matched by the chromosomal divergence (Bezy *et al.*, 1977). According to Gould *et al.* (1974, 1977) in a joint multivariate morphometric and general investigation of the land snail *Cerion,* the coordinated application of biochemical genetics and multivariate statistics can resolve many issues in the systematics of highly variable organisms. Warner (1976) investigated geographical variation, based on ten characters, in the plains woodrat and found that the distribution of three chromosomal morphs was consistent with their undergoing the type of speciation put forward by White (1973) and which he termed stasipatric speciation.

Atchley *et al.* (1981) found that morphological changes in rat- and mouse-skull measurements failed to accord with the genetic differences observed. They used 18 skull and skeletal characters subsumed into principal components, which were rotated orthogonally. They also noted that when a size vector is isolated, it does not automatically represent isometric growth. An important theoretical point has been made by Lin (1978), to the effect that the vectors isolated in multivariate selection experiments are virtually selection indices (cf. Lande, 1979). It is common practice in multivariate investigations to explore the variation in question in terms of vectors which, as Leamy (1977) points out, confound phenotypic and genotypic variation. He recommends that, where practicable, separate analyses should be conducted on the genotypic and phenotypic variation (see also Cheverud, 1982c).

An important matter which is rarely addressed is the extent to which multivariate morphometric changes confer selective advantage or disadvantage on the organism in question. The classic instance is that of Bumpus' sparrows, where, it was claimed, the largest and the smallest individuals failed to survive a severe frost. This phenomenon is sometimes called centripetal selection. Johnstone et al. (1972) have re-examined Bumpus' data with the six morphometric measurements available. They find that some of Bumpus' main conclusions are supported, and also that among those individuals that failed to survive were males whose form tended to that of females, and vice versa. This finding suggests that phenotypically intersexual individuals were maladapted.

The ferment of ideas on the evolutionary relationships of hybrids has stimulated many investigations. Neff and Smith (1979) examined hybrid fishes of the genera Lotropis and Notropis. They concluded that principal component analysis is superior to linear discriminant functions for this purpose. Shaklee and Tamaru (1981) used 31 morphological characters to separate two Hawaiian bone-fish species with the aid of a stepwise discriminant analysis. The cyprinid fishes of the genera Phoxinus and Semitolus were found by Legendre (1970) to produce fertile hybrids in some crosses between species: this finding, together with the establishment of the chromosome number of the species involved, suggested that S. margarita be transferred to the genus Phoxinus. Colless (1980) has compared the taxonomic arrangements of 16 populations of the fish Manidia using morphometric and allozyme data. He found that when both are ordinated on a principal component representation, there is good agreement.

The effects of secondary contact between separated populations have been evaluated by Schueler and Rising (1976) as including both character convergence, including mimicry, and character divergence. The degree of natural hybridization attained was assessed using discriminant function scores. It is particularly interesting to see how selection maintains the broad integrity of two species even though some gene-flow may be taking place across the zone of hybridization. Rohwer and Kilgore (1973) deduced that such a process is taking place in the zone of overlap between two species of arid-land foxes, as indicated by 14 skull characters that they measured, and combined in a principal component analysis.

Vogt and McPherson (1972) suggested a multivariate technique for separating closely related species, using quantitative differences; under favourable conditions this can be used to distinguish hybrids. Both discriminant functions and principal components have been used by Rohwer (1972) to show that hybridizing meadow-larks, with their strong premating isolation from song-patterns, maintain a clear distinction between the species.

Atchley (1974) and Atchley and Cheney (1974) explored morphometric relationships within the *viatica* group of morabine grasshoppers of Australia, detailed cytogenetic studies of which are among the more significant contributions to the theory of speciation. These authors refuted the view that chromosomal rearrangements, which accompany 92% of all speciations in the animal kingdom, are not usually accompanied by corresponding amounts of phenotypic divergence. Atchley (1981) has summarized the complex relations between morphometrics and cytogenetics in the parthenogenetic grasshopper *Warramaba virgo*, by comparing C-banding patterns and morphometrics. The morabine grasshoppers are wingless and relatively immobile, and the concept of panmixis in their populations makes little sense; their reluctance to move from the immediate area where they were hatched brings into play mechanisms of speciation unthinkable by those accustomed to highly mobile animals. Moreover, the morabines have no premating isolation mechanisms, so that intraspecific and even intergeneric matings are easy and apparently frequent.

The extensive cytogenetic and morphometric investigations of the morabine grasshoppers, and of the *Jaera* complex of isopod crustaceans, are revealing facts about evolution which ill consort with older concepts. There has been a fruitful marriage between morphometric and cytogenetic studies on *Jaera albifrons*, which exists all down the coastline of North Europe, from Sweden to Spain, exhibiting a cline in chromosome number along the way (Prunus and Lefèbvre, 1971).

Because of the prevalence of the belief that speciation requires some substantial measure of geographical separation, studies are now being conducted to examine the rate at which species so separated diverge from their original sympatric conditions. Lessios (1981) looked at three genera of sea-urchins on either side of the Isthmus of Panama, known to have separated the Atlantic from the Pacific between two and six million years ago. He found that although at least 21 characters, incorporated into canonical variates, indicate that considerable divergence has by now taken place, the divergence between those populations which are still sympatric, i.e. on the same side of the isthmus, was even greater. Lessios attributed this extra divergence to the need for habitat separation between closely related congeneric species. This argument ignores the fact that, without the barrier, all the populations would have remained sympatric, so that if anything, the allopatric separation of the past few million years has served only to reduce divergence in the characters analysed that would otherwise have occurred. It might be more realistic to say that whether the species were allopatric or sympatric has had very little to do with their divergence one way or the other, particularly since Rees (1970), who incorporated ten cranial characters into canonical variates, found an increased morphological divergence after some 4000 years of separation of deer on either side of the Straits of Mackinac.

Pimentel (1981) investigated a hybrid swarm of sand verbenas using multivariate techniques, and found that principal coordinates using Gower's metric, together with non-metric multidimensional scaling, worked well. Difficulties were, however, encountered when binary data were employed. The introgressive replacement of *Clarkia speciosa polyantha* by *C. nitens* was investigated by Bloom (1976), who used canonical variates. He found that the morphological changes consequent upon introgression did not accord well with the chromosomal changes. Nine intermediate populations were studied involving 12 metric characters and eight flower colours. The geographically intermediate populations on either side of the north-south boundary between the species are very similar to reference populations of *polyantha*. This boundary is marked by a complex system of chromosomal rearrangements due to the hybridization.

At an apparently intraspecific level, Tantravahi (1971) has examined the chromosomes of the *Tripsacum lanceolatum* complex in the light of the variation that they show when assessed by multivariate analyses.

Thorpe (1979) and Thorpe and McCarthy (1978) have made considerable use of multivariate analysis for unravelling complicated evolutionary situations in snakes. In one of these studies, it was found that the African House Snake, *Boaedon fuliginosus*, occurs in two morphs, previously thought to be distinct species. The multivariate morphometric analyses show that in West Africa, the two morphs behave as if they were two separate species (named *B. fuliginosus* and *B. lineatus*) but in the rest of Africa, these two forms appear to be conspecific morphs of a single species. Thorpe and McCarthy (1978) believe that there is evidence of a "character-shift" and that this species-complex might possibly pose interesting problems for the concepts of "gene-pool cohesion" and "phenetic species". The rôle of polymorphism in speciation is a significant one, eminently amenable to multivariate elucidation. Reyment (1982c) has considered certain implications with respect to speciation in Cretaceous ostracods.

Lande (1979) has made a valuable contribution to the possibilities of applying the theory and methodology of quantitative genetics to evolution in the phenotype. In that paper, he has particularly taken up the question of multivariate evolution, with special emphasis on the analysis of multivariate allometry. Lande's results are taken up on p. 121 in the section on differential growth.

Cheverud (1982a, 1982b) has given the subject of genetic, phenotypic, environmental and morphological integration in macaques close attention. Phenotypic, genetic and environmental correlations among 48 cranial characters were used to compare patterns of phenotypic, genetic and environmental integration with each other and with a theoretically derived pattern based on cranial development and function. Cluster analyses of the principal component loadings for traits derived from all three matrices were performed in order to compare empirical clusters with theoretically derived clusters. The pattern of phenotypic integration was found to match the functional pattern, while environmental clusters differentiate between the gross neurocranial and facial functional units. The genetic clusters differentiate between functional subunits within the neurocranial and facial units, but do not efficiently delineate these two gross units themselves. The overall similarity of the phenotypic, genetic and environmental correlation patterns was found to be low, which differs from the opinions expressed by Lande (1979) on this type of relationship.

12
The Analysis of Size and Shape

Some of the more recent studies on size and shape of both classical and non-classical stamp will now be briefly reviewed. Cuzin-Roudy and Laval (1975) have taken up the "classical interpretation" of allometric relationships by canonical variates. In their study of the heteropteran *Notonecta maculata* Fabricius, interpretable results were obtained for the analysis of post-embryonic development whereby canonical variates could be related to general growth, sexual differences, and the effects of a juvenile hormone.

Cheetham and Lorenz (1976) utilized the standard size-shape interpretation of principal component analysis for studying growth in bryozoan zooids. They rotated the first two principal component axes in the manner of factor-analysts. They used a rather complicated method of "vector analysis" for producing the variables used in the principal component study. Size, i.e. the area within the autozooidal outline, accounts for one third of the variation and tends to vary less within colonies than does shape. The portion of the shape variability independent of size could be partitioned into three components. One of these is connected with the asymmetry of the habitus. The second shape-component is associated with elongation and distal inflation; these two variables seem to be less influenced by the microenvironment than is asymmetry. The third component was given a rather tentative interpretation in that it was thought to indicate that distal inflation is slightly more sensitive to the microenvironment than is elongation.

A novel approach to the problem of size and shape variation has been presented by Stützle *et al.* (1980). This study was concerned with modelling growth in human height in connexion with which methods entirely different from the concept of allometric growth have been developed. In fact these authors seem to be totally unaware of the concept of differential growth as developed by biologists. Stützle and coworkers frame their concept in terms of shape-invariance (as opposed to the idea of growth-invariance in some classical allometric work) and, using non-linear regression, they came up with a rather interesting method of analysis. For certain kinds of studies involving ecophenotypic variation in the morphometry of some organism this approach would seem to be of value. At the present stage of its development, however, more work on the biological significance of the mathematical operations is necessary.

ALLOMETRY AND GENETICS

Lande (1979), who has pioneered the application of the principles of quantitative genetics in palaeontology, has taken up the question of multivariate allometry in the wider context of the genetical interpretation of allometry in the size of the brain. He observes that methods of multivariate analysis, functional analysis and optimality criteria do not account for dynamical constraints imposed by the pattern of genetic variation within populations. The expression of phenotypic variation often does not suggest any clear mechanism connecting growth patterns or adult variation to inter-specific evolution, as exposed by quantitative-genetic methods, even when natural selection operates on body-size alone. When there is individual variation in development, no necessary correspondence exists between ontogenetic and adult variation in a population.

Evolutionary allometry can arise from natural selection on more than one character. Multivariate selection can, however, also produce inter-specific patterns other than allometric. Thorpe and Leamy (1983) have studied size and shape variation in house mice by multivariate methods in a genetic context.

Essential concepts developed by Lande for studying multivariate evolution in the mean phenotype are couched in terms of a modified gradient system, the selection gradient, the Malthusian mean fitness, which is a generalized genetic distance of Mahalanobis' type, formed from the mean changes in the phenotype and the genetic covariance matrix, and the selection index, which has the form of a linear discriminant function.

With respect to the brain-body-allometry question, Lande finds that at various taxonomic levels, brain and body-weights tend to follow the usual allometric equation. Data from selection experiments in mice indicate that (1) the short-term differentiation of brain- and body-sizes in closely related mammals result either from directional selection on body-size with changes in brain-sizes (largely a genetically correlated response) or from random genetic drift; (2) during the long-term allometric differentiation within most mammalian orders there has been more net directional selection on brain-sizes than on body-sizes.

Reyment (1982b) has applied Lande's method of analysis to phenotypic evolution in the Late Cretaceous foraminiferal species *Afrobolivina afra*. Two species of Cretaceous ostracods have been studied by the same approach (Reyment, 1982a, c) with particular emphasis on evolution in the multivariate phenotype.

OTHER METHODS FOR ANALYSING SIZE AND SHAPE RELATIONSHIPS

The principal component (and factor analytical) interpretation of size and shape variation and multivariate allometry reigned supreme for two decades and this was the situation at the time the first edition of "Multivariate Morphometrics" went to press. Few were really satisfied with aspects of the principal component interpretation of multivariate differential growth. Attempts at providing a fully rounded generalization of Huxley's bivariate allometry equation have met various difficulties, the most serious of which concern adequate definitions of size and shape. Jolicoeur's (1963b) generalization is intuitively attractive but runs into trouble in that the entire differential growth relationship must be summarized in the first logarithmic principal component. In this respect, it requires a model of the one-factor kind, such as proposed by Hopkins (1966). Mosimann's (1970) solution, in which he defines specific size and shape variables, seems to provide a more general approach to the problem. The approach of Humphries *et al.* (1981) is addressed to a semi-graphical multivariate procedure for "multivariate discrimination" in which emphasis lies with the separation of groups (of, say, related species) on the grounds of differentiation in shape. In this respect, the method seems to be successful for two groups, provided the first two principal components summarize most of the differences. It can rapidly become onerous if many groups are involved.

From the point of view of a graphical appraisal of the role of shape differentiation in relationships between species, this method is an advance, but it does not contribute much towards improving our analytical capabilities for the detailed study of differential growth in one-organ systems, such as is necessary in the treatment of multivariate quantitative genetic problems (Lande, 1979).

Mosimann (1970) brought into clearer focus some of the concepts of size and shape in relation to differential growth. If we, with Sprent (1972), consider three measurements on the human body: head (H), body (B) and total length (L), we can suggest, as a measure of shape, the ratio H/L, for example, which represents the proportion of total length that is head-height. Proportion is a convenient measure of shape; however, a ratio such as H/B also designates a shape relationship. There are various measures of total size; for example, $H + B$ is such a measure, as are also H and B each on their own. Mosimann (1970) generalized these ideas to p dimensions and defines "shape vectors" and "size variables" associated with a set of p distances between specified points. He referred to such a vector of distances as a "measurement function".

If these distances are measured between corresponding points on two individuals, the measurement functions may be denoted as x_1 and x_2. Mosimann's definition of equality of shape is expressed as $x_1 = cx_2$ for some positive scalar c. Sameness of shape is therefore dependent on the components of the vectors x_1 and x_2 as well as the number of dimensions involved. Thus, if two individuals have the same shape, then every shape vector of the first is equal to the corresponding shape vector of the second. A standard size-variable in Mosimann's terminology is any real-valued function $g(x)$ that is positive for all positive values of x_i and such that $g(ax) = ag(x)$ for all positive values of a and x_i; a is some constant. Typical standard size-variables are the characters measured themselves (the x_i), their sum and their geometric mean. Typical shape vectors considered by Mosimann are proportions of various kinds.

One of the keystones of Mosimann's analysis is the role of isometry. He defines isometry as independence between a shape-vector and a given size-variable. This concept helps in avoiding some of the difficulties associated with Jolicoeur's (1963b, d) definition of isometry when there is not even a straight-line relationship between logarithms (see also Humphries *et al.*, 1981). Sprent (1972) observes that isometry is rare in the bivariate case and must be even less common in the multivariate case.

An advantage of Mosimann's (1970) treatment of the problem in relation to the established interpretation based on a principal component analysis is that it does away with arbitrary decisions about what are shape variables and what are size measures. If one principal component is said to be a size-component, and all other components are interpreted as shape-components, then, *per definitionem*, the so-called shape-components are all independent of the size-component and can tell us nothing about how shape depends on size. As all know who have attempted such analyses, these other components are likely to be a mixture of size and shape-indicators. The consequences of this reasoning have been drawn in full by the recent work of Bookstein and his collaborators (see below).

Mosimann and James (1979) have taken up the problem expounded by Mosimann (1970) in relation to the analysis of size and shape variation in North American blackbirds. The choice of the size variable is shown to be important and exact statistical tests are provided under a multivariate log-normal assumption. According to Mosimann and James, the standard approach to allometric analysis by means of the classical allometry equation and direct generalizations of it, such as employed by Gould (1966), Spielman (1973) and Thorpe (1976), can at best only reflect the mean trend of shape with some size variables. They believe the method used in their paper can determine visually and geometrically meaningful variables of size and shape which permit the direct analysis of the association of one with the other. They analyse relationships in Florida red-winged blackbirds and find that bill-depth and bill-shape, but not bill-length, display interesting covariation across Florida in a manner suggestive of variation found in Darwin's celebrated finches. It is also consistent with size-trends observed in a variety of bird species in eastern North America.

Mosimann and Malley (1979) give a precise account of Mosimann's analysis of size and shape variation in relation to some other methods and interpreted in a geometrical context.

Reyment (1982d) has applied Mosimann's method to the analysis of size-shape relationships in Middle Cretaceous ammonites of the genera *Subprionocyclus* and *Reesidites* from Japan.

Siegel and Benson (1982) discussed the problem of matching two dimensionless polygons with the end in view of analysing the comparative morphology of animal skeletons. The method is based on resistant fitting techniques, featuring an algorithm for median resistant fitting, a procedure which is a kind of robust regression and which seems to provide ready identification of parts with similar shapes. The results obtained for such disparate remains as ostracods (comparing male and female carapaces in phylogenetic work) and primate skulls seem encouraging.

An early worker in the field of shape variation was Thompson (1942). His graphical technique for representing shape homologies and affine relationships has recently been taken up quantitatively by several workers in a series of publications.

Bookstein (1978) formalized the ideas involved in Thompson's method of depicting shape variation. In this monographic treatment of the analysis of shape, Bookstein lays claim to reforming the subject of morphometrics, which he redefines as the measurements of shapes, their variation and change, within the framework of modern applied geometry. A shape change is defined as a map of one shape onto another which sends arcs (or surface patches) smoothly onto arcs, and corners (or edges) onto corners; then it sends landmarks' onto landmarks (Bookstein, 1978, pp. 8–9). He takes issue with Mosimann's (1970) treatment of the subject as this is in terms of "shape-variables" which are ratios of sums of distance measurements among "landmarks". Bookstein insists that the problem should be defined as being the "extraction of a finite set of measures for shapes already given", thus allowing for the recognition of divers irregularities of form, which the conventional construction of shape variables from distances does not permit. Standard methods of multivariate analysis applied to size and shape changes are given poor marks by Bookstein (1978, pp. 31–32), in a discussion not free of polemic overtones, because such analyses do not depict biological forms and are therefore claimed by him to be not much better than worthless. Doubtlessly, some of Bookstein's criticisms must be considered carefully, particularly with respect to the inadequacy of the usual system of selecting variables and the desirability of considering this problem in the light of Bookstein's (1978, p. 90) biorthogonal grids. Bookstein (1978, p. 63) projects his methodology into a plea for "*geometric* morphometrics" in which multivariate statistics enters solely to compress the information extracted from a delicate analysis of actual curved form".

Rather curiously, Bookstein and co-workers (Humphries *et al.*, 1981), following on Bookstein's rejection of linear shape representations as a useful analytical tool, propose a modified principal components approach for quantifying shape differences among populations independent of size, exemplified by species of the fish *Coregonus* from Lake Superior. In cases in which the first two principal components confound size and shape, "size-effects" are eliminated from one axis with what are termed "shear-coefficients", derived from the regression of general size on principal components centred by group. The general size-factor is estimated by the principal axis of the within-group covariance matrix of the logarithmically transformed data. It is suggested that residuals from the regression of general size on the transformed axes approximate a shape-discriminating factor which is uncorrelated with size within groups and which shows up the inter-populational shape differences contained in the first two principal components. The rather heterogeneous, somewhat arbitrary and partly graphical, technique advocated by Humphries *et al.* (1981), clearly worthy of consideration, is interesting in the present connotation of the analysis of size and shape because it brings out with almost blinding clarity the real difficulties involved in producing an unchallengeable treatment of differential growth. In contrast to Mosimann's ideas, they prefer to approach a shape-discrimination study via a factor of the Jolicoeur type, ratios are not acceptable, and there are many other objections (Humphries *et al.*, 1981, pp. 291–292).

The analyses of Humphries *et al.* of species of *Cyprinodon* from Lake Chichancanab, Mexico, three species of minnows and five allopatric populations of ciscoes (genus *Coregonus*) from Lake Superior illustrate the usefulness of the methodology in certain circumstances. However, if the variation is not restricted to the first two principal components, the method may fail. This would be the case for the crab data in Chapter 3.

Of particular significance for the work of all morphometricians is the emphasis Bookstein and his co-workers place on a proper consideration of the selection of diagnostic characters for measurement.

Essentially the same idea as that expounded in the above work is expressed in the study by Brower and Veinus (1978) of the Silurian eurypterid *Eurypteris remipes* and the Silurian-Devonian brachiopod *Dicoelosia varica*. Brower and Veinus analyse characters obtained from Thompson-grids that are the coordinates of points which are homologous and, or, topographically constant relative to orthogonal axes. The resulting data are then analysed by means of Jolicoeur's (1963b) model for multivariate allometry, just as was done by Humphries *et al.* (1981). However, Brower and Veinus do not attempt to partition size-effects from shape-effects. A further study in the same vein is that of Brower *et al.* (1979) in which the Early Permian amphibian *Diplocaulus magnicornis* and fossil man are analysed, using variables obtained from point-coordinates. Another application of Thompson-grids is that of Andrews *et al.* (1974).

Cherry *et al.* (1982) have compared body shapes for 184 taxa of frogs, lizards and mammals using eight linear traits from the major parts of the body. Of the metrics used by them, the Mahalanobis' generalized distance was claimed to be the least adequate for quantifying overall differences in shape. The most satisfactory results were said to be yielded by the simple measures known as the Manhattan distance and the Canberra metric (Cherry *et al.*, 1982, p.916). We find these conclusions rather surprising.

GRAPHICAL DISPLAYS

We append here a few comments on the role of graphical displays in morphometrical work. One of the main interests attaching to a multivariate analysis concerns the graphical presentation of the results obtained. Everitt (1978) has written a book entirely devoted to this topic to which we refer the interested reader.

In most cases, the graphical representation will consist of the plots of scores in the planes of interest. Such plots can be augmented by the inclusion of figures of the organisms represented by the points, as, for example, was done by Reyment (1973) for ammonites.

Andrews (1972) proposed a method for obtaining visual representations of multivariate data whereby each multivariate point is mapped into a function of a particular form. Each of the p-dimensional observations, which defines a trigonometric function, is plotted so that the set of points will appear as a group of lines drawn across the page. An example of the use of this method in palaeontology is given by Reyment and Neufville (1974). The technique can be usefully employed in conjunction with principal components and canonical variates.

We can also mention the problem of the horseshoe (or Guttman) effect in plots of principal component scores and principal coordinates for displays arising from dichotomous data (and highly correlated continuous data for some association measures). Williamson (1978) has described a way to avoid this nuisance.

13
The Measurement of Environmental Variation

In many animals and plants, the environment in which growth takes place determines to some extent the final form adopted by the organism. This is often referred to as the norm of reaction (cf. plants, bryozoans, corals). It is remarkable how little this environmental variation has been studied, no doubt partly as a reaction against "variety hunting" of the last century, when any variation which could be established was named and described. Environmental variation is, nevertheless, of interest in its own right, and may attain magnitudes comparable with specific or even generic variation. Examples are the effects of strong winds and grazing pressures on plants (Whitehead, 1959), the variation of locusts when reared in crowded habitats, the changes in shape being essentially the same in a wide range of species and genera (Blackith, 1962), and variation in Recent brachiopods (McCammon, 1970). Two kinds of variation were identified in the brachiopods, one of which was due to physical constraints arising from overcrowding or growth between dead shells and rocks; this type of variation resulted in asymmetry. The second kind of variation encountered was seen in the principal components of the dispersion matrix for five characters: length, width, thickness, beak angle of pedicle valve, and the ratio of the area of the two sections, termed the asymmetry ratio. The first principal component was found to represent size variation, having all its coefficients positive. The next two components represent shape variation of different kinds. Using these components as the axes of a chart on which the positions of the individual brachiopods of a different species were plotted, it was concluded that the costate species, even though not closely related to one another, exhibited one kind of shape change, whereas the non-costate species exhibits another. As in research into the patterns of variation in locusts and grasshoppers, we have the curious result that species within the same genus may show quite distinct patterns of variation, yet much less closely related species, sometimes allotted to distinct families, may show identical patterns. Here, apparently, is a widespread type of variation which has been neglected but for which the necessary techniques of study are now available in morphometric analyses. Yet the analysis may not be easy because variation of form may be strikingly distinct from the variation in colour (Blackith and Roberts, 1958) so that similar colour forms are widely disparate morphometrically.

Campbell and Dearn (1980) studied morphological variation between and within three closely related species of grasshopper. They found that a certain combination or pattern of characters, related to the shape and size of the eye, differed markedly between species, while being relatively invariant within species, with other patterns of characters exhibiting obvious altitudinal variation. Such a result appears to be relatively widespread; work by Campbell and colleagues on such diverse things as blowflies, bats (Campbell and Kitchener, 1980), kangaroo paws (Hopper and Campbell, 1977), and eucalypts (Hopper, Campbell and Moran, 1982) all show geographical variation in certain combinations of the characters but with the conjunction of characters effecting discrimination between species being relatively invariant within and between populations of the same species.

These results would appear to support the proposal by Carson (1975) that there are two different systems of genetic variability in bisexual species—an "open" variability system showing intraspecific variation, and a "closed" variability system determining species-specific characteristics and showing no variation within species.

The anostracan brine-shrimp *Artemia salina* can live in a wide range of environments, and is tolerant of variations in salinity. For about a hundred years now it has been known that the body proportions of this branchiopod are correlated with salinity. Gilchrist (1960) analysed brine-shrimps from water of two different salinities, and from three widely separated localities. She measured the total length, abdominal length, abdominal width and the length of the caudal furca. Later, Reyment (1966a) treated these data by means of principal component and canonical variate analyses.

The principal components, whether extracted from the dispersion matrix of the untransformed data or from the dispersion matrix of the logarithms of the data, did not alter appreciably with salinity or provenance. The general consistency of these components was of the same order as Reyment's earlier analyses of fossil and Recent ostracods had uncovered. In agreement with Gilchrist, Reyment concluded that the variability is partly of genetic and partly of environmental origin.

Rather more informative was a canonical variate analysis carried out on six variables, the original four together with the length of the prosoma and

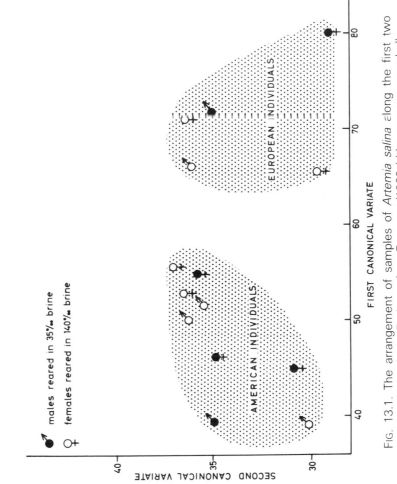

FIG. 13.1. The arrangement of samples of *Artemia salina* along the first two canonical variates. Redrawn from Reyment (1966a) (the axes are unscaled).

the number of setae per furca. Figure 13.1 shows the arrangement of the various samples along the first two canonical variates. Material of American origin tends to have low values of the first canonical variate, in which the length of the abdomen is important, but total length much less so. The width of the abdomen and the furcal length, though about half as important as the abdominal length, act in opposite senses. Material of Mediterranean origin tends to cluster along those parts of the first canonical variate where the score is highest. It is interesting, and rather strange, that differences of salinity affect the shapes of brine-shrimps much less than differences of geographical origin. There is, moreover, evidence of sexual dimorphism which expresses itself in the fact that males have consistently lower scores along the first canonical variate axis than have females, just as brine-shrimps from 3.5 per cent sea-water have consistently lower scores than those from 14 per cent sea-water.

A species of the living freshwater ostracod genus *Cypridopsis* has also been found to be influenced by the environment. The shape of the carapace differs slightly according to whether the water in which it lives is relatively rich in lime, stagnant, or fresh. Three characters were measured on the carapace by Reyment and Brännström (1962), to wit, length, height and breadth. Both generalized distances and canonical variates were computed. Adults from a normal freshwater environment score most highly along the first canonical variate, which accounts for 99.8% of the variation, in conjunction with the second variate; the canonical vector is (0.785, 0.428, −0.449).

As shown in Fig. 13.2, there is a regular pattern for adults (which are growth-stage IX) and individuals of the last larval stage (which belong to growth-stage VIII) by which ostracods grown in fresh water score more highly, not only along the first canonical variate but also along the second, in comparison with the individuals reared in lime-rich, or stagnant water. Thus the effects of "contaminating" the water are almost identical for the two contaminants and form an axis of variation which lies obliquely across the first two canonical variates.

Changes of form at different geographical localities make up one of the most frequently encountered kinds of environmental variation. Sokal and Rinkel (1963) remarked that "It may occasion some surprise that the statistical problems involved in the analysis of geographic variation have been almost untouched in spite of the amount of emphasis that has been given to this subject as a cornerstone of the New Systematics. With the exception of the valuable papers of Pimentel (1958, 1959), directions for statistical analysis of such data have been either erroneous or lacking altogether". This was something of an overstatement, since Burla and Kälin (1957) had clearly demonstrated how to combine seven morphometric

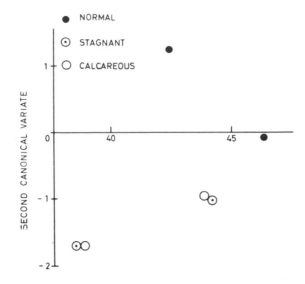

FIRST CANONICAL VARIATE

FIG. 13.2. Plot of the first two canonical variates for *Cypridopsis vidua* from "normal", "lime-rich" (calcareous) and "stagnant" (axes unscaled) laboratory environments. Redrawn from Reyment and Brännström (1962).

characters of *Drosophila* into a discriminant function contrasting two geographically separated populations: nevertheless, the complaint was broadly justified. No doubt this state of affairs arose because of the need to handle a multiplicity of characters simultaneously. Sokal and Rinkel used "factor analysis" to reduce this multiplicity of geographically modified attributes to a smaller number based on the limited capacity of the organism to bring into play more than a few independent patterns of growth, much as Jolicoeur had earlier tackled geographic variation in the wolf (Jolicoeur, 1959, 1963c).

Other ways of investigating geographical variation have been framed in terms of trend-surface analysis and by partitioning the area in question into definable categories; Gabriel and Sokal (1969) discuss these avenues of approach for univariate and multivariate analyses.

Sometimes a species living in two different places shows a distinct pattern of growth at each. Malmgren (1970) measured the bryozoan species *Floridina brydonei* from the Danian of Denmark, and made a principal component analysis of the covariance matrix of the four measurements: length of zooecium, width of zooecium, and length and width of the aperture, taken on two samples. The sample from Ny Klostergaard (northwestern Jutland) contained 100 specimens and that from Klintholm (eastern Funen) contained 89.

The first principal components for each sample are essentially size vectors with all the coefficients positive. For the sample from Ny Klostergaard, the first vector was determined to represent 51.7% of the total variation: this vector is (0.155, 0.576, 0.466, 0.653). For the Klintholm sample, the first vector (0.150, 0.828, 0.191, 0.505) accounts for 60.7% of the total variation. These vectors, which are mutually inclined at 23°, differ significantly from each other in orientation. The two samples, however, do not differ significantly in the directions of the remaining principal components, which represent essentially shape factors.

The principal components reflect the direction of evolutionary or environmental change under geographical isolation; canonical variates can be used to construct a diagram of the cross-section of this process actually attained by the organisms at any given moment. For instance, Delany and Healy (1964) measured ten characters on each of 156 specimens of the long-tailed field-mouse from eight islands, and two points on the mainland, of northwest Scotland. The first two canonical variates were used as the axes of a chart showing much overlap between the various populations. Indeed, the concept of subspeciation previously applied to these mice was shown to be dubious. It is interesting that the larger mice from the island of Rhum are discriminated along the second canonical variate axis and not along the first, although this situation would probably have been altered had more of the groups been involved in substantial size variation. It cannot be stressed too often that the order in which a particular biologically meaningful vector occurs in a multivariate analysis depends on the selection of the original material for study as well as on the intrinsic morphometric properties of the material. The foregoing study has been extended by Delany and Whittaker (1969) who combined nine skull characters to obtain both canonical variates and generalized distances.

Environmental variation in the morphology of animals is often thought of as adaptive, although the justification for this point of view is sometimes tenuous. Fujii (1969) has explored the possibility of classifying animals into a hierarchy according to their ecological characteristics rather than their morphological ones. He employed 11 attributes such as developmental time, longevity, weight, and number of eggs per female, and obtained a hierarchy which distinguished the two species of bean weevil that he used, at low degrees of similarity, followed by clustering of the Japanese strains together, and the American strains together, before the remaining strains entered the clusters. Fujii also notes the possible advantages of representing the groups of weevils, as expressed by their ecological characteristics, on a multi-dimensional ordination system rather than in a hierarchy.

Just as geographical variation among related organisms may involve only a limited number of common dimensions of variation, so may other types of variation, such as sexual dimorphism, have a similar effect. Geographical variation in grasshoppers has been studied by Blackith and Roberts (1958) and by Blackith and Kevan (1967), using multivariate methods of analysis.

Discrimination between closely related taxa may be rendered less accurate because of superimposed geographical variation. When Rempe and Bühler (1969) were distinguishing the mandibles of the water-vole *Neomys fodiens fodiens* from those of *N. anomalus milleri* by means of the three characters: mandible-length, height of the lower jaw, and length of the tooth-row, geographical variation was observed to interfere with the comparisons along the discriminant function $(-1.00, 2.58, 2.58)$. However, once a fourth character, the minimum height of the lower jaw, was included in a canonical variate analysis, this interference was effectively suppressed. It seems from the data published by these authors that the geographical variation in mandible-shape takes the form of an essentially north-south cline.

Palaeontologists are frequently faced with the problem of material which varies not only from place to place but from time to time in a geological sequence. There may be variation not only in the form of the individual organisms, but also in their relative abundances (Reyment, 1963a). The potential causes of such variation are many and various; among them are post-mortem sorting by currents of water; changes in the salinity of the aquatic environment; differential growth leading to different shapes for individuals of the same species but of different sizes; changes in climate leading to warmer or colder environments, and, in some cases, crowding effects when sessile animals are bunched together because the available resting sites (rocks etc. on the sea-floor) are in short supply.

Reyment (1971b) has partitioned the changes of morphology in the ostracod carapace into a "size" component, based on the first principal component, and a "shape factor", based on the second such component, which he considers to be, in part, a measure of the genetically determined variation, as opposed to the environmentally controlled ecophenotypic variation, reflected in a broad sense by the first principal component. Of course, these partitions of the total variation (when there are only three variables measured) can only be interpreted in this way in a very general sense.

Reyment (1971b) used measurements on the carapace of Paleocene ostracods sampled at different levels in bore-hole cores in a clay-shale for his comparisons and found that the responses of several species are correlated such that they grow larger or smaller on the average when the environment becomes slightly more favourable or slightly less optimal. In this way, some approach to the ecology of past environments may be made, and for this purpose those fossil forms which extend over a wide range of strata are most valuable, although such forms are often studied less precisely because they are of less use as zone fossils. Ecophenotypic variation in microfossils has been used by Reyment (1980b) as the basis of a morphometrical logging procedure.

This method of biostratigraphical analysis is analogous to the electrical and radioactive logging methods of the petroleum industry. Under certain conditions, the electrical logs (self potential, resistivity and redox—for an account of scientific logging, see Pirson, 1977) provide a measure of the palaeo-environment and can be used as part of the process of identifying ecophenotypic variation and genetic tracking of minor shifts in the environment as opposed to evolutionary changes of ecological significance, as determinable by the application of the criteria of Lande (1976). The quantitative genetical aspects of this kind of analysis are considered by Reyment (1982c) in a paper investigating the relationships between morphometrical variation through time, environmental factors of several kinds, physical and chemical, and evolutionary shifts in the morphology of fossil micro-organisms.

A few words must be said concerning the concept of size-invariant discriminant functions and generalized distances. Burnaby (1966a) raised the question of size-invariance arising from his interest in "pure" taxonomic comparisons, unsullied by arbitrary size compositions of samples arising from varying stages of growth, ecophenotypic differences, and polymorphisms, including sexual dimorphism. By means of a transformation within complementary subspaces, he produced a solution of this problem which is intellectually satisfying for many situations. Rao (1966) presented a somewhat more general view of growth invariance (he encountered the problem through being a referee for Burnaby's paper). Subsequently, Gower (1976) gave a general treatment of the subject including explicit computational details for a wide range of situations. More recently, Humphries *et al.* (1981) have questioned the generality of the Burnaby–Rao–Gower procedure for growth invariance. Despite their comments, there is little doubt that the concept of growth invariance is often useful and necessary in taxonomic work.

Reyment and Banfield (1976) applied the Burnaby–Gower solution to the analysis of planktic foraminifers from the Danian of southern Sweden; the results obtained are useful and accord well with the overall multivariate interpretation of the data (Malmgren, 1974).

Another kind of ecological problem is posed by nearness relationships such as occur, for example, in a forest, between various species of corals in a reef, epiphytes, and sedimentational environments. Reyment and Banfield (1981) analysed the multivariate asymmetric relationships generated by nearest unlike depositional environments (among six categories) in the Mississippi Delta. The mathematical method employed may be thought of as an asymmetric analogue of principal component analysis in which principal planes replace principal axes. Presumably, some of the phytosociological cases considered in Chapter 14 could be amenable to analysis by this method.

14
Multivariate Studies in Ecology

From the measure of the amount of variation in an organism, according to its environment, to a measure of the features of the environment itself, is a modest step conceptually, but one that requires some consideration at the biological level. If we leave aside the few remaining vestiges of the idea that morphological characters could be divided conceptually into adaptive and non-adaptive ones, and few ideas seem more dated and inappropriate to current ways of thinking about evolutionary problems, the ensemble of morphological characters can be considered as the direct result of environmental pressures on the genotype. We cannot have any such confidence when we come to consider the ensemble of environmental features that we are capable of appreciating, for in all but a few instances, we know next to nothing of the extent to which changes in any one of the features influence an organism living in the habitat. For this reason, the association of environmental features into vectors in terms of changes in the abundance, behaviour, or growth of the organism is even more tentative than in morphological studies, but this statement holds true, essentially, for all synecological investigations.

Buzas (1967) used canonical variate analysis to uncover the features determining, or appearing to determine, the abundance of 45 species of foraminifers in eight areas of the lagoon or open sea bed, off the Texas coast. Bays on the lagoon were shown to have different abundances of foraminifers when the living populations are considered but not when the dead population is estimated. Three stations of the open ocean at which the samples were taken at depths of 0–30 m gave patterns of abundance which clustered together closely on a two-dimensional plot of the first and second canonical variates, and were well separated from the deep-water station samples. The first canonical variate seems to reflect depth differences faithfully, whereas the second canonical variate seem to express ecological differences other than depth and associated features of the environment. Along this variate, populations from the bays seem more like the deep-water samples than do those from the shallow sea, whereas on the first canonical variate, there is a steady progression from bay samples to shallow water samples to deeper water samples. The samples from the moderately deep oceanic water seem to be aberrant when assessed along the second and third variates. Evidently, factors other than depth are of consequence to foraminifers and Buzas suggested that canonical variate analysis should help ecologists to distinguish between such basically distinct combinations of factors as those associated with depth and those not primarily associated with depth.

Cassie and Michael (1968) have applied principal component and canonical correlation analyses to the study of the invertebrate fauna of a silty intertidal mud-flat in New Zealand. They were able to discern four communities of species by means of the principal component analysis. The community into which any given species falls is determined by the principal component to which it contributes the highest "factor loading" (score along the principal component). The distribution of the various communities seemed to be determined by the coarseness of the underlying sediments, as expressed by a linear function of the eight fractions into which the particle size-range of the sediment had been separated. This linear function proved to be about twice as effective as the simple arithmetic mean particle size, and its more widespread use may help to reduce some of the loss of information associated with the use of the mean particle size for sediments where the grain size has a skewed distribution.

Cassie (1962) utilized principal components to distinguish a group of four species of plankton, the behaviour of which was mainly determined by the mixing of two water masses in an estuary, and a smaller group of two species whose behaviour did not appear to be governed by this factor.

Pelto (1954) used the fact that in an association which is virtually a pure stand, the entropy of mixing of the constituent elements is low relative to that of an association in which many elements are roughly equally represented. Thus high entropy delimits the zone of intergradation between two associations or communities (Howarth and Murray, 1969). This concept has been employed by McCammon (1966) to provide a criterion by which principal components could be rotated to positions such that the entropy of the system is minimized, when the components might be expected to link the centres of associations. More recently, McCammon (1968) has compared this approach with several others, and shows that his "multiple component analysis" brings about groupings of the elements of Bahamian sediments which have a lower intra-group variance than other candidate methods, including rotated principal components, although the improvement on the principal component solution was not large. McCammon initiates his principal component analysis by first clustering the data by means of the unweighted pair-group method of Sokal and Sneath (1963). The number of distinct facies is then estimated from the resulting dendrogram, thus allotting the original data (variables of the facies) into submatrices in accordance with the decision taken as to the number of distinct facies present. Principal components are then taken for each of the submatrices, and the scores of the samples computed and converted into a facies-map showing the distribution of the facies over the area sampled. The hypothesis is, essentially, that to each of these principal components corresponds a particular type of environment or facies. Each sample is then classified according to the component it most closely resembles, and then allocated to the appropriate environment.

As an example of the use of multivariate statistical techniques in an ecological problem, we offer the case of rotifers living in a series of inland waters in Europe, which have been examined in detail by Dr A. Nauwerck. Four characters of the rotifers, the length and breadth of the body and the lengths of the anterior and posterior spines, were measured on samples of the order of 50 individuals from ten different inland waters, of which three, the Wolfgangsee in Germany, the Vierwaldstättersee in Switzerland, and Lago Maggiore in northern Italy, were chosen for more detailed morphometric analysis. We are indebted to Dr Nauwerck for permitting us to make use of this material.

As a first step, a canonical variate analysis was run on the measurements of *Kellicottia longispina* and *Keratella cochlearis* from the three localities to see whether the variation of form from lake to lake was consistent from one species to another. Figure 14.1 shows the analyses for each species, which in fact display a remarkable degree of parallel variation: the first canonical variate consistently distinguishes material from the Wolfgangsee from that taken in the other two lakes, whereas the second canonical variate distinguishes the Vierwaldstättersee material from the remainder. Both canonical vectors correspond to significant roots, a statement which translates itself into Fig. 14.2 which shows that there is a substantial, but not complete, degree of separation between the three sets of points relating to *Kellicottia,* so that the reality of the ecological differences between the forms of the rotifers from the three lakes cannot be doubted.

A second canonical variate analysis was performed on the combined samples of *Kellicottia* and *Keratella* from the three localities, the results of which are displayed in Fig. 14.2. The extent to which the second canonical variate separates the material from the three lakes is unaltered, but there is a curious reversal of the direction of the separation. In the case of *Kellicottia,* the specimens from Lago Maggiore have the smallest negative scores,

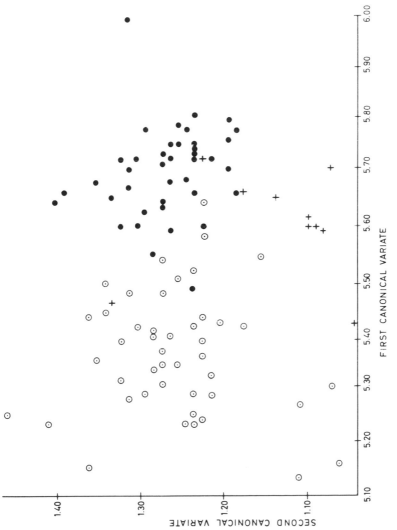

FIG. 14.1. Canonical variate chart showing the relative morphometric distances between individual rotifers of the species *Kellicottia longispina* from the Wolfgangsee (open circles), the Vierwaldstättersee (crosses) and Lago di Maggiore (closed circles). Data of Dr A. Nauwerck (axes unscaled).

FIRST CANONICAL VARIATE

SECOND CANONICAL VARIATE

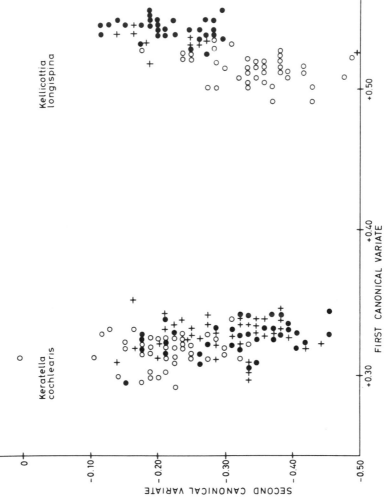

FIG. 14.2. Canonical variate chart of *Kellicottia longispina* and *Keratella cochlearis*, based on four morphometric characters. Symbols and locations as in Fig. 14.1. Data of Dr A. Nauwerck (axes unscaled).

whereas in material of *Keratella,* they have the largest. The clear separation between the two species, entirely on the basis of the first canonical variate, is apparent. A third canonical variate separates the localities but the physical interpretation of this is not clear.

The seven remaining lakes for which samples were available are more or less intermediate between the three selected for a detailed analysis, so that by studying the physical properties of the three exemplar lakes, we may gain some clues as to the reasons for the variation in shape of the rotifers. The first canonical variate seems to correspond fairly closely to the pH of the water (Wolfgangsee, 7.78; Vierwaldstättersee, 7.47; Lago Maggiore, 7.14) but it has not proved practicable to reify the nature of the second canonical variate, which does not seem to depend at all closely on temperature, or on conductivity. The changes of form of organisms living in a lake may, of course, respond to shifts in trace-element concentrations or in other factors which are unlikely to be measured in the absence of some definite reason for so doing.

We feel that we should underline the significance of the rotifer analysis for quantitative interpretations of palaeontological observations. It is often speculated on as to whether ecological differences in form and ecophenotypic variation in series of a fossil species can be made the basis of a meaningful analysis, as suitable comparative material is difficult to find for Recent organisms. Nauwerck's data provide at least one good set of comparative material of value to palaeontologists.

Cassie (1967), in a principal component analysis of the plankton of Lago Maggiore, has shown that there is no "winter" component of the major groups of plankton ceding to a "summer" grouping: the larvae of *Kellicottia* were found to occur in Cassie's group 1 (a twelve-species group including *Eudiaptomus vulgaris, Cyclops* larvae and other rotifers) throughout the year.

Hendrickson (1979) discussed the use of discrete multivariate data in ecology, with interesting examples concerning industrial melanism in moths, the sizes of turtles at high latitudes, and the growth form of pines near the tree-line.

One rarely considered matter is that of the scale on which field studies are conducted. Hengeveld and Hogeweg (1979) have described the effects of moving from the scale of the Netherlands to that of Europe as a whole when assessing the distributional patterns of Carabidae (ground beetles).

Not only the organisms living in a particular environment, but that environment itself, can be examined for significant contributions to elements of the synoptic, multivariate, measures of climatic response. Hocker (1956) has used discriminant functions to distinguish the meteorological factors of consequence to the establishment of *Pinus taeda.*

Laurec *et al.* (1979) emphasize the essentially dual nature of ordination experiments in ecology, with, in general, two-way tables in which the rows are observations whereas the columns contain the variables derived from the set of taxa. They considered these first in the context of principal component analysis, and then of correspondence analysis, finally using principal coordinate analysis as an ordination technique. They illustrated their arguments with an example in which the annual cycle of phytoplankton at Paluel is followed within the framework of axes of variation which display both the species involved and either the number of samples in which it occurs, or the abundance of the species.

Steinhorst (1979) has provided a useful account of the problem of assessing niche overlap by traditional techniques. He compared niche overlap with that of communities or habitats and commended a multivariate approach involving the volume of intersection of the hyperellipsoids relating to each species in multivariate space, the breadth of the niche of each species being defined by the volume of the hyperellipsoid itself, which is proportional to the square root of the determinant of the covariance matrix of the organisms and of the resources available to them.

Dillon (1980) analysed the relationships between patterns of distribution of two species of land-snails by principal components. The study was constructed so as to comprise "predictor variables", elevation of the sampling sites, the angle of slope of the terrain, the aspect of the slope, the percentage of vegetable cover and the nature of the substrate. The plot of the scores for the first two principal components is interesting because it gives a good picture of snail-sites in relation to vegetational cover (which dominates the first principal component) and elevation and slope-orientation, which dominate the second principal component.

Gittins (1979) has summarized the application of canonical variate analysis and canonical correlation to the study of vegetational distribution patterns in a very comprehensive account of this type of quantitative ecological analysis. This work gives a useful and timely discussion of the problems and general methodology generated by the use of frequency data for the distribution of species in connexion with standard methods of multivariate analysis.

Erez and Gill (1977), using a principal component analysis with rotation, studied 64 biogenic constituents at a depth of 40 m in the Rad Burka of the Gulf of Eilat, a marine environment characterized by dunes, a swash-zone, dead-reef platforms, sand channels, a living coral reef, a *Cymodocea* "lawn", reef knolls, and a *Halophila* "lawn". The multivariate analysis ordinated the samples into three classes: shallow near-shore, the upper slope, and the lower slope zones. This is an example of a palaeoecologically oriented study of a Recent environment.

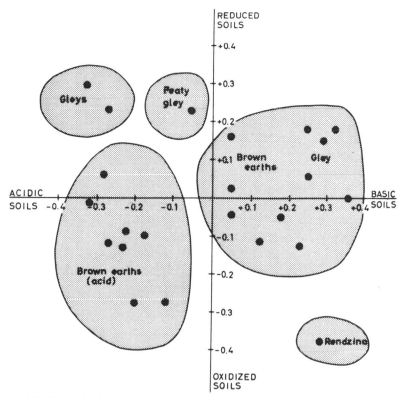

FIG. 14.3. A principal coordinate clustering of Glamorganshire soil profiles based on mean horizon similarities: distances between the points approximate $(2(1 - S))^{1/2}$ where S is the measure of similarity. Redrawn from Rayner (1966).

In his book on classification techniques, Gordon (1981) has, in one of the chapters, provided a useful and concise account of the application of multivariate methods to ecological problems.

As an example of correspondence analysis applied to the study of the palaeoecology of microfossils, we cite a paper by Cugny and Rey (1975). Their analysis succeeded in extracting "factors" descriptive of the distributions of foraminifers and algae, energy of the palaeo-environment, depth and temperature. (We note here that in some of the recent French literature, the term "factor analysis" is applied to Benzécri's (1973) correspondence analysis.)

Rayner (1966, 1969) has compared soils in Glamorganshire, Wales, by means of dendrograms obtained from a numerical taxonomic approach, and also by an ordination using principal coordinates. A difficulty arises in such work, because the same type of horizon does not always occur at the same depth in the soil profile, yet the horizons cannot always be unambiguously classified from a knowledge of their physical properties. Attempting to sort horizons on the basis of their similarities does not lead to a conclusive result and there are substantial differences between the dendrograms constructed for the similarity of the most alike pairs of horizons, and on the average of the similarities of the best matched pair of horizons.

The principal coordinate approach gave a clear interpretation along the two axes tentatively reified as reflecting acidity and redox potential: the major soil types involved were distinctly separated along these two axes of variation. Figure 14.3 shows these groups. Although the picture yielded by the principal coordinate method was not unlike that given by the principal components, it did represent an improvement in terms of the ease with which the results could be interpreted. Multivariate pedology has come a long way since the pioneering work of Cox and Martin (1937). Much of this improvement stems from the recognition of orthogonal axes of variation, and in particular, those arising from multivariate analyses of the data give the experimenter the same intellectual and practical control over the experimental material that the analysis of variance provides in the univariate case, with the added advantage of affording multivariate vectorial representations of the processes at work in the soil. This approach comes out strongly from Holland's (1969) work on soil nutrients, in which he combined the variables into principal components and showed how their orthogonality was of conceptual value.

Gökcen and Özkaya (1981) ordinated turbidites using canonical variates and found clear separation between olistostromes, proximal turbidites and distal turbidites when seven size distribution characters of the sediments were taken into account.

Usher et al. (1982) considered that some analysts have fallen into the trap of neglecting uncommon species in animal associations or plant communities. They liken these rarer species to the oil in an engine, slight in mass relative to the whole, but vital in function. They find that a principal coordinate analysis of Antarctic ecosystems reveals clear groupings dependent on the moss or lichen cover.

15
The Distribution of Plants and Animals

This chapter is in two sections, respectively, botanical and zoological. It deals with the comparison of floral and faunal elements and the associations into which they might be thought to fall. These examples of "abstract morphometrics" are broadly complementary.

PHYTOSOCIOLOGICAL STUDIES

Many of the problems with which phytosociologists have to deal are essentially morphometrical; but it is a moot point whether phytosociological studies as a whole are properly to be subsumed under the heading "morphometrics". In general, the presence or absence of plant species in a quadrat constitutes a "character" of that quadrat with the aid of which (together with the corresponding information about many other species) quadrats can be clustered or otherwise investigated. Nevertheless, there are special concepts in phytosociology, such as "fidelity" and "indicators", which can be incorporated with what we ordinarily treat as morphometric studies only at the risk of taking us far from topics familiar to the authors of this work. We content ourselves, then, with reminding readers that there are some excellent reviews of the topic which it would be foolish to try to summarize here. Outstanding are those by Dagnelie (1965a, b), who, after presenting the basic theory, goes on to give an example of factor analysis applied to a matrix of correlations between the various ecological variables measured. A factor reflecting the influence of physical variates was extracted, as was one reflecting the distribution of the plants which was not accounted for in these terms.

Detailed accounts of multivariate methods have been written by van Groenewoud (1965), Gittins (1969, 1979) and Orlóci (1978, 1979) with phytosociologists particularly in mind.

Orlóci (1968b) has commented that the application of information analysis in phytosociological work is rapidly gaining momentum, and it may be that for special purposes information analysis is more suitable than are multivariate methods. However, there are constant swings of the pendulum of fashion between the entropy-information approach to classification and the more strictly multivariate approach, and it is unwise to ignore the potentialities of either.

The very idea of ordination stems from the essentially multivariate approach to phytosociology of Goodall (1954a, b) whose work has been of great influence in this field. More than a quarter of a century later it is hard to appreciate the hostility with which Goodall's early papers in phytosociology were met.

Although there has been much valuable work done using analytical methods constructed particularly with phytosociological investigations in mind (Williams and Lambert, 1959) there is no real evidence that it is necessary to leave the general framework of multivariate techniques as we have outlined it in this book. Harberd (1962) has used the generalized distance and canonical variate techniques quite successfully to examine plant communities, even with qualitative data. Whilst no one could deny that certain problems may require special analytical methods, it seems open to question whether the enormous number of special methods devised for particular fields of study is at all necessary; in fact, the construction of such methods often reflects a restricted knowledge of what is going on outside the special field.

Gittins (1965a, b) has shown that there are benefits in applying both Q-type and R-type analyses to phytosociological observations, and has made the valuable recommendation that ecological and environmental variates should be included in the analysis to help in the reification of the various axes of variation that can be extracted. He also points out the advantages to be gained from the very detailed analyses possible when the monothetic association analysis can be applied both in the direct and inverse forms (the nodal analysis of Lambert and Williams, 1962). Despite our general preferences for methods of analysis that fall into the framework of multivariate techniques outlined in this book, and a suspicion of special techniques invented to solve particular problems, there can be little doubt that the intensive study of definite aspects of a problem by workers deeply familiar with it and who have learnt to exploit the potentialities of the method of their choice, will often produce more meaningful results than the casual application of even the most powerful analytical techniques by workers whose grasp of the problem, and perhaps also of the technique for its solution, is imperfect. Jeffers (1967b) has pointed out that when multivariate methods have been discarded as failing to contribute to the solution of a problem, an interpretation by a more experienced worker will often show that the initial use of the technique was poorly judged.

Orlóci (1979) makes the point that most of the multivariate methods traditionally employed by phytosociologists, and indeed by ecologists in general, involve assumptions about linearity which are often inappropriate when dealing with systems characterized by nonlinearity. He goes on to commend techniques such as nonmetric multidimensional scaling as a way round this difficulty.

In general, phytosociological research has exploited the battery of special techniques which it has invented with flair and persistence, and one might think that the study of animal communities would be further advanced if it had been prosecuted with equal devotion. As always happens, there is a tension between the advantages of maintaining special techniques in terms of which groups of workers have learnt to conceptualize, and the advantages of using general methods of analysis that most workers can discuss in the reasonable hope of being mutually intelligible.

That general methods can be applied to phytosociological problems to good effect has been shown by Orlóci (1966), who made a principal component analysis of the frequencies with which certain plants occurred in nine district habitats within a "dune-and-slack" vegetation complex. The ordination that was produced by the first two of these components is shown in Fig. 15.1: a reasonably clearcut separation of the habitats is evident. The efficiencies of principal components, and of two other methods of ordination, were compared, and principal components proved to be materially more efficient in accounting for the interstand distances, whether raw frequencies transformed by the arcsine square root transformation, or simple presence or absence data were involved.

To provide for views to the contrary, we must quote Ivimey-Cook and Proctor (1967) who found that a varimax rotation of principal component axes gave a more informative picture of floristic relationships within an east Devon heath than did the unrotated principal component analysis. A principal coordinate analysis might have been a more appropriate multivariate method to compare with the varimax principal component analysis, but one can see the point that the authors were trying to make, that the component analysis revealed axes oriented along such polarities as abundance-scarcity, soil moisture-dryness and base status, whereas the subsequent rotation reoriented these axes to polarities based on floristic relationships in which the experimenters were more immediately interested.

On the other hand, Proctor (1967) found that ordinations of some British

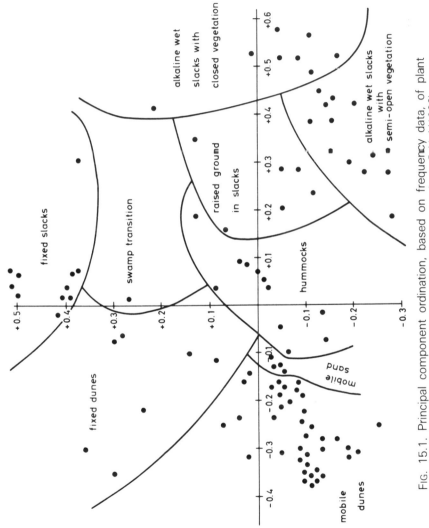

FIG. 15.1. Principal component ordination, based on frequency data, of plant habitats in "dune-and-slack" vegetation. Redrawn from Orlóci (1966).

liverworts gave readily interpretable polarities. The first axis, as usual in phytosociological studies analysed with principal components, reflected abundance (analogous to the "size" component of zoologists) whereas the second axis contrasted the mountainous northwestern regions with the southeastern lowlands. A third axis was polarized between those areas with oceanic and montane plant distributions. Noy-Meir and Austin (1970) note that ordinations may well be non-linear: rotation of the principal component axes would then lead to serious difficulties.

Phytosociologists currently emphasize the contrasts between the classification of a community (Orlóci, 1967) and its ordination, the former being achieved by various techniques such as association analysis and the latter by more obviously multivariate methods such as principal component analysis (Greig-Smith, 1964) and principal coordinate analysis. However, the objectives and resources of the experimenter are usually mixed, rather than confined to one of these approaches. It is understandable that differences of emphasis should exist. For instance, in two papers on the analysis of data from tropical rain forests, Webb et al. (1967) consider that a classification should be applied first, and ordination invoked only if the classification proved unprofitable: they commend Gower's ordination procedure (Gower, 1967) in such cases. Writing from a different standpoint, Greig-Smith et al. (1967) make the point that classification is more likely to prove satisfactory at high levels of vegetational variation and ordination at low levels: they found that ordination within the forest types revealed correlations between composition and environment not otherwise apparent.

Ordinations of grassland vegetation and associated soil attributes are discussed by Gittins (1969) and by Ferrari et al. (1957), whereas the soil characters are ordinated with beechwood vegetation by Barkham and Norris (1970).

A valuable comparison of the possible ways of examining floral assemblages has been published by Moore et al. (1970); this study shows that the classical techniques associated with the Montpellier and Uppsala schools constitute a compromise between a thoroughgoing ordination and a thoroughgoing classification. In this way much of the information which is lost when one or other extreme approach is adopted can be retained, and this preservation of information is much to be desired; it corresponds to a considerable extent with the practice of reification (identification of the physical, biological or geological meaning of the axes of variation) that we have recommended elsewhere in this book, since reification amounts to writing back on to an ordination the information that would have been preserved in a classification, or conversely. Biometricians from several disciplines are moving towards this less partisan and more broadly-based view of the classification versus ordination controversy, as Clunies-Ross (1969) has noted for some methods of handling market research surveys.

Nevertheless, the comments that Moore *et al.* (1970) make on the time taken to compute the various sets of results apply to work on a computer of an obsolete generation; so far as the principal component analysis is concerned, the time seems to be one or two orders of magnitude greater than would now be needed; computer time is rarely a limiting restraint of the choice of multivariate methods except for the special case of polythetic divisive classification. It is interesting that Moore *et al.* (1970) found a number of types of variation represented along each of their component axes, amounting to what we have termed symbatic components of variation.

THE QUANTITATIVE COMPARISON OF FAUNAL AND FLORAL ELEMENTS

A wide variety of measures of association is now available in the literature for the purpose of assessing how similar two assemblages may be, in terms of the animals or plants that live in them. Readers will find various similarity indices such as the Jaccard coefficient, Sorensen's QS, Mountford's I and many others. Over 60 of such measures are considered, for binary data only, by Lamont and Grant (1979). After setting aside those unsuitable for ordination purposes, these authors compare 21 measures, and note the profound effects of choice of measure on the results obtained. They argue that the best method of measuring biotic distance is one which reflects the ecological distance along an environmental gradient, but then note that we rarely know enough to define this ideal measure.

If S_i denotes the number of species found in site i and S_j the corresponding number found at site j, then Lamont and Grant recommend the simple absolute index $S_i + S_j - 2S_{ij}$ where S_{ij} is the number of species common to both sites. They also recommend the Euclidean distance $(S_i + S_j - 2)^{1/2}$.

Many of these coefficients, which are basically a count of the number of species common to both assemblages, can be adjusted to allow for the total number of species present in each assemblage. Now, there are certain occasions when such an adjustment is desirable, and certain occasions when it is not; the choice is illustrated by the following examples.

Consider the faunal assemblages in two habitats of which one (A) contains 40 species and the other (B) 10, all the ten in B being included in the forty in A, the remaining thirty of the species of A being absent from B. The question now arises as to why these 30 species are absent from B, to which there are two broad categories of answer. Either the 30 species do not occur in B because that assemblage inhabits a region inimical to the 30 species, i.e. the absence affords positive information about the nature of B relative to A, or the 30 species do occur in B at such low densities that they do not happen to have been found in the samples taken, i.e. the absence tells us little or nothing about the differences between A and B. We assume, for simplicity, that the sampling effort and sampling efficacy are comparable for the two habitats. If an adjustment is made to the measure of similarity, to allow for the different numbers of species in the two assemblages, the implication is that absences are telling us little or nothing; if we consider the absences to be informative, we shall not want to make adjustments of this kind. Presences and absences may, in fact, provide distinct, even contradictory, evidence about ecological relationships, as Field (1969) has shown, in a study in which joint absences were sometimes treated on the same footing as joint presences and sometimes distinguished from them.

In many practical situations our knowledge of the reasons for absences will be inadequate for a firm decision on the matter. Nevertheless, in zoogeographical investigations, and much ecological work, one might think that absences from faunal assemblages are too useful to jettison, since in such cases we would not usually attribute absences to the vagaries of sampling variation. A history of some of the various coefficients invented (and reinvented) in zoogeographical work is given by Udvardy (1969) and Lamont and Grant (1979).

One group of coefficients of similarity (or, more appropriately, dissimilarity) not mentioned in Udvardy's review is the distance measurement.

Matsakis (1964) used a distance coefficient which contains the sums of the dissimilarities between assemblages each standardized by the number of species "at risk" averaged over the two habitats compared. The important advance made by Matsakis is to concentrate attention on the dissimilarities rather than on the similarities, since we learn how habitats differ by studying the faunal elements which do not occur in one of them but do occur in others. Species common to, or absent from, both of two habitats tell us nothing about the comparison of these habitats; this is a source of vagueness when attention is focused on similarity indices. Matsakis' coefficient has the disadvantage of not being additive, in the sense that a comparison between habitats A and B cannot be directly assessed against that between B and C, and to that extent is less suitable for zoogeographic work; it has the property that two habitats each of which is almost barren, and hence with very few species, but barren for different reasons, for instance one a saline marsh, the other an arid desert, will appear to be quite dissimilar on Matsakis' computation, but quite similar on that using a matching coefficient. Again, it is a matter of deciding what kind of information one wishes to preserve in the calculation which is one reason why no finality in the matter of coefficients of similarity is to be expected. Matsakis' coefficient has been used to good effect by Cassagnau and Matsakis (1966) in studies of collembolan ecology.

Goodman (1972) has compared a variety of distance metrics, including the generalized distance, Sokal's distance, principal components and the chord-distance of Cavalli-Sforza and Edwards, in biological usage. None was wholly satisfactory, but he recommended a combination of the generalized distance and principal components as a working technique.

It seems to be generally agreed that the elements of a fauna are less closely associated into communities than is often thought to be the case with the floral elements. However, for many purposes the quantitative comparison of faunal elements is of interest: for instance, the effects of pollution, or of agricultural practices, on the animal associations may need to be assessed. Where these elements are scored as being present or absent, on a 0, 1 basis, the problems involved are similar to those met with in numerical taxonomy, each habitat being regarded as an "organism" with the various faunal elements as its "characters".

According to Baroni-Urbani and Buser (1976) none of the metrics mentioned above is wholly satisfactory when used with dichotomized data. They recommend the use of a metric which they have designed to range between 0 and 1 or between -1 and $+1$ and which otherwise satisfies all the criteria they specify.

ZOOGEOGRAPHY

Much evolutionary theory has been based on the concept of panmixis, the proposition that an individual may mate with any other individual of the appropriate sex within the same population. Patently, for the vast majority of the less spectacular organisms such as soil-dwelling invertebrates, benthic foraminifers and endobiontic ostracods, such a proposition makes little sense. Such organisms move only a few metres from the spot where they were hatched during their lifetimes. Even on a time-scale of interest to a palaeontologist, the spread of such populations may be of the same order of magnitude as continental drift.

In 1973 Sneath and McKenzie published an important paper showing that it was practicable to project a three-dimensional model of the distribution of an organism world-wide on to a sphere. They then went on to rotate the sphere so as to match the distribution of the animals (and/or plants) with present-day geographical features. If the organisms had reached their present positions as a result of their intrinsic mobility, the distribution of the organisms should then match their present geographic distribution.

However, if the organisms have been carried to their present positions by continental drift, the existing geographic distribution will no longer match the distribution as calculated from the dissimilarities of the faunal or floral components of the assemblages of organism. Two parts of what was originally a single population, carried asunder by drifting tectonic plates, will have a dissimilarity corresponding to their original physical separation, and not that corresponding to the separation augmented by drift. Genetic drift is also potentially relevant in this context.

In this way, a fairly clear distinction can be drawn between organisms whose present distribution reflects passive dispersal on a tectonic plate and those who have arrived by active dispersal.

Sneath and McKenzie (1973) give examples using gingkos and ostracods. Blackith and Blackith (1975) looked at the world distribution of Collembola and noted that when a three-dimensional principal coordinate model of their distribution based on faunal dissimilarity was projected onto a sphere, only about one-third of the sphere was covered, suggesting a substantial contribution from continental drift. This view was later confirmed by a more detailed analysis of the peri-Arctic distribution of the same group. On the hypothesis of a Gondwanaland origin for the apterygote insects, the species that have reached the Arctic should represent the end-products of the radiative process. In that case the peri-Arctic fauna should be very diverse. If, on the other hand, Collembola are transported by, for instance, storms carrying airborne soil, we should expect the same faunas to be more similar. In fact, the peri-Arctic Collembola are very diverse with low similarity coefficients between the sites chosen (Blackith, 1981). This finding may do something to redress the memory of Jeannel (1943) who was pilloried academically for making the same suggestion when continental drift was widely supposed to be a figment of a deranged imagination.

The distribution of dinoflagellates around the coasts of the British Isles has been studied by Holligan et al. (1980) using correspondence analysis. They were able to show that the occurrence of the four communities of dinoflagellates that they recognized was related to a variety of environmental parameters, and that the most important factor in determining the relative abundance of individual species was the vertical stability of the water column.

Today's heretic is tomorrow's scientist: Dale (1979) has commented that vegetation classification has been a paradise of individuality, eccentricity, heresy, anomalies, hobbies and perhaps humour. Both phytosociology and zoogeography are entering a more balanced, professional phase, greatly aided in this by the discipline of multivariate analysis, but we hope that however exasperating it may be to have the wheel re-invented time after time, we shall never quite lose the initiatives that strike out along genuinely new directions.

16
Coordinated Applications of Multivariate Analysis

There are so many ways of looking at multivariate problems that any one line of attack may be inadequate. When all calculations had to be done by hand there were obvious reasons for limiting the number of ways in which a given body of data was analysed; with computers taking over the work, there are advantages in the examination of as many aspects of the problem as seem reasonably likely to help the experimenter to gain new insights into his material. Indeed, Gower (1975a) has formalized the approach to the question by developing statistical methods of comparing different multivariate analyses of the same data. A commonly used *ad hoc* combination of analyses involves the utilization of principal component analysis to examine the variation within certain groups in conjunction with a canonical variate analysis to examine the variation between the groups, with or without the addition of a principal coordinate study.

If this particular combination of techniques is employed, there will be some interest in knowing how principal components and canonical variates are related: Rouvier (1966) gives a derivation of the expression for the canonical variates in terms of the principal components of the dispersion matrix pooled within groups. Other pertinent references are Digby and Gower (1981), Campbell (1980), Campbell and Reyment (1980) and Campbell and Atchley (1981). Gower (1972) has discussed how measures of taxonomic distance can be examined by a battery of multivariate methods, with particular reference to the analysis of hominid data.

It is most undesirable to make comparisons between methods of analysis without a reasonable exploration of the interpretative capacities of each; this is all the more true when the methods are closely related, amounting almost to rearrangements of the same calculation. It is by no means rare to see that one or other method is considered by a particular author to be in some respect superior to another when the capacities of the method considered as being inferior have been inadequately explored or understood. This is a transient phase of incomplete dissemination of the basic theories, comparable to that occurring more than a quarter of a century ago in univariate analysis, when one could find alleged comparisons of the effectiveness of the *t*-test and the one-way analysis of variance in the biological literature. Unfortunately, the transient phase is still with us almost as manifestly as when attention was drawn to the situation in the first edition. For example, a great deal of unnecessary polemics are devoted to the question of method-superiority by LeMaitre (1982).

For reasons such as this, we have included a series of coordinated multivariate analyses, using more than one analytical method, in the hope that these exercises will provide the reader with a starting point capable of adaptation to his particular interests.

AN UPPER CRETACEOUS BELEMNITE SPECIES

The belemnite species *Actinocamax verus* from nine localities in the Upper Cretaceous of the Soviet Union (Russian Platform) was measured in respect of four characters: the length and breadth of the rostrum, the maximum width of the rostrum at right angles to the preceding measurement, and the asymmetry of the alveolar scar. The main results obtained by Reyment and Naidin (1962) are now presented.

Principal Component Analysis

Each of eight samples was subjected to a principal component analysis; it was found that eight sets of latent vectors could be grouped into two sets; one set comprised of the four samples from the Campanian localities, and one from the Santonian, and the other set the remaining samples from the Santonian localities, and one from the Upper Coniacian; an analysis of the pooled data for the entire set of eight samples was also carried out to provide further comparison. In both sets, the first principal component results from a "size vector" similar to that for the pooled data which is (0.96, 0.13, 0.20, 0.12). Reyment and Naidin used the covariance matrix for their calculations; with the hindsight of experience, it would have been preferable to have used the correlation matrix.

The second principal component differs considerably between the two sets. In the first set, it is a "shape vector" with most of the variation in the fourth dimension (alveolar scar). In the second set, it is likewise a "shape vector", but a distinct one, with most of the variation in the third (width of rostrum) dimension. Reyment and Naidin interpreted these differences as being evidence for a measure of subspeciation during the period intervening between the (older) Late Coniacian and the (younger) Campanian times. This interpretation now seems less plausible than a simpler one, namely, that the second and third principal components of the first set are, in fact, the third and second components of the second set. This interpretation implies that essentially the same changes in shape are occurring in both sets of samples. The "size vectors" of both sets of samples are essentially parallel. It transpires that a switch in the order of latent roots and vectors is not uncommon for fossil material and such has been observed among fossil ostracods (Reyment, 1966a).

Generalized Distance Models

Generalized distances computed on the basis of four attributes gave an unsatisfactory three-dimensional model in which some of the distances had to be "bent" drastically in order to make them fit. This is probably due to the four-dimensional nature of the "true" model, though heterogeneity of the dispersion matrices in the original samples may have some effect on the results. A test for multivariate normality, supported by a suite of univariate tests, led to the exclusion of the Upper Coniacian sample. Despite this step, there was still considerable heterogeneity and the generalized distance model was only marginally improved. If a canonical variate analysis had been tried at the time, it would have shown that all four canonical roots were both statistically and biologically significant.

Reyment and Naidin concluded that the distortions in the dispersion matrices stemmed largely from post-mortem sorting of the rostra, owing to the winnowing effects of bottom currents. Christensen (1974) also drew attention to this source of inaccuracy in statistical studies of belemnites in his multivariate analysis of English material of *Actinocamax plenus*.

GENETIC CLINES IN THE BODY SHAPE OF FRUIT-FLIES

Five characters of *Drosophilia obscura*, to wit, the length of the thorax, width of the head, wing length and width, and the length of the fore-tibia, were measured on samples from 12 localities ranging from Scotland to Israel. Misra (1966) commented that there will in general be a continuum of "normal" shape changes, such as Teissier (1960) has discussed, assessable in terms of some multivariate extension of the allometry equation, together with a number of dimensions of variation representing deviations from this "normal" continuum.

Misra used a "factor" model (actually, a principal components calculation applied to a correlation matrix with the unities of the leading diagonal replaced by "communalities" estimated from the averages of the elements of the corresponding columns) the merits of which are not unchallengeable. However, it seems unlikely that the size factor, which he isolated as a vector corresponding to the greatest latent root, when so amended is seriously different from the first latent vector.

The general "size factor" estimated in this manner was found to be remarkably close to a vector $(1, 1, 1, 1, 1)$ representing isometric growth in *Drosophila*; the two vectors make an angle of only $1°$ $50'$. There is an apparently uniform north-south cline in the size of the fruit-flies expressed in the "size factor", a population from Edinburgh having a mean score along the vector of 90.4, whereas a population from Qiryat Anavim, Israel, has a score of 83.11, with the remaining populations falling neatly between these limits.

The data were then reanalysed by means of canonical variates, using an unweighted between-groups matrix. About 88% of the total variation was accounted for in terms of the first three canonical variates. The variate corresponding to the largest root is something of a size vector, but it differs from that extracted from the modified correlation matrix in that it reflects relative, rather than absolute, size changes, with the thorax and head growing less rapidly than the other dimensions of the body. There is a general north-south cline reflected in this variate.

The size and shape interpretations tentatively used in the present connexion are to be taken as being very approximate descriptions of the differential growth relationships in the material. Work over the last decade (for example, Mosimann, 1970; Bookstein, 1978; Siegel and Benson, 1982) has shown that the problem is more complex than was realized by Teissier (1938) and Jolicoeur and Mosimann (1960) when the size and shape interpretation of latent vectors was first put forward.

The next two variates do not display any clear-cut geographical trends. The second variate shows a contrast of wing-area and length of the fore-tibia, with extremes which are geographically quite close (Lunz, Austria and Switzerland); the third variate illustrates an attenuation of the wings which again has geographically close populations showing extreme shapes; Pavia has the least, Formia the most, attenuation, both are in Italy. But in Central Europe, altitudes and associated climatic factors vary greatly, so that physical contrasts among the populations of fruit-flies might well be essentially geographical, but not obviously so, unless examined with a knowledge of the altitude at which they were obtained.

Misra (1966) compared the morphological changes in *Drosophila robusta*, published by Stalker and Carson (1947), with those found in *D. subobscura*. It appears that in the North American populations of *robusta*, size does not follow a north-south cline, but that the smallest individuals were bred in the warmest localities. There is thus every likelihood that the cline found in *subobscura* operates through the mediation of temperature.

MORPHOMETRIC CHANGES IN CRETACEOUS HEART-URCHINS

The heart-urchins of the English chalk appear to form a virtually continuous series of measurable forms; the fact that there are no substantial breaks in the fossil record makes it all the more difficult to distinguish between genuine species and ecophenotypic variants. Nichols (1959) measured four characters on the test: the length, breadth and height and the number of tube-feet on eight samples of *Micraster*. He identified these samples as *M. corbovis*, *M. cortestudinarium*, *M. coranguinum*, *M. senonensis*, "intermediates" between *coranguinum* and *senonensis*, *M. glyphus*, *M. stolleyi*, "intermediates" between *glyphus* and *stolleyi*. Reyment and Ramdén (1970) have analysed Nichols' data using principal components, principal coordinates (employing Gower's measure of association), and canonical variates. We note that a later analysis of the same data has been made by Ottestad (1975), but from a purely methodological aspect.

Principal Component Analysis

In all the samples, the first component accounts for at least 89% of the total variation and constitutes a general "size" component with all the coefficients of about the same size and with the same signs. This is a very common situation in the analysis of organisms with calcareous shells.

Principal Coordinate Analysis

Reyment and Ramdén also subjected the echinoid measurements for the individual specimens, irrespective of taxonomic status, to a principal coordinate analysis. The latent roots were observed to fall off rapidly being successively 42.96, 21.67, 6.69 and 3.56. The first two roots account for most of the variation.

By calculating the positions of each individual along the first two principal coordinate vectors, as is illustrated in Fig. 16.1, we can plot the locations of the measured individuals of *M. coranguinum*, *M. cortestudinarium* and *M. corbovis*. These three species evidently correspond to the three clusters of points on the chart, although there is some overlap. An interesting feature of the graph is that the individual points are arranged in a Y-shaped

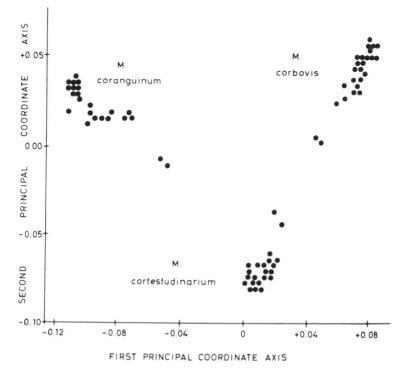

FIG. 16.1. Principal coordinate analysis of the genus *Micraster* showing the morphometric relationship between two species and an ancestral one. Data from
Nichols (1959).

"super-cluster", which may be interpreted as evidence of morphological
connexions between *corbovis* and *coranguinum* via *cortestudinarium*. It is
significant that morphological intermediates seem to occur as points which
straggle between the main clusters.

Canonical Variate Analysis

Figure 16.2 shows the arrangement of the eight samples on a chart the axes
of which are the first two canonical variates. The first canonical variate,
accounting for 75% of the total variation, is one in which the number of
tube-feet moves in one direction whereas the three other characters move in
the other. The second canonical variate, accounting for 20% of the total
variation, reflects a contrast between the breadth counterpoised against
variation in height and number of tube-feet. There does not seem to be a

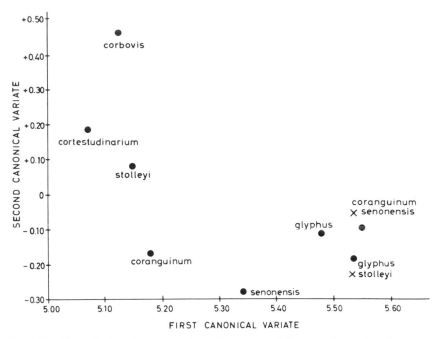

FIG. 16.2. Canonical variate analysis of the genus *Micraster* based on four morphometric characters. Data from Nichols (1959) (axes unscaled).

pure "size" canonical variate, which may be seen as an indication of the
marked sensitivity of the shape of the echinoid test to ecological pressures.

Examination of the plot of canonical variate means in Fig. 16.2 does not
seem to reveal any relationship between the burrowing habit of a form and
the position occupied in the chart. In fact, alternate members of the arc
composed by the means are representatives of deep burrowers, less deep
burrowers and non-burrowers. The most striking finding of this analysis is
that those samples which are putative "passage forms" between two species
do not occupy the positions that would be expected according to this
hypothesis. The *coranguinum-senonensis* intermediates are like neither of
the end members. The *glyphus-stolleyi* "intermediates" are very close to
glyphus, but not at all intermediate between that species and *stolleyi.*

The investigation of Kermack (1954), who used a kind of allometric
analysis to sort out the confused situation in these echinoids, has indicated
that there are only two main lineages. Modern methods of evolutionary
biology can be expected to cast light onto a situation which is still far from
reaching satisfactory elucidation.

MOORS AND CHRISTIANS: COORDINATED MULTIVARIATE ANALYSIS OF DATA ON BLOOD GROUPS

It is widely accepted that data on frequencies of blood groups offer considerable potential for the interpretation of patterns of human migration (Mourant et al., 1976, 1978). In some cases, the frequencies of a single gene may vary significantly from one population to another and thus supply adequate comparative information. In most situations, however, differences between populations can only be identified at the multivariate level. An analysis of the situation in the Iberian Peninsula has recently been given by Reyment (1983).

The colourful history of the Iberian Peninsular has long fascinated historians, linguists and anthropologists. Few have not heard of the wonders of the Alhambra and the marvels of Moorish medieval culture, from which stems much of our Western learning. But who were the Moors? Are the present-day inhabitants of Morocco their direct descendants? And if almost 300 000 forcefully christianized descendants (the Moriscos) of the Moors were expelled from their native land early in the 17th century, what happened to the estimated 2–4 000 000 Hispano-Moslems of the heyday of Al-Andalus?

Modern Spanish historical research recognizes that most of the Spanish Moslems were converted Iberians, known as Muwallads (Medina, 1980, 1981) and the North African (Berber) and Arab elements in the population of Medieval Spain were never great. In addition, Spain and Portugal had a large Jewish (Sephardim) population of which a high proportion was absorbed by the Christian community as a result of forced conversions over a long period, the last phases of which were vastly accelerated by the attentions of the Holy Inquisition. Even here, however, it seems as though the majority of the Sephardim descended from proselytes (Medina, 1980, Chapter 13).

During the 11th and 12th centuries, the eastern regions of Moslem Spain (Aragon and the Levante) were in the hands of imported slave warriors (Slavs, French, Catalonians, northern Spaniards and black Africans—the saqaliba). The breakdown of the central power structure, due largely to the desire for Home Rule on the part of native Andalusians, led to the development of petty kingdoms or taifas. Thus, the largest part of Spain was under the rule of Muwallad dynasties, whereas those of the southern coasts were governed by clans of Arab, Berber or Syrian origin.

Generally, then, the people of the Iberian Peninsula during the 8th to 15th centuries are to be seen mainly as being homogeneous ethnically but partitioned into three religious castes (Castro, 1965) with boundaries that were far from rigid.

At the time of the capitulation of the Kingdom of Granada, only Granada and parts of Aragon, Murcia and Valencia retained Arabic-speaking populations, the remaining Moorish elements either having been assimilated imperceptibly, or remaining nominal Moslems (Mudéjars), greatly facilitated by their linguistic homogeneity with the Christians (most Hispano-Moslems spoke Romance, a kind of Proto-Castilian, as their mother-tongue, in common with the non-Moslems).

The act of expulsion of 1609 was only partly successful, largely owing to its arbitrariness, and many Moriscos returned clandestinely to Spain rather than withstand the perils and persecution they encountered in North Africa. By 1618 the hopeless task of tracking down returned Moriscos and the people who had kept themselves in hiding, many masquerading as Gypsies, was abandoned by the authorities. Thus a tragic chapter in the history of Spain came to an end to the relief of even the "Old Christians", not a few of whom had been caught up in the deportation frenzy, the victims of denunciations of jealous neighbours or vindictive rivals.

Let us now see what the genes can tell us. On the basis of the foregoing, we should expect (1) that the populations of the eastern coast of Spain should show deviations with respect to the B and cDe frequencies (traces of the slave taifas with a black African element)—the pre-Moslem history of the region has a record of Semitic colonization, which cannot be separated from the effects of later events; (2) the population of southern Portugal can be expected to display identifiable "Moorish characteristics", if such exist, as there have been no appreciable dislocations in the population of the region since the 13th century; (3) the ubiquitous distribution of descendants of the *conversos* (Christianized Jews) should have some influence on the blood-group frequencies of Spain and Portugal, unless most of the converts descended from proselytised natives.

All data derive from the compilations of human blood groups in the tables published by Mourant *et al.* (1976, 1978). Coverage is understandably incomplete, particularly for the less commonly assessed genes, but the ABO system is well documented for western Europe and North Africa. Where applicable, the log-ratio transformation has been used (Aitchison, 1982).

Canonical Variate Analysis—ABO-system

Figure 16.3 shows the projections of the individual observational vectors onto the plane of the first and second canonical variates for material from Spain, Italy, Algeria, Tunisia, Libya, Italy and Morocco. The Spanish samples form a group to the right with one Tunisian sample (denoted W, the Wahabitic Moslems from the island of Djerba, adherents of an ascetic sect)

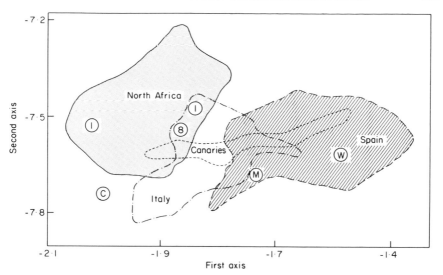

FIG. 16.3. Projection of the individual sample observations onto the plane of the axes for the first and second canonical variates for the ABO blood polymorphisms. The key for the outlying observations is: I = Sicilian samples; 8 = Valencia, M = Morocco, W = Wahabites from Djerba Island, Tunisia; C = Canary Islands (from Reyment, 1983).

centrally embedded in the Spanish field. A Moroccan sample falls among the Spanish observations and one Spanish observation (Valencia) lies among North Africans. Sicilian samples (Sicily has a North African populational background) fall in the North African field as well.

Correspondence Analysis of the Combination ABO:cDe

Figure 16.4 shows the scores for the third axis of the correspondence analysis for these four variables plotted on a map of the western Mediterranean. It will be seen that the scores for the eastern region of Spain agree well with those for Berbers of the northern Maghreb. The score for the Algarve (southernmost Portugal) lies within this range as do those for Sicily and the Canary Islands. Sephardic Jews sampled in Israel display the highest positive scores of all, whereas Moroccan Jews fall within the general range occupied by Berbers (Jewish values are enclosed in brackets). The scores for north-western Spain and northern Portugal are similar and differ essentially from those yielded by eastern Spain, southern Portugal and North Africa. Of interest is the value for Lebanese Moslems of the Shia sect in that it is close to the values found for the northwestern Iberian Peninsula (and to those for northern Europe).

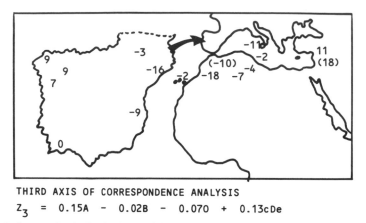

THIRD AXIS OF CORRESPONDENCE ANALYSIS

$$z_3 = 0.15A - 0.02B - 0.070 + 0.13cDe$$

FIG. 16.4. Scores for individual populations for the third axis in the correspondence analysis of A, B, O and cDe in relation to geographical locations. The values for Jews are given in brackets (from Reyment, 1983).

Principal Coordinate Analysis of the ABO-system and Rh Complexes CDe, cDE, cDe and cde

The plot of the first two principal coordinates for the seven variables of interest is shown in Fig. 16.5. The differences between northwestern Iberia and southern Portugal-Aragon are emphasized in this projection. Interesting features are the positions of Sephardic Jews in relation to northern Portugal, an area of *converso* concentration, and the suggestive connexion between Aragon, southern Portugal, Valencia and Berbers of Mouley Idriss. Basques were included in this analysis and it will be seen that they lie apart, differentiated by the strongly Rh negative properties and in particular, frequencies of *cde* (note received from Dr A. E. Mourant). The lines drawn between points in this illustration are a minimum spanning tree, in this case, that of Prim (1957).

Analysis

The most striking feature of the analyses is the fact that the populations of Aragon and the eastern provinces of Spain deviate rather markedly from those of northwestern Spain. There is, also, a certain gradational tendency for southern Spain with respect to the distribution of ABO-frequencies. The mass assimilation of a substantial portion of the originally considerable Jewish population has left surprisingly little trace in the blood polymorphisms analysed, which offers support for the belief that the Jews were mainly proselytes. As far as can be expressed by the data available, the frequencies of the blood polymorphisms for Galicia and Leon do not differ greatly from those occurring in Northern Europe.

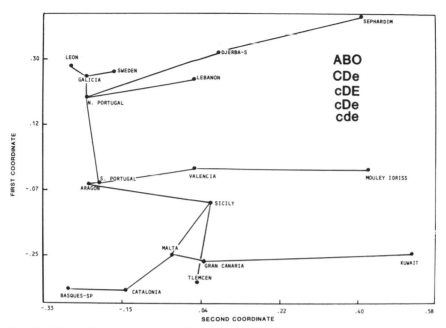

FIG. 16.5. Plot of the principal coordinate analysis for the first two coordinates for the polymorphisms A, B, O, CDe, cDE, and cde. The Prim minimum spanning tree for the points is shown: the connexion between Sicily and Berbers (Tlemcen) passes, but does not cut, the line between Malta and Gran Canaria (from Reyment, 1983).

The multivariate statistical analyses of blood polymorphisms indicate the existence of what might be termed a Moorish substrate in some populations of the Iberian Peninsula insofar as the original populations were characterized by exotic elements. Other western Mediterranean areas with a similar past to Spain and Portugal (Sicily, Malta) behave analogously in the analyses, as do the data for the Canary Islands.

Related analyses have been published by Carmelli and Cavalli-Sforza (1979) and Piazza *et al.* (1981). The former paper deals with discriminant function and canonical variate analyses of four markers (ABO, MN, Rh, Hp) for Jewish populations. Here it was confirmed that Jewish minorities tend to resemble the non-Jewish majority of a country, a fact known from standard univariate work, but the new information yielded by the multivariate approach is that the Rh frequencies tend to be differentiated in the direction of a Mediterranean origin. The paper by Piazza *et al.* (1981) uses principal component analysis and canonical variates for constructing gene-frequency maps for aboriginal populations of the world (using 39 independent alleles from 10 loci). These authors claim that some gene frequencies are correlated with distance from the equator which they interpret as evidence of the selective effects of climate.

17
Some Applications of Multivariate Morphometrics in Palaeontology

BIOSTRATIGRAPHY

Biostratigraphy is an important branch of applied palaeontology. Its main goal is the application of fossils, and hence of evolutionary sequences, to the subdivision and characterization of time-sequences and their data in geological terms. As practised in qualitative connexions, the chronological subdivisions are made by the identification of particular zonal indices, which are normally limited to short time-ranges. Quantitative biostratigraphy is not really a new subject—its roots go back to the days of Lyell in the last century. However, as conceived today, there is a vast difference between the aims of the originators of the idea and the methodology now employed. Quantitative biostratigraphy makes use of the evolutionary sequences of particularly microfossils preserved in the sediment and studies, by morphometric methods, changes of shape and size of critical forms over time. In other words, the evolution of the phenotype is put to practical use.

The method of analysis advocated here employs a kind of profile treatment of multivariate data coupled with specific tests for agreement in sequences.

The practical application of morphometrically oriented biostratigraphy is clearly of importance in the petroleum industry in the detailed analysis of fossil sequences in boreholes. A well produced study of long-ranging forms and their morphological variation (which contrasts sharply with the aims of traditional biostratigraphy) provides an excellent basis upon which to found a detailed, fine-stratigraphical analysis of boreholes. Inasmuch as the fossil organisms, usually ostracods, foraminifers and coccoliths, will have reacted to environmental effects (ecophenotypic reactions) by genetic tracking of slight fluctuations, the possibilities of correlations between boreholes in different basins, or in widely differing parts of a sedimentary basin, are not great, but the method is valuable for work in restricted areas, such as a single sedimentary basin.

Reyment (1980b) has developed the topic of morphometric biostratigraphy so as to unite the analysis of biological variation through time with quantifiable factors of the palaeo-environment. The interested reader is referred to that text for details of the methods involved.

Hazel (1977) has summarized the use of multivariate analysis in biostratigraphical work, in particular with respect to stratigraphy based on assemblage zones (i.e. zones composed of a set of species). He draws examples from various parts of the fossil record, including the Palaeozoic, Mesozoic and Cenozoic, with emphasis on benthic invertebrates. The groups given special consideration in this study are trilobites, and Cretaceous and Pliocene ostracods. Hazel recommends the use of principal components and principal coordinates, combined with cluster analysis, in order to display graphically the affinities between zones. Although the approach utilized by Hazel is not based on morphometrics as we interpret the subject, the work is cited here for its general interest to workers in multivariate analysis.

PALAEOECOLOGY

In a natural way, some aspects of palaeoecology follow on from the foregoing section. As an example taken from palaeobotany, we briefly cite some work by Birks and Peglar (1980).

Five collections of modern *Picea glauca* pollen, four of *P. mariana* pollen and two of *P. rubens* pollen from the Late Quaternary of eastern North America were examined in an attempt to distinguish between these species. Three characters were measured on individual grains of each collection, namely, the sculpture of the furrow-membrane, the presence of an undulating margin to the cap, and irregular reticulation in the sacci of which the latter two characteristics are typical of the pollen of *P. rubens*. Seven size variables were also measured for each grain. It was found that no simple combination of morphological and size criteria provides effective discrimination between *P. glauca* and *P. mariana*.

Linear discriminant function analysis, applied to *P. glauca* and *P. mariana* for size variation in the continuous criteria, was used to allocate individual grains, with 91.5% being correctly identified. The same size variables were

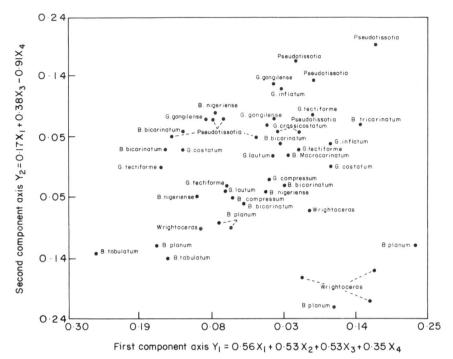

FIG. 17.1. All specimens projected onto the plane of the first two principal components. G. *Gombeoceras; B. Bauchioceras.* Note component for $x_2 \simeq 0$ in the second latent vector. After Reyment (1979a).

measured on fossil *Picea* from two Late Wisconsin localities in Minnesota and one Holocene sequence in Labrador. The individual fossil pollens were assigned to either *P. glauca* or *P. mariana* by means of the linear discriminant function.

Care was taken to justify the mathematical assumptions of the analysis by testing for multivariate normality and homogeneity of covariance matrices. It is virtually impossible to distinguish between *P. glauca* and *P. mariana* pollen without the use of size-statistics and discriminant analysis. Birks and Peglar believe that they have demonstrated the importance of the linear discriminant function for Quaternary palaeoecology.

PALAEONTOLOGICAL TAXONOMY

As an example of the usefulness of multivariate morphometrics for unravelling a knotty taxonomical situation, we offer the analysis by Reyment

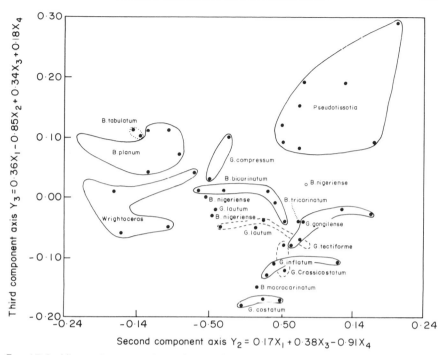

FIG. 17.2. All specimens projected onto the plane of the second and third principal components. *G. Gombeoceras; B. Bauchioceras.* Note component for $x_2 \simeq 0$ in the second latent vector. After Reyment (1979a).

(1979a) of relationships in closely related ammonites of the Turonian (Cretaceous) of northern Nigeria.

Principal component analysis was used to study the variational morphology of the shell of these ammonites with particular attention directed towards the recognition of homeomorphy. The four most commonly measured dimensions of ammonite shells are the maximum diameter of the conch (x_1), the maximum width of the last whorl (x_2), the maximum height of the last whorl (x_3) and the maximum umbilical diameter (x_4). These are not ideal by any means, but we shall continue none the less. These were analysed for the species mentioned in Fig. 17.1 for a total of 51 samples.

In order to stabilize the analysis with respect to size differences, the logarithms of the observations were used and the computations made on the correlation matrix. The first principal component is isometric with respect to x_1, x_2 and x_3, these variables being locked in an allometric relationship with x_4. The plot of the first and second principal components, shown in Fig. 17.1, contrasts general morphological variation in all four variables with a whorl-

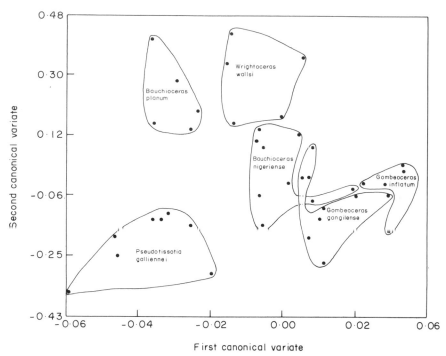

FIG. 17.3. Plot of the first two canonical variates for the six groups of species (axes unscaled). After Reyment (1979a).

height and umbilical-diameter relationship. Several significant patterns are evident in this figure.

1. The morphological closeness of *Wrightoceras* and *Bauchioceras* through *B. planum* and *B. tabulatum*.
2. The specimens of *Pseudotissotia galliennei* lie in the upper third of the chart.
3. The locations of the individuals in Fig. 17.1 suggest several simplifications in the taxonomy of *Bauchioceras* and *Gombeoceras*. The use of sub-specific names for various morphological types is not permissible, for there are neither space nor time differences involved. Some variants may be ecophenotypic, whereas others may be genuine species.

Support for the above interpretations is offered by the projection of the data points into the space of the second and third principal components, shown in Fig. 17.2. The isolated position of all specimens of *P. galliennei* is clearly shown, as is also the close liaison between most *Bauchioceras* and *Gombeoceras*. *Wrightoceras*, and the highly compressed, weakly

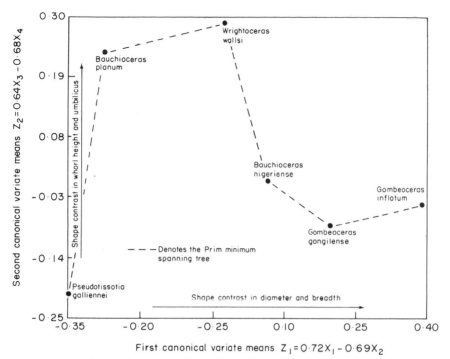

FIG. 17.4. Plot of the canonical variate means for six species projected onto the plane of the first two axes of canonical variates. The minimum spanning tree (Prim network) between means is shown. Note the components for x_3 and x_4 in the first canonical vector and x_1 and x_2 in the second canonical $\simeq 0$. After Reyment (1979a).

ornamented forms of *Bauchioceras*, to wit, *B. planum* and *B. tabulatum*, occupy a field characterized by an identical inverse relationship between variation in the umbilical diameter and the maximum whorl-height.

The species groups indicated by the foregoing analyses (see Fig. 17.3) were then analysed further by canonical variate analysis. Some caution is required here, as the groupings were erected from an examination of the data. A canonical variate may then tend to reproduce the same configuration of clusters. The plot of the first and second canonical variates, shown in Fig. 17.3, shows *B. nigeriense*, *G. gongilense* and *G. inflatum* to be closely located. *W. wallsi* and *B. planum* are relatively near but all are far from *P. galliennei*. This impression is confirmed by the Prim minimum spanning tree shown in Fig. 17.4 for the six pairs of first and second canonical variate means. The close morphological relationships between the groups of *Gombeoceras* and *Bauchioceras* are clearly manifested. Particularly informative is the link with *P. galliennei*, which lies right at the end of the tree. The

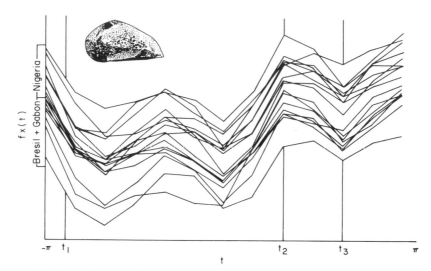

F<small>IG</small>. 17.5. Graphical multivariate analysis of the Turonian ostracod *Brachycythere sapucariensis* Krömmelbein. This Fig. gives a comparison between samples obtained from eastern Nigeria, western Gabon and Brazil (Sergipe). The curves are very similar and thus show that all individuals belong to the same species. The Brazilian and Gabonese individuals are closer than either of these groups are to the Nigerian material. This indicates that the biological population of the species was not dissociated in Early Turonian time. t_1, t_2, t_3 denote locations in the fig. where the individuals are exceptionally similar (after Reyment and Neufville, 1974).

multivariate morphometrical analysis of this difficult group led to the conclusion (Reyment, 1979a) that *Bauchioceras* and *Gombeoceras* are both vascoceratids, and not phylogenetically separate, as was once thought, and that *Bauchioceras* and *Pseudotissotia* are homeomorphs, a conclusion also supported by the differing ontogeny of species of the two genera.

An alternative hypothesis requiring testing, but needing more material and more diagnostic variables, is that we have in fact no more than two highly variable species *B. nigeriense* and *G. gaongilense*, possibly assignable to the same genus. The present multivariate analysis offers suggestions in this direction.

PALAEOBIOGEOGRAPHY AND VISUALIZATION OF MULTIVARIATE DATA

Andrews (1972) method of plotting multivariate data is taxonomically useful. His plots, as Neff and Marcus (1980) note, are a graphical method for systematically scanning an *m*-dimensional space in order to plot scores for the objects analysed as continuous functions of the directions chosen in the space. If the data arc graphed on an arbitrary initial axis through the origin, one can then rotate the axis slightly, to produce new scores. Repetition of the rotation gives a quasi-continuous line representing each object. Objects that are unusual in some respects stand out from the cord-like cluster of lines for similar objects. Gnanadesikan (1977) has suggested a method which scans the space more thoroughly.

Wilkinson (1978) described multivariate graphical techniques for representing computer-made classifications of organisms. Everitt and Nicholls (1975) have presented a review of the state of the art in representing multivariate data by means of graphical techniques and this review has been expanded into a book by Everitt (1978). Reyment (1978b) offers a graphical display of growth-free variation in a foraminiferal species.

As an example of the application of Andrews' (1972) technique to a palaeontological problem we offer the representation of variation in the Turonian (Cretaceous) ostracod species *Brachycythere sapucariensis* in relation to a drift hypothesis for the origin of the South Atlantic Ocean (Fig. 17.5).

Fourier analysis can be employed in connexion with the visualization of shapes and Younker and Ehrlich (1977) recommend harmonic amplitudes as multivariate shape-descriptors.

18
Recent Applications of Multivariate Analysis

Until about 1960, we could be reasonably sure of knowing about most applications of multivariate statistical analysis to biological problems: this had ceased to be true when the first edition of this book was written, and is now a vain hope. However, there is the need for less experienced workers to see what applications have been made, and with what techniques, in fields sufficiently like their own for them to be able to identify with the problems involved.

Naturally, since many of the earlier applications had something of the pioneering spirit about them, the techniques were not always wisely chosen, nor were they applied optimally. This comment applies with particular force to techniques selected because they happened to be available (often as a computer program) and known to the scientist responsible, more appropriate ones going by default.

We have classified the applications by the subject to which the techniques were applied, as classification by method would not help his own interests. Many arbitrary decisions had to be made: the biogeography of North American flies could, for example, be considered under geography or entomology. The reader should be alert to the limits of cross-indexing.

BIOGEOGRAPHY, PALAEONTOLOGY AND GEOLOGY

There have been a number of palaeontological investigations aided by multivariate techniques, of which just a few can be listed here. Malmgren (1972) used principal component analysis to show that dextrally coiled specimens of *"Globorotalia" pseudobulloides*, a foraminifer of the Cretaceous-Tertiary boundary, are significantly thinner than the sinistrally coiled specimens. Reyment (1973) used principal coordinates to illustrate the essentially two-dimensional variation of ammonoid shells of various inflations and umbilical widths. He related these axes of variation, essentially shifts in width and changes in coiling, to buoyancy.

The inter-relationships of the genera of vascoceratid ammonites were restudied by Reyment (1978a, 1979a) in an investigation using principal components and principal coordinates with the end in view of establishing an evolutionarily based pattern of taxonomic connexions. Reyment (1971a) studied dimorphism in *Benueites,* an ammonite genus of the Turonian (Cretaceous), by a battery of multivariate techniques.

Reyment and Neufville (1974) looked at the ostracod species *Brachycythere sapucariensis* from the Turonian of northeastern Brazil and West Africa and deduced that these populations were in genetic contact until about early Turonian time, at least, after which they were separated by continental drift, since the two West African areas of distribution (Nigeria and Gabon) are less alike than either is to the Brazilian (Fig. 17.5).

Vascoceratid ammonites of the genera *Vascoceras* and *Fallotites* have been analysed by Berthou, Brower and Reyment (1975) with the aid of principal components and other multivariate techniques.

A *Q*-mode analysis of 2800 fossil families in 91 Metazoan classes led Sepkoski (1981) to consider that there were three major axes of variation, one corresponding to a trilobite-dominated Cambrian fauna, the second to a brachiopod-dominated late Palaeozoic fauna, and the third to a mollusc-dominated Mesozoic fauna of modern aspect.

Campbell and Reyment (1980) offer a method of detecting atypical values in highly variable foraminiferal material, based on probability plots of the generalized statistical distances for each sample member. Canonical correlations were used by Reyment (1980b) to associate a "biolog" of the morphometric pattern of ecophenotypic variation of an organism with the "ecolog", consisting of scores on the principal axes of the matrix describing its environment. The same author (Reyment, 1978a) has prepared biologs from the growth-invariant discriminant function scores (cf. Burnaby, 1966a) analogous to physical borehole logs, and (Reyment, 1978b) gives details of the preparation of growth-invariant measures by "slotting" (cf. Gordon and Reyment, 1979).

Igneous differentiation and metamorphism are identified as the processes controlling the geochemistry of the Bourlamaque batholith in Canada by David, Campaglio and Darling (1974) who employed correspondence analysis with an unusually detailed explanation for this method. The same technique had, however, been introduced earlier into geology and oceanography by Melguen (1972, 1973) for the study of marine sedimentary facies, with good results. Mahé (1974) used correspondence analysis for a taxonomic evaluation of lemurs.

Geographic variation in ribbed frogs was represented by Pauken and Metter (1971) using the generalized statistical distance, and principal component analysis. Reyment and Banfield (1976) applied Gower's (1976) work on growth-free canonical variates to fossil foraminifers to compare species observed at different periods.

Another application of multivariate analysis has been the determination of the marine or terrestrial origins of Cretaceous sediments (Reyment, Berthou and Moberg, 1976). Both canonical correlations and principal coordinates were techniques utilized by Reyment (1976) to investigate the association between fossil organisms and geochemical components of the host-sediment. Ivert (1980) carried out a similar type of analysis for the same borehole from western Nigeria. Morever, multivariate analysis of discrete variables of the ostracod carapace was found to be a useful aid in biostratigraphical studies because there is a shift in time of the specific phenotypes (Reyment, Hayami and Carbonnel, 1977).

Bryant (1977) and, later, Bryant and Turner (1978) have studied the adaptation of the housefly and facefly to life in the United States of America, after introduction from Europe. They consider size-changes to be mainly related to additive genetic factors, whereas shape-changes are connected with non-additive, i.e. interlocus, interactions. They used principal components and discriminant functions, and observed that the vectors of the two species under morphometric analysis were remarkably similar. Similarly, Niles (1973) looked at adaptive variation in the horned larks of the United States and Skeel and Carbyn (1977) examined the relationships of wolves in Canadian national parks, using principal components and non-metric multidimensional scaling.

Kennedy and Schnell (1978) produced graphs showing the geographical variation of kangaroo rats plotted on principal component axes.

As the house sparrow spread across New Zealand it underwent morphometric changes amounting to clinal variation, as demonstrated by Baker (1980). He measured 16 characters on 13 samples totalling 791 adult birds, spanning their range in New Zealand; he also regarded the evidence for the Kluge-Kerfoort phenomenon, that inter-locality differences are positively correlated with intra-locality variance.

Atchley (1971) explored geographical variation in pupae of the biting midge *Culicoides* and found that the changes of shape in the pupae run parallel with those in the adults.

If this account of geographical applications seems heavily biased towards biogeography, it is partly because excellent reviews in human geography are now available (Mather and Openshaw, 1974; Clark *et al.* 1974). The former paper notes acerbically that in the social and behavioural sciences, the authors of many articles have demonstrated little more than their ability to leap onto a band-wagon and write copiously, even while it is in motion. Both papers distinguish factor-analytical techniques from the main corpus of multivariate analysis in the sense expounded in this book; failure to establish this distinction leads to unnecessary confusion for students entering the field, and for some more experienced practitioners as well. Mather and Openshaw refer to the habit of collecting data from ready sources such as census returns, and picking over them as if they were cans on a rubbish tip. This approach to multivariate analysis tends to obscure the rationale behind the procedure adopted. On the other hand, exploratory investigations are doubtless important.

The difficulty of striking a reasonable balance between exploration and the need to match technique with problem has led to a healthy back-lash of scepticism exemplified in a paper by Scott (1975). However, although scepticism may be a valuable corrective for excessive optimism, such as is bound to be generated in the early "heroic" stages of applying a novel set of techniques, real advances are also to be found.

Karr and James (1975) have successfully employed canonical correlations to link the environmental with the morphometric parameters of the organisms living in an assemblage or community. Finally in this section we note that the whole topic of biogeographical variation and its impact on racial affinities has been discussed by Thorpe (1976).

THE ANALYSIS OF ASSEMBLAGES OF ORGANISMS

There is a sense in which many of the applications mentioned elsewhere in this book could be deemed to fall into this category. However, there are several, often large-scale applications in which quite complicated assemblages are explored to discover, if possible, the factors which make them what they are.

The work of Erez and Gill (1977) and Cugny and Rey (1975) on relating the distribution of micro-organisms to environmental factors has already been reviewed (p. 142). These studies are predated by Reyment's (1963a) application of maximum likelihood factor analysis to distributional patterns of Paleocene ostracods and the reification of the results with respect to environmental implications.

The application of most of the techniques described in this book to the analysis of assemblages has been discussed by Hazel (1977), with particular emphasis on biostratigraphical relevance (see p. 167). He noted that when data matrices are large, univariate techniques tend to become cumbersome to such an extent that attempts to reduce the volume of the comparisons leads to bias: multivariate methods help to overcome this bias by greatly reducing the need to "trim" data matrices to make them accessible. However, some of the limitations Hazel records, in respect of the problems of running Q- and R-mode analyses together, can be obviated by the use of correspondence analysis, which he did not mention.

Correspondence analysis and principal components were applied to the data from a Jurassic plant-bed by Spicer and Hill (1979), and the general problem of assessing how similar two or more assemblages, of organisms or of environmental parameters, may be has been taken up by Green (1980).

Canonical variate analysis of X-ray spectrophotometric data from 34 species of aquatic invertebrates allowed Calaprice, McSheffrey and Lapi (1971) to conclude that such organisms may possess characteristic elemental compositions that are, in effect, species-specific, and that each species may be identified on the basis of its chemical composition, with a few errors. The development of populations of phytoplankton in lakes has long been an ecological puzzle, since on a superficial view, one might expect a small number of species to predominate, instead of the richness of diversity usually encountered. Lefèbvre, Laurent and Louis (1972) have followed the development of the phytoplankton in Lake Annecy with the aid of correspondence analysis.

Cassie (1972) has continued his studies of the mudbank fauna and sediments along the coastline of New Zealand. It has proven practicable to define areas of the South Pacific by means of principal components to assess the assemblages of foraminifers (Malmgren and Kennett, 1973). These assemblages are separated by convergence zones in the oceanic water. This work is just one example of the wealth of detail deriving from the Deep Sea Drilling Project (DSDP), one of the major international scientific endeavours of our time.

Kaesler, Mulvany and Kornicker (1977) used multidimensional scaling of myodocopid ostracods for the same purpose, as that just mentioned, in order to locate the Antarctic convergence.

An unusual approach to assemblage evaluation by means of correspondence analysis is that of Bonnet (1974), who discovered that the assemblage of testaceous amoebae in French soils differs according to the suitability of those soils for truffle growing, among other reasons for the variation in the assemblages. One can, therefore, distinguish likely truffle-bearing soils by their amoebae. Bonnet, Cassagnau and Deharveng (1976) have looked at the Collembola of high woodlands in the French Pyrenees, before and after felling of the trees; they observed that the biotic equilibrium is disrupted within four years of the felling, so that the collembolan assemblage can act as an indicator of ecological damage.

O'Brien and Rock (1978) found that canonical variates and principal components, applied to the distribution of chaetognaths in Galway Bay, Ireland, divided the assemblages of these animals into one characteristic of the neritic area of the bay and one typical of the oceanic area. The discriminatory capability of canonical variates helped Wharfe (1977) to explore the benthic invertebrates of the Medway estuary in England.

Bonnet and Rey (1973) probed the ecological conditions under which a marine Cretaceous assemblage can have lived: they deduced that the beds clustered according to whether they had been deposited in poorly or richly oxygenated seas, were land-locked or open, or were within reefs or outside them. The age of the deposit and the depth of the water were also major factors showing up in the principal components and the axes resulting from a correspondence analysis. Once one is in a position to assess similarity between assemblages, one can begin to speak of their converging in evolutionary time and use multivariate statistics to establish this convergence (Karr and James, 1975), an observation which can be put to practical use in "ecostratigraphy".

A principle originally due to McArthur states, in effect, that the survivors of the battle for resources in any given habitat constitute the assemblage which leaves as little as possible of the available resources unutilized. Obviously, such a statement forms a word-model which requires verification in any particular instance, but in its light we can think of assemblages as communities rather than simply aggregations of competing species. To this end we need techniques which will encompass the entirety of what may be a numerous assemblage of species, influenced by many environmental factors. In other words, a highly multivariate approach is dictated by theory and expediency. Boucot (1978) has, in general terms, viewed the field from the palaeontological point of view.

The relative abundances of 13 species of foraminifers from Christchurch Harbour, England, were assessed in terms of the physicochemical properties of the sediments in which they were found. These sediments and the bottom water were compared for dissolved oxygen, chlorinity, magnesium and calcium contents, temperature, pH and depth of water, as well as the mud, silt, sand and gravel contents. Howarth and Murray (1969) then applied a variety of multivariate methods to these data in order to unveil the relationships involved.

ANTHROPOLOGY, ETHOLOGY AND LINGUISTIC APPLICATIONS

Two fields in which a variety of multivariate techniques flourish are physical anthropology and psychometry. In the case of psychometry, the flood of factor-analytic studies continues unabated, far too numerous to permit review here. A welcome trend is that some of the more controversial factor-techniques are now being supplanted or, at least, shored up by other multivariate methods. For instance, Worsley (1981) conducted a factor analysis of 59 female and 79 male secondary school students in respect of 28 personal rating tests, and backed this up with a discriminant function analysis.

In the field of physical anthropology, quartimax rotated factor analysis has been supported by discriminant analysis in studies of adolescent growth by Largo *et al.* (1978). Suzanne and Sharma (1978) employed generalized distances in their analysis of Punjabi head measurements, whilst Marcellino *et al.* (1978) looked at size and shape differences among six Amer-Indian tribes. They found that the generalized distances correlate well with linguistic attributes, but not with geographical differentiation. The range of application of multivariate methods to anthropological investigations has been discussed by Corrucini (1975, 1978) and by Campbell (1978, 1980).

Dermatoglyphics figure in two papers. Leguebe and Vrydagh (1981) have uncovered a striking clustering of world human populations, more or less independent of sex, based on digital ridge-counts, ordinated with principal component analyses. Reed and Young (1979) also made use of principal components to sort out dermatoglyphs and link them with individual variables.

Roberts *et al.* (1981) investigated the regions of Cumbria in northwestern England, and with the aid of principal components, discovered that the six regions differ genetically, the south and central regions in particular. The central region shows traces of its ancient Norse population, according to these authors. The effects of migration and biological differentiation in the west of Ireland are documented and analysed by Relethford *et al.* (1980).

Two useful papers on the multivariate interpretation of human skeletal remains have been published by van Vark (1974, 1975), and Wolpoff (1976) has considered the special case of measurements on teeth. Petit-Maire and Ponge (1979) studied primate cranial morphology by means of correspondence analysis obtaining results of considerable classificatory significance. The pelvic girdle and femur are other parts of the anatomy to which multivariate methods have been applied by McHenry and Corrucini (1975, 1978).

Morgan (1981) reviewed the use of several clustering methods, including non-metric and multidimensional scaling, in order to cluster English dialects, the ways in which people confuse letters acoustically, and the elderly living at home, respectively. He drew attention to the usefulness of Andrews' curves for plots of high-dimensional data (Andrews, 1972).

Reyment (1981b) noted that the shape of the cranium of travelling clans in Middle Sweden differs from some European Gipsies as well as from the settled people of that country, pointing out, however, that this result could be ascribable to the small sample size for travellers. Correspondence analysis was used to show that European Gipsy dialects with preserved inflections differ markedly from the language of Gipsies, such as those of England and Spain, in which all grammatical structure has been lost and that there is a definite order of priority for the loss of basic words of Indian origin connected with the transition from uninflected language. There are no essential differences between the vocabulary of the secret language of Swedish and Norwegian travellers and that of Andalusian gipsies *(Chipicallí)*, or of Anglo-Romany (which does not include Welsh *Romanès*, which is inflected).

Two multivariate studies on Australian aborigines have come to hand. Fingerprints and other traits have been used by Parsons and White (1973) to evaluate differences between the tribes; they found that the aborigines are heterogeneous when assessed by allele-frequency traits, but not with respect to morphological criteria. The view that the tribes are physically homogeneous needs revision in the light of this work in which good agreement between allele-frequency traits and dermatoglyphics was obtained. The tribes are, of course, highly diverse culturally and with a differing history of arrival in Australia, and cultural traits tend to go hand-in-hand with the allele-frequency ones. Tasman Brown (1973) has explored multivariate techniques for the study of the aboriginal Australian skull.

Brothwell, Healy and Harvey (1972) have looked closely at facial patterns in the people of Tristan da Cunha and the Ainu of Japan; they used canonical variates as their main technique. The generalized statistical distance was utilized by Spielman (1973) to probe size and shape variation in Yanomama Indians.

Canonical variate analysis was employed by Margetts, Campbell and Armstrong (1981) to investigate the dietary patterns of native-born Australians in relation to the eating habits of immigrants of British or Italian parentage.

Habbema (1976) developed a Bayesian discriminant procedure for identifying the 46 chromosomes of the human complement and Rightmire (1976) has made good use of multidimensional scaling to investigate human diversity south of the Sahara.

A functional analysis of the proportions of primate long-bones is reported by Jouffroy and Lessertisseur (1978) who relied mainly on correspondence analysis as did Mahé (1974) in his study of the 33 craniometric variables he measured on 175 lemurs, a group which plays an important rôle as a distant ancestor of the primates. Lavelle (1977) considered the relation pertaining between tooth and long-bone size, with canonical correlations as the preferred method of analysis. Most of Chapter 7 of Bookstein (1978) consists of a thoughtful consideration of ways of comparing the crania of various primates, with a full literature review (see also p. 124).

Oxnard (1973) summarized his researches into the relations between multivariate appreciations of shape and the functions of the primate skeleton. He drew attention to a matter dealt with in more detail elsewhere (Ashton, 1981) that the traditional relationships of the fossil primates to man may well stand in need of revision; this is a wide topic, many aspects of which transcend the sphere of multivariate applications. Indeed, Oxnard and Yang (1981) explicitly reach out beyond the scope of multivariate statistics, as considered in this book, to a study of cancellous plates inside the lumbar vertebrae by harmonizing Fourier transformations of the pattern met by taking cross-sections of the bones with the stress patterns in the bones under the more stressful tasks that the animal of interest may have to tackle in its everyday existence. Healy and Tanner (1981) have given a contemporary survey of the relations between size and shape to growth and form with echoes of Bookstein's (1978) attempt to bring Thompson's transformation grids within the ambit of multivariate morphometrics. We refer the reader to the section on size and shape (Chapter 12) where the tangled web of concepts is, we trust, to some extent unravelled but mention Pilbeam's and Gould's (1974) review in passing.

Because of the ferment of interest currently directed towards the study of primate relationships, studies such as that of Jacobshagen (1979) on variation in the orang-utan are doubly welcome and should be extended to the primates and anthropoids as a whole.

The use of principal components in physical anthropology has been reviewed by Andrews and Williams (1973), and Corrucini (1978) has considered the effects of allometry on the evolutionary course of hominoids. The locomotor adaptations in the genus *Cercopithecus* are described by Manaster (1979).

There are several subtle problems in multivariate anthropology which are given further consideration by Campbell (1978). Gould (1975) discussed the scaling problems relating to studies of the primate brain; Lande (1979) has investigated the relation of brain to body-size in a multivariate allometric context, and Healy and Tanner (1981) have reviewed size and shape in relation to growth.

Discriminant functions have been used to define obesity in children of 7–11 years of age by Jayakar *et al.* (1978). This stage would normally be followed by the adolescent growth-spurt, which presents particular technical difficulties for those studying growth. Smoothing spline functions are employed by Largo *et al.* (1978) and out of their work grew an independent analysis of the size and shape problem with the objective of removing shape differences from analyses of size (Stützle *et al.* 1980). Whatever the problems involved in removing shape differences at the statistical level, they pale into insignificance before the real-life problems of young people who are unable to slim. Worsley (1978) has described their reactions to themselves and to others, both slim and obese, illustrating his approach by means of a principal component analysis.

A very full account of the use of multivariate methods in ethological investigations is available in Colgan (1978) so that more detailed consideration here would be superfluous.

ORNITHOLOGY

Thirteen morphological measurements were used to distinguish the two subspecies of the white-tailed Black Cockatoo in Western Australia via a discriminant function (Campbell and Saunders, 1976). Madsen (1977) performed a canonical variate analysis of three characters measured on 25 strains of chickens in order to elucidate the growth curves of the animals.

Rochford (1983) has measured a variety of characters of the woodcock in the hope of assisting in the sexing of the birds, but with limited success; normally, the woodcock can only be sexed during the short period of the year when the gonads are active—at other times even these features regress.

Principal components in conjunction with minimum spanning trees were used for studying the Lari by Schnell (1970) and the former technique was applied to evolution in the house sparrow by Johnstone (1973). Mosimann and James (1979) applied the work of Mosimann (1970) to the multivariate allometric analysis of growth in Florida Redwing Blackbirds (see also p. 123). Multivariate analyses of the house sparrow by means of principal component analysis have been published by Johnstone *et al.* (1972).

HERPETOLOGY

Incipient speciation in the ringed snake *Natrix natrix* was revealed when Thorpe (1975) collected 750 specimens and subjected 52 of their characters to a canonical variate analysis. This work developed into an analysis of racial differentiation, the greater part of the species complex aggregating into "eastern" and "western" forms meeting along a zone of hybridization (Thorpe, 1979). In that study, canonical variate analysis provided tight, well delineated clusters which were easy to interpret, whereas principal coordinates yielded clusters which were more dispersed and required precise knowledge of the locality from which the sample was collected before full interpretation was practicable: inasmuch as neither approach is overwhelmingly advantageous, Thorpe recommended that both be employed. We note that where well defined groups are available at the outset, a canonical variate approach is preferable.

The systematics of the Black and Lined House Snakes, within the *Boaedon fuliginosus* species complex, was analysed by Thorpe and McCarthy (1978) in a paper of fundamental importance for discussions concerning the evolution of polymorphism as a speciation mechanism. These writers found that these snakes in West Africa present all the hallmarks of being distinct species. In the rest of Africa, however, they seem to be conspecific, a fact which poses interesting questions about the cohesion of the gene pool and related matters.

Jackson, Ingram and Campbell (1976) used canonical variate analysis to demonstrate that patterns of ornamentation in snakes are more closely associated with the behaviour of the snakes than to the appearance of the habitat in which they live. Irregularly banded and blotched patterns are connected with snakes whose response to predators is that of active defence, whereas striped and unicoloured snakes tend to respond by flight rather than defence.

ENTOMOLOGY (ACRIDOLOGY)

The need to appreciate the effects of crowding on locusts stimulated one of the earliest applications of multivariate analysis in biology, because locusts which are crowded give rise to offspring which have a different shape from those born to isolated mothers. Such studies continue: Lauga (1977) used 18 morphological characters to evaluate changes in new-born hatchlings of the migratory locust as the parents were crowded. Whereas increasing the osmotic pressure of solutions in which the eggs are incubated reduces the hatchling size, but leaves its shape unchanged, treating crowded eggs with acetone solutions of the contents of solitary eggs changed the shape of hatchlings issuing from them to that of solitary locusts.

Gillett (1978) used discriminant functions to show that some modification of behaviour, colour and morphology ensued when crowded locusts were reared at different temperatures and photo-periods. de Gregorio and Lauga (1981) have confirmed that pyrgomorphid grasshoppers exhibit only two contrasts of form; *Zonocerus variegatus* reared during the wet season in Togo differ in size and shape from those reared during the dry season.

Lauga (1974) recommended principal component analysis and discriminant functions for such grasshopper investigations as those just outlined. He obtained poor results using correspondence analysis.

Key (1982) was able to separate the eumastacid genus *Geckomima* into two distinct groups of species by means of a principal coordinate analysis using 19 morphometric characters, and computing the axes of variation based on the Manhattan distance metric.

In evaluating the apparent phylogenetic links between the various subfamilies of the Eumastacidae, which seem to have evolved during the Mesozoic Era, Blackith (1973) utilized eleven morphometric measurements in a principal coordinate analysis. Campbell and Dearn (1980) used 12 such measurements to investigate populations of *Praxibulus* sp. and *Kosciuscola* sp. at various altitudes in southeastern Australia, employing canonical variate analysis. They concluded that the observed character displacement could not have been an artefact of clinal variation, evident in allopatric populations, and being carried over to the sympatric populations.

GENERAL ENTOMOLOGY

Inasmuch as insects constitute about three fourths of the animal species of the world, it comes as no surprise that there are several entomological applications to report.

Most of these are morphological, as in the studies of geographical variation in water-beetles by Zimmerman and Ludwig (1975), but there are important ecological applications as well. Bonnet *et al.* (1976) analysed the collembolan fauna of the High Pyrenees using correspondence analysis (see p. 178) to good effect.

In what must be one of the most comprehensive applications of principal component and principal coordinate analysis yet attempted for taxonomic ends, Davies and Boratynski (1979) looked at 24 species of male Coccoidea (scale insects) of the family Diaspididae and compared subsets of the 101 characters they had investigated. In these insects, the males are dramatically different from the wingless females so that the two sexes offer virtually two separate classifications necessarily based on the same phylogeny. Ordinations were based on subsets of about 25 characters selected in various ways. Some of the subsets could reproduce classifications based on the reference set containing four times as many characters. Such subsets included: the set selected subjectively by a taxonomist with long experience in this group; the set obtained by choosing those characters with high absolute values on the first five latent vectors of a standardized correlation matrix of characters;

and the set yielded by picking one character from each of the characters formed by single linkage of simple matching coefficients between the characters. These authors stressed the value of using the characters of males and of immature forms of both sexes to supplement the customary taxonomies based on the degenerate and specialized females.

Post-embryonic growth in a cockroach (Blattidae) was discussed in some detail by Davies and Brown (1972), mainly with principal component analyses; these authors provide a thoughtful account of differential growth patterns in insects. Canonical variate analysis enabled Cuzin-Roudy and Laval (1975) to undertake a similar investigation, this time on the water-bug *Notonecta maculata*.

The divergence of two sibling species of Lepidoptera has been studied by Guillaumin and Lefèbvre (1974) using the generalized statistical distance. Lane (1981) noted that whereas wing-shading patterns are of taxonomic value, they have rarely been used quantitatively, and he employed both principal components and principal coordinates to assimilate these patterns into the assemblage of characters by means of which a difficult group, of biting midges of veterinary interest, may be identified.

The ecology of small arthropods (Collembola and Oribatidae) living in moss on a huge boulder in a forest in France has been examined by Bonnet *et al.* (1970; 1975) by correspondence analysis.

Wilkinson (1974) used the add-a-point facility of principal coordinate analysis to answer some questions about the minimization of the number of characters exploited for the classification of the Lepidoptera. Usher (1979) looked at the value of the white spots on the wings of some Lepidoptera with a view to including them in a principal coordinate analysis in order to evaluate the taxonomic relationships of the *Neptis* butterflies. See also C. Wilkinson (1970).

Louis and Lefèbvre (1971) applied canonical variate analysis to sort out the races of the honey-bee. Rohlf and Sokal (1972) made use of a form of factor analysis with rotation of the axes to isolate and identify the morphogenetic influences that interact to give the observed correlations between characters of two species of flies. Again, Rohlf (1972) compared multidimensional scaling, principal component and principal coordinate analyses: he found that principal coordinates respond better than principal components when data are missing, but that multidimensional scaling is better than either in this respect. In behavioural studies on closely related Hawaiian species of *Drosophila*, Ringo and Hodosh (1978) put canonical variates to good use, whereas Rohlf and Sokal (1972) chose "factor analysis" for their work on two dipteran species.

Sakai (1978) deduces a Gondwanaland origin for the Dermaptera from an analysis of 20 environmental variables determining, or associated with, the distribution of the world's earwigs.

MOLLUSCS

The pulmonate gastropod *Cerion* has long been regarded as an array of seven allopatric taxa: from this view, Gould and Paull (1977) dissent. They believe, as a result of canonical variate analyses of 19 variables measured on 20 specimens from each of 23 samples, that each island in the Caribbean, from Hispaniola to the Virgin Islands, has a *Cerion* population with a distinct morphology. The first canonical array takes up 59% of the variation between populations, arranging them in perfect geographical order, from finely and copiously ribbed shells in the east to more cylindrical, apically pointed shells with fewer but stronger ribs in the west.

Dillon (1980) applied canonical variate and principal component analyses to the distributional data for two land-snails in Arizona (see p. 142). Phillips *et al.* (1973) enlisted principal components and canonical variates to examine variation in the whelk *Dicathais*. A highly variable species of the genus was found to respond morphologically to selective pressures at each locality investigated.

CRUSTACEA

Campbell and Mahon (1974) used canonical variates to compare two species of rock crab (genus *Leptograpsus*): they showed that the colour forms, blue and orange, do in fact represent two distinct species (see also pp. 13–28).

Riska (1981) explored morphological variation in the horseshoe crab, finding that the differences between localities were not merely an extension of differences within each locality, when these differences were represented along discriminant functions. As shown in Chapter 15, analyses comparing the results of applying several multivariate techniques to the same set of data are frequently especially instructive. Holmes (1975) used principal coordinates, principal components and canonical variates to ordinate the species of amphipod crustaceans in the genera *Gammarus* and *Marinogammarus*.

Variation in the ostracods has been studied by Reyment and his co-workers in several papers (for example, Reyment, 1973, 1974, 1982b, c, Reyment and Reyment, 1981). Principles of phenotypic evolution in fossils can be conveniently exemplified by means of ostracods. Reyment (1982f) has probed a case of "speciation caught in the act" in Moroccan Upper Cretaceous ostracods by means of linear and quadratic discriminant functions. Here, the mechanism promoting the speciation event may have been environmentally cued polymorphism in relation to ecological changes arising from a eustatic change of sea-level. Laval (1975) used several multivariate techniques to explore the influence of restricted food on growth of the amphipod *Phronima sedentaria*.

NEMATOLOGY

There have been at least two multivariate studies of free-living nematodes, a particularly difficult group whose taxonomists need all the help they can get. Townshend and Blackith (1975) used both principal component and canonical variate analyses in their studies of the effects of fungal diet on growth in *Aphelenchus avenae,* whereas Blackith and Blackith (1976) examined the inter-relationships between the shapes of species in the genus *Tylenchus* and related genera by means of principal coordinates.

Moss and Webster (1970) reviewed work to that year on the multivariate approach to nematology.

PROTOZOOLOGY

Gates, Powelson and Berger (1974) employed both principal components and canonical variates to examine the relationship between the various non-interbreeding, but morphologically almost identical, clones of *Paramecium aurelia.* Moreover, Berger (1978) has recommended appropriate multivariate methods to resolve taxonomic problems in the ciliophores. Gates and Berger (1974) resolved the differences in form of three strains of ciliates with the aid of principal components and canonical variates.

The investigation of incipient speciation mechanisms is arousing interest among protozoologists, as in other fields. Berger and Hatzidimitriou (1978) have elucidated such mechanisms in a sexually reproducing species of ciliate *Ancistrum mytili,* which inhabits species of the mussels *Modiolus* and *Mytilus* in North America. Three sympatric forms were confirmed by principal component analysis of 20 continuous and one discrete attribute (a procedure which immediately invites caution). The sympatric forms differ in somatic size and in the number of kineties. An example of the application of canonical variate analysis to *Paramecium* is given by Powelson *et al.* (1975) and the same technique proved successful in separating two species of microfilariae (Campbell, 1976).

Gates and Berger (1974) reported on the biometry of three strains of the ciliate *Tetrahymena pyriformis,* and in the same year, Gates *et al.* (1974) discussed the possibility of deciding to which clone (virtually a species) samples of *Paramecium aurelia* belong.

Turning to the Foraminifera, many aspects of their morphometrics have been investigated by Reyment (1978b, 1980b, 1982b), Reyment and Banfield (1976) and Campbell and Reyment (1978, 1980). Such studies are invaluable in constructing biologs, characterizations of sedimentary horizons intended for comparison with resistivity, or other physical parameters.

Reyment (1982e) has exploited a wide range of multivariate statistical techniques in his detailed study of phenotypic evolution in the Upper Cretaceous foraminiferan species *Afrobolivina afra*. Here, modern genetic theory is merged with the multivariate analyses. Phenotypic variation in the planktonic foraminiferan species *Globigerina pachyderma* was examined by Malmgren and Kennett (1972) by principal component analysis.

OTHER ZOOLOGICAL APPLICATIONS

The perennial attraction of discovering links between form and function is evident in McMahon's (1975) paper on allometry in ungulate long-bones. Dodson (1975, 1976) published two papers on relative growth in *Alligator* and *Protoceratops* respectively. The distribution and relationships between two species of Canadian bat have been described by Eger and Peterson (1979) putting both principal component analysis and discriminant analysis to use, and in another North American application, Elder and Hayden (1977) consider the determination of Canidae from Missouri.

The morphological relationships between the grey wolves of Canadian National Parks were explored by Skeel and Carbyn (1977) by a battery of techniques, including multi-dimensional scaling, discriminant functions and principal components. The last technique was involved in work on the dugong skull reported on by Spain *et al.* (1976), who were concerned to remove sexual dimorphic and size variation from their appreciation of skull-form. Jameson and Richmond (1971) studied parallelism and convergence in the frog *Hyla* by principal components and discriminant functions, whereas there is an unusual use of canonical correlations by Calhoon and Jameson (1970) in an attempt to associate features of the weather in southern California with size-variation in *Hyla regilla*.

Bernard (1981) has recommended multivariate methods for examining growth in fish and in a paper considering a new species of *Myotis*, Bogan (1978) plotted not only the scores of the organisms on canonical variate axes, but also the vectors of the original measurements, an idea that might be followed up.

Relationships between the genera and species of the feather mite are the subject of a publication by Moss *et al.* (1977) in which multivariate methods are put to use. Turning to the problem of phytoseiid mites, Hansell *et al.* (1980) applied operational point homology by Cartesian transformations to elucidate shifts in shape.

Many fossil and living species of bryozoans were investigated by Cheetham and Lorenz (1976); these authors employed the standard dichotomy of size and shape brought to light by a form of principal component analysis, but they went even further and partitioned the shape component into three parts, one associated with asymmetry of the outline, the second associated with elongation and distal inflation, and the third was tentatively identified as indicating that distal inflation is somewhat more sensitive to environmental fluctuations than is elongation. Cook (1977) put principal coordinates to good use in exploring species relationships in a bryozoan genus. Brood (1972), in his exhaustive monograph of Swedish Cretaceous bryozoans, has made frequent use of standard methods of multivariate analysis in his taxonomic descriptions.

Reyment (1971, 1973, 1978a) has continued to work on post-mortem floating of cephalopod shells and its influence on the palaeobiogeography of the group; he has shown that the twin patterns of growth, degree of coiling and greater width are determining features in supporting positive buoyancy.

BOTANY

The plants of the sites explored during the investigations forming part of the International Biological Programme were linked to the abiotic properties of the sites in a principal component analysis by French (1981), just as Heal *et al.* (1974) had looked at the abiotic conditions influencing the decomposition of strips of cotton buried in tundra biome soils. Moore (1975) investigated the growth and fruiting of apple trees by principal components: as a result it became apparent that an increase in stem dry-weight is controlled by the root-stock, whereas the scion is largely responsible for controlling the general trend of growth of the tree as a whole with time.

Riggins *et al.* (1978) showed that the dwarf lupin in the USA is a single, highly variable, species without detectable subspecific differentiation. Décamps and Bonnet (1976) compared the plantules of Ranunculaceae using correspondence analysis, principal components and canonical variates, and found that the classifications produced from these techniques agree well with those already established from consideration of the adult plants.

Multivariate discriminant analysis was used by Garten (1978) to explore the mineral composition of the leaves of nine species of plants, with results suggesting that each species survives because it occupies a restricted niche in multivariate space defined by the elements as axes. In a like manner, the discriminant analysis of 12 fatty acids from strains of the gliding bacterium *Simonsiella* correctly allocated them to their source, which was the oral cavity of sheep, dogs, cats and man (Jenkins *et al.*, 1977).

The topic of variation in the grass *Poa annua* was explored using canonical variates by Ellis *et al.* (1971). Variation in the growth of Corsican Pine stands in southern Britain suggested, when analysed by principal components, that low minimum winter temperatures are more influential than warmth during the growing season (Fourt *et al.*, 1971) and also that soil total phosphate, together with pH values, are important for growth potential.

The phylogenetic relationships of 12 species of kangaroo paw have been clarified by Hopper and Campbell (1977) using canonical variate analysis. Mastroguiseppe *et al.* (1970) employed stepwise discriminants to evaluate relationships among recent and fossil Ginkgo wood, claiming that its analysis brought out details not otherwise readily observable.

The possibility of using aerial photographs, taken on colour film, to detect diseased trees, regarding the image densities on red, green, and blue layers of the colour film as three variables, was explored by Stiteler (1979) who also examined differences in the concentration of nutrients in the foliage of sugar maple trees on good and on poor sites.

In a survey of the theory and uses of canonical correlation and variate analyses, Gittins (1979) has investigated soil-vegetational relationships in the tropical rain forest, the dynamic status of such a forest, the structure of grassland vegetation in North Wales, the nitrogen nutrition of eight grass species, and herbivore-environment relationships in the Ruwenzori National Park in Uganda, each example being used to illustrate different aspects of the multivariate techniques. This article alone contains some 200 references, by no means all botanically oriented, with remarkably little overlap with those given in this book, an eloquent testimony to the wide-ranging diversity of multivariate applications in current science.

Backman (1980) exploited several multivariate methods in his monographic study of the taxonomy and biostratigraphy of Neogene coccoliths from the North Atlantic realm.

Palynology

Non-metric multidimensional scaling was used by Birks (1973) to represent surface pollen spectra. Quaternary pollen diagrams were compared using the same approach. The work by Birks and Peglar on *Picea* pollen has already been discussed on p. 167.

Birks and Saanisto (1975) noted that when Finnish pollen data for three periods (4000, 6000 and 8000 years ago) were subjected to a principal component analysis, no less than 50% of the total variation could be related to major geographic patterns within each period of time. These patterns emphasize the distinction between south and central Finland and the extreme north of the country.

A map of scores along the first principal components shows the geographical form of covariance relationship between pollen types with high scores in northern Lappland as compared with the negative scores of assemblages in central and southern Finland. The first principal component thus reflects the distinction between birch-dominated assemblages of the north and the pine-dominated assemblages of the south.

Pennington and Sackin (1975) investigated Late Devensian pollen and chemical analyses and found that the two sets of assessments were closely related. Mahé (1974), in an overview of the uses of correspondence analysis, gave preliminary details of an investigation of saxifrage pollen grains. Gordon and Birks (1974) consider ways to compare pollen diagrams. A valuable reference for Quaternary palaeoecology is Birks and Birks (1980).

NON-BIOLOGICAL APPLICATIONS

Correspondence analysis has caught the public imagination in France rather more than such techniques normally do and the method has been explained in the popular press. French popular attitudes to economic difficulties have been analysed by this method by de Virieu (1975). Diday and Lebart (1977) have given a pocket-book type of account of how the necessary calculations are to be done. With all this public acclamation of Benzécri's (1973) method, it is good to remember that the idea was first developed by H. O. Hirschfeld (later Hartley) in Cambridge in 1935.

Possible strategies for improving public transport in Dublin, and assessing the preferences and intentions of users, were evaluated with the aid of linear discriminant functions by Markham and O'Farrell (1974).

In the course of a more general exposition of the theory and uses of correspondence analysis, Hill (1974) reanalysed the Münsinger-Rain incidence matrix. This is a table specifying the occurrence of 70 types of artefact in 59 tombs. Correspondence analysis yields almost the same results as had been obtained earlier by Kendall (1971). Hill's paper is of importance for those interested in making use of correspondence analysis.

Anderson (1971) explained how to apply multivariate methods to soil samples. Discriminant functions help to sort out geochemical data from various parts of the United Kingdom (Castillo-Munoz and Howarth, 1976). Smith et al. (1982) have utilized discrimination and allocation techniques to separate and identify gossans with the help of geochemical analyses. The paper by Hodson et al. (1966) remains a classic of its kind for anyone concerned with archaeological studies: in this case, an ordination of brooches by computer, by an anatomist, and by four archaeologists were compared.

The use of multiple measurements in meteorology was explained by Craddock (1965) from the standpoint of long-range weather-forecasting.

Bibliography

Numbers in square brackets indicate the page on which the reference is to be found.

Adanson, M. (1759). "Histoire naturelle du Sénégal: Coquillages. Avec la relation abrégée d'un voyage en ce pays pendant les années 1749–1753", 461 pp. Bauche, Paris. [7]

Adanson, M. (1763). "Familles des plantes", Vol. 1, 515 pp. Vincent, Paris. [7]

Aitchison, J. (1982). The statistical analysis of compositional data. *J. Roy. Statist. Soc.* B **44**, 139–160. [162]

Aitchison, J., Habbema, J. D. F. and Kay, J. W. (1977). A critical comparison of two methods of statistical discrimination. *Appl. Statist.* **26**, 15–25. [46]

Albrecht, F. O. and Blackith, R. E. (1957). Phase and moulting polymorphism in locusts. *Evolution.* **11**, 166–177. [72]

Albrecht, G. H. (1978). Some comments on the use of ratios. *Syst. Zool.* **27**, 67–71. [30]

Amtmann, E. (1965). Biometrische Untersuchungen zur introgressiven Hybridization der Waldmaus (*Apodemus sylvaticus* L., 1785) und der Gelbhalsmaus (*Apodemus tauricus* Pallas, 1811). *Z. zool. Syst. Evol.* **3**, 103–156. [88, 112]

Anderson, A. J. B. (1971). Numerical examination of multivariate soil samples. *Int. Inst. Math. Geol. J.* **3**, 1–14. [192]

Anderson, T. W. (1958). "An Introduction to Multivariate Statistical Analysis", 374 pp. John Wiley and Sons, New York.

Anderson, T. W. (1963). Asymptotic theory for principal components analysis. *Ann. Math. Statist.* **34**, 122–148.

Anderson, T. W. and Bahadur, R. R. (1962). Two sample comparisons of dispersion matrices for alternatives of immediate specificity. *Ann. Math. Statist.* **33**, 420–431.

Andrews, D. F. (1972). Plots of high-dimensional data. *Biometrics.* **28**, 125–136. [126, 172, 173, 181]

Andrews, H. E., Brower, J. C., Gould, S. J. and Reyment, R. A. (1974). Growth and variation in *Eurypterus remipes* DeKay. *Bull. Geol. Inst. Univ. Uppsala*, NS **4**, 81–114. [126]

Andrews, P. and Williams, D. B. (1973). The use of principal components analysis in physical anthropology. *Am. J. Phys. Anthropol.* **39**, 291–303. [182]

Arber, A. (1950). "The Natural Philosophy of Plant Form", 247 pp. University Press, Cambridge. [5]

Ashton, E. H. (1981). The Australopithecinae: their biometrical study. *In* "Perspectives in Primate Biology", Eds Ashton, E. H. and Holmes, R. L., pp. 67–126. Academic Press, London. [182]

Ashton, E. H., Healy, M. J. R. and Lipton, S. (1957). The descriptive use of discriminant functions in physical anthropology. *Proc. R. Soc.* B. **146**, 552–572. [64]

Ashton, E. H. Healy, M. J. R., Oxnard, C. E. and Spence, T. F. (1965). The combination of locomotor features of the primate shoulder girdle by canonical analysis. *J. Zool.* **147**, 406–429. [62, 63]

Atchley, W. R. (1971). A statistical analysis of geographic variation in the pupae of three species of *Culicoides* (Diptera: Ceratopogonidae). *Evolution* **25**, 51–74. [176]

193

[115] Atchley, W. R. (1972). The chromosome karyotype in estimation of lineage relationships. *Syst. Zool.* **21**, 199–209.

[117] Atchley, W. R. (1974). Morphometric differentiation in chromosomally characterized parapatric races of morabine grasshoppers. *Aust. J. Zool.* **22**, 25–37.

Atchley, W. R. (1981). Chromosomal evolution and morphometric variability in the thelytokous insect *Warramaba virgo* (Key). *In* "Evolution and Speciation: Essays in Honor of M. J. D. White", Eds Atchley, W. R. and Woodruff, D., pp. 371–397.
[115, 117] Cambridge University Press.

[30] Atchley, W. R. and Anderson, D. (1978). Ratios and the statistical analysis of biological data. *Syst. Zool.* **27**, 71–77.

Atchley, W. R. and Cheney, J. (1974). Morphometric differentiation in the *viatica* group of morabine grasshoppers (Orthoptera, Eumastacidae). *Syst. Zool.* **23**,
[117] 400–415.

[30] Atchley, W. R., Gaskins, C. T. and Anderson, D. (1976). Statistical properties of ratios, I. Empirical results. *Syst. Zool.* **25**, 137–148.

Atchley, W. R., Rutledge, J. J. and Cowley, D. E. (1981). Genetic components of size and shape. II Multivariate covariance patterns in the Rat and Mouse skull. *Evolution* **35**, 1037–1055.

Atchley, W. R., Rutledge, J. J. and Cowley, D. E. (1982). A multivariate statistical analysis of direct and correlated response to selection in the rat. *Evolution* **36**, 677–698.

[191] Backman, J. (1980). Miocene-Pliocene nannofossils and sedimentation rates in the Hatton-Rockall Basin, NE Atlantic Ocean. *Stockh. Contr. Geol.* **36**, 1–91.

[98] Baer, J. L. (1969). Paleoecology of cyclic sediments of the Lower Green River Formation, Central Utah. *Brigham Young Univ. Geol. Studies* **16** (I), 3–95.

[111] Bailey, D. W. (1956). A comparison of genetic and environmental principal components of morphogenesis in mice. *Growth* **20**, 63–74.

[176] Baker, A. J. (1980). Morphometric differentiation in New Zealand populations of the House Sparrow, (*Passer domesticus*). *Evolution* **33**, 638–653.

Baker, G. A. (1954). Organoleptic ratings and analytical data for wines analysed into orthogonal factors. *Food Res.* **19**, 575–580.

Baker, R. J., Atchley, W. R. and McDaniel, V. R. (1972). Karyology and morphometrics of Peters' Tent-making Bat, *Uroderma bilobatum* Peters (Chiroptera,
[115] Phyllostamatidae). *Syst. Zool.* **21**, 414–429.

Barkham, J. P. and Norris, J. M. (1970). Multivariate procedures in an investigation of vegetation and soil relations of two beech woodlands, Cotswold Hills, England.
[149] *Ecology* **51**, 630–639.

Baroni-Urbani, C. and Buser, M. W. (1976). Similarity of binary data. *Syst. Zool.*
[152] **24**, 251–259.

Barraclough, R. M. and Blackith, R. E. (1962). Morphometric relationships in the
[30] genus *Ditylenchus*. *Nematologica* **8**, 51–58.

Bartlett, M. S. (1965). Multivariate analysis. *In* "Theoretical and Mathematical Biology", Eds Waterman, T. H. and Morowitz, H. J., pp. 202–224. Blaisdell
[37] Publishing Co., New York.

Bennett, B. M. (1951). Note on a solution of the generalised Behrens-Fisher problem. *Ann. Inst. Statist. Math. Tokyo* **2**, 87–90.

Benzécri, J. P. (1973). "L'Analyse des Données. 2, L'Analyse des Correspon-
[86, 192] dances", 619 pp. Dunod, Paris.

Berger, J. (1978). Quantification of ciliophoran species descriptions: an appeal to
[188] reason. *Trans. Am. Micros. Soc.* **97**, 121–126.

Berger, J. and Hatzidimitriou, G. (1978). Multivariate morphometric analysis of demic variation in *Ancistrum mytili* (Ciliophora: Scuticociliatidae) commensal in two mytilid pelecypods. *Protistologica* **14**, 133–153. [188]

Bergson, H. (1911). "Creative evolution", 425 pp. Macmillan, London. [6]

Bernard, D. R. (1981). Multivariate analysis as a means of comparing growth in fish. *Can. J. Fish Aquatic Sci.* **38**, 233–236. [189]

Berthou, P. Y., Brower, J. C. and Reyment, R. A. (1975). Morphometrical study of Choffat's vascoceratids from Portugal. *Bull. Geol. Inst. Univ. Uppsala N. Ser.* **6**, 73–83. [175]

Bezy, R. L., Gorman, G. C., Kim, Y. J. and Wright, J. W. (1977). Chromosomal and genetic divergence in the fossorial lizards of the family Anniellidae. *Syst. Zool.* **26**, 57–71. [115]

Birks, H. J. B. (1973). Modern pollen rain studies in some arctic and alpine environments. *In* "Quarternary Plant Ecology", Eds Birks, H. J. B. and West, R. G., pp. 143–168. Blackwells, Oxford. [191]

Birks, H. J. B. and Birks, H. H. (1980). 'Quaternary Palaeoecology", Edward Arnold, London. [192]

Birks, H. J. B. and Peglar, S. M. (1980). Identification of *Picea* pollen of Late Quaternary age in eastern North America: a numerical approach. *Can. J. Bot.* **58**, 2043–2058. [167]

Birks, H. J. B. and Saarnisto, M. (1975). Isopollen maps and principal components analysis of Finnish pollen data for 4000, 6000 and 8000 years ago. *Boreas* **4**, 77–96. [191]

Blackith, R. E. (1957). Polymorphism in some Australian locusts and grasshoppers. *Biometrics* **13**, 183–196.

Blackith, R. E. (1958). An analysis of polymorphism in social wasps. *Insectes Sociaux* **5**, 263–272. [38, 110]

Blackith, R. E. (1960). A synthesis of multivariate techniques to distinguish patterns of growth in grasshoppers. *Biometrics* **16**, 28–40. [108, 111]

Blackith, R. E. (1962). L'identité des manifestations phasaires chez les acridiens migrateurs. *Colloques Int. Cent. Natn. Rech. Scient.* **114**, 299–310. [51,52,110,126]

Blackith, R. E. (1963). A multivariate analysis of Latin elegiac verse. *Language and Speech* **6**, 196–205. [94]

Blackith, R. E. (1965). Morphometrics. *In* "Theoretical and Mathematical Biology", Eds Waterman, T. H. and Morowitz, H. J., pp. 225–249, Blaisdell Publishing Co., New York. [66]

Blackith, R. E. (1973). Aperçu sur l'évolution mésozoique des eumasticides. *Acrida* **2**, V-XVIII. [185]

Blackith, R. E. (1981). The Collembola of the tundra biome sites: a zoogeographical synthesis. *In* "Tundra Ecosystems: a Comparative Analysis", Eds Bliss, L. C., Heal, O. W. and Moore, J. J. (The International Biological Programme 25), pp. 541–545, Cambridge University Press. [153]

Blackith, R. E. and Albrecht, F. O. (1959). Morphometric differences between the eye-stripe polymorphs of the Red Locust. *Scient. J. R. Col. Sci.* **27**, 13–27. [60]

Blackith, R. E. and Blackith, R. M. (1969). Variation of shape and of discrete anatomical characters in the morabine grasshoppers. *Aust. J. Zool.* **17**, 697–718. [60]

Blackith, R. E. and Blackith, R. M. (1975). Zoogeographical and ecological determinants of collembolan distribution. *Proc. R. Irish Acad.* **75** (B), 345–368. [153]

Blackith, R. M. and Blackith, R. E. (1976). A multivariate study of *Tylenchus* Bastian 1865 (Nematoda, Tylenchidae) and some related genera. *Nematologica* **22**, 235–259. [188]

Blackith, R. E., Davies, R. G. and Moy, E. A. (1963). A biometric analysis of the development of *Dysdercus fasciatus* Sign. *Growth* **27**, 317–334.

[58, 59, 111, 112, 134]
Blackith, R. E. and Kevan, D. K. McE. (1967). A study of the genus *Chrotogonus* (Orthoptera). VIII. Patterns of variation in external morphology. *Evolution* **21**, 76–84.

[71, 110, 129, 133]
Blackith, R. E. and Roberts, M. I. (1958). Farbenpolymorphismus bei einigen Feldheuschrecken. *Z. Vererb Lehre* **89**, 328–337.

[29]
Blackwelder, R. E. (1964). Phyletic and phenetic versus omnispective classification. *In* "Phenetic and Phylogenetic classification", Eds Heywood, V. H. and McNeill, J., pp. 17–28, Systematics Assn. Publ. No. 6, London.

[88]
Blair, C. A., Blackith, R. E. and Boratynski, K. L. (1964). Variation in *Coccus hesperidum* L. (Homoptera; Coccidae). *Proc. R. Ent. Soc. Lond.* (A) **39**, 129–134.

Bliss, C. I. (1958). Periodic regression in biology and climatology. *Bull. Conn. Agric. Exp. Stn.* 615 pp.

[118]
Bloom, W. L. (1976). Multivariate analysis of the introgressive replacement of *Clarkia nitens* by *Clarkia speciosa polyantha* (Onagraceae). *Evolution* **30**, 412–424.

[6]
Bocquet, C. (1953). Recherches sur le polymorphism naturel des *Jaera marina* Fabr. (Isopodes asellotes). Essai de systématique évolutive. *Archs Zool. Éxp. Gén.* **90**, 107–450.

[113]
Bocquet, C. and Solignac, M. (1969). Etude morphologique des hybrides expérimentaux entre *Jaera albifrons* et *Jaera (albifrons) praehirsuta* (Isopoda: Asellota). *Archs Zool. Éxp. Gén.* **110**, 435–452.

[189]
Bogan, M. A. (1978). A new species of *Myotis* from the Islas Tres Marias, Nayarit, Mexico, with comments on variation in *Myotis nigricans*. *J. Mammalogy* **59**, 519–530.

[178]
Bonnet, L. (1974). Quelques aspects du peuplement thécamoebien des sols de truffières. *Protistologica* **10**, 281–291.

[186]
Bonnet, L., Cassagnau, P. and de Izarra, D. -C. (1970). Etude écologique des collemboles muscicoles du Sidobre (Tarn). II. Modèle mathématique de la distribution des espèces sur un rocher. *Bull. Soc. Hist. Nat. Toulouse* **106**, 127–145.

[178, 185]
Bonnet, L., Cassagnau, P. and Deharveng, L. (1976). Un exemple de rupture de l'équilibre biocénotique par déboisement: les peuplements de Collemboles édaphiques du Piau d'Engaly (Hautes-Pyrénées). *Rev. Ecol. Biol. Sol.* **13**, 337–351.

[186]
Bonnet, L., Cassagnau, P. and Travé, J. (1975). L'Ecologie des arthropodes muscicoles à la lumière de l'analyse des correspondances: Collemboles et Oribatides du Sidobre (Tarn, France). *Oecologia (Berl.)* **21**, 359–373.

[179]
Bonnet, L. and Rey, J. (1973). Modèles mathématiques des facteurs paléoécologiques dans quelques gisements du Crétacé inférieur d'Estremadura. *Comun. Serv. Geol. Portugal* **56**, 421–450.

[2, 5, 31, 124, 125, 157]
Bookstein, F. L. (1978). "The measurement of biological shape and shape change". Lecture notes in biomathematics No. 24, 191 pp. Springer-Verlag, New York.

[101]
Boratynski, K. and Davies, R. G. (1971). The taxonomic value of male Coccoidea (Homoptera) with an application and evaluation of some numerical techniques. *Biol. J. Linn. Soc. Lond.* **3**, 57–102.

[179]
Boucot, A. J. (1978). Community evolution and rates of cladogenesis. *Evol. Biol.* **11**, 545–655.

Bouroche, J. -M. and Saporta, G. (1980). "L'analyse des données". (Que sais-je?) No. 1854, 127 pp. Presses Universitaires de France, Paris.

[114]
Brieger, F. G., Vencovsky, R. and Paker, I. U. (1963). Distancias filogenéticas no genero *Cattleya. Cienc. Cult.* **15**, 187–188.

Brood, K. (1972). Cyclostomatous Bryozoa from the Upper Cretaceous and Danian in Scandinavia. *Stockh. Contr. Geol.* **26**, 1–464. [190]

Brothwell, D. R., Healy, M. J. R. and Harvey, R. G. (1972). Canonical analysis of facial variation. *J. Biosoc. Sci.* **4**, 175–185. [181]

Brower, J. C., Cubitt, J. M., Veinus, J. and Morton, M. (1979). Principal components analysis, and point coordinates in the study of multivariate allometry. *In* "Geomathematical and Petrophysical Studies in Sedimentology", Eds Gill, D. and Merriam, D. F. *Computers and Geology*, **3**, 245–266. [126]

Brower, J. C. and Veinus, J. (1978). Multivariate analysis of allometry using point coordinates. *J. Paleontology* **52**, 1037–1053. [31, 126]

Brown, T. (1973). "Morphology of the Australian Skull Studied by Multivariate Analysis". Australian aboriginal studies No. 49. [181]

Brožek, J. and Keys, A. (1951). Evaluation of leanness–fatness: norms and interrelationships. *Br. J. Nutr.* **5**, 194–206.

Bryant, E. H. (1977). Morphometric adaptation of the housefly (*Musca domestica* L.) in the United States. *Evolution* **31**, 580–596. [176]

Bryant, E. H. and Turner, C. R. (1978). Comparative morphometric adaptation of the housefly and the face-fly in the United States. *Evolution* **32**, 759–770. [176]

Burla, H. and Kälin, A. (1957). Biometrischer Vergleich zweier Populationen von *Drosophila obscura*. *Revue Suisse Zool.* **64**, 246–252. [131]

Burnaby, T. P. (1966a). Growth-invariant discriminant functions and generalised distances. *Biometrics* **22**, 96–110. [31, 135]

Burnaby, T. P. (1966b). Distribution-free quadratic discriminant functions in palaeontology. *In* "Computer Applications in the Earth Sciences", pp. 70–77, State Geol. Survey, Kansas. [33, 50]

Buzas, M. A. (1967). An application of canonical analysis as a method for comparing faunal areas. *J. Anim. Ecol.* **36**, 563–577. [136]

Cacoullos, T. (Ed.) (1973). "Discriminant Analysis and Applications", 434 pp. Academic Press, New York. [44]

Cain, A. J. (1958). Logic and memory in Linnaeus system of taxonomy. *Proc. Linn. Soc. Lond.* **170**, 185–217. [4]

Cain, A. J. (1959). Taxonomic concepts. *Ibis* **101**, 302–318. [7]

Cain, A. J. (1962). The evolution of taxonomic principles. *In* "Microbial Classification", Eds Ainsworth, G. C. and Sneath, P. H. A., pp. 1–13. University Press, Cambridge. [4]

Calaprice, J. R., McSheffrey, H. M. and Lapi, L. A. (1971). Radioisotope X-ray fluorescence spectrometry in aquatic biology: a review. *J. Fish. Res. Bd Can.* **28**, 1538–1594. [178]

Calhoon, R. E. and Jameson, D. L. (1970). Canonical correlation between variation in weather and variation in size in the Pacific Tree Frog, *Hyla regilla*, in Southern California. *Copeia* **1970**, 124–134. [189]

Campbell, B. (1962). The systematics of man. *Nature*. **194**, 225–232. [73]

Campbell, N. A. (1976). A multivariate approach to variation in microfilariae: examination of the species *Wuchereria lewisi* and demes of the species *W. bancrofti*. *Aust. J. Zool.* **24**, 105–114. [37, 188]

Campbell, N. A. (1978). Multivariate analysis in biological anthropology: some further considerations. *J. Human Evolution* **7**, 197–203. [180, 183]

Campbell, N. A. (1979). Canonical Variate Analysis: Some Practical Aspects. Thesis, Imperial College, 242 pp. University of London. [20, 46]

Campbell, N. A. (1980). On the study of the Border cave remains: statistical comments. *Current Anthropology* **21**, 532–535. [46]

Campbell, N. A. (1981). Graphical comparison of covariance matrices. *Aust. J. Statist.* **23**, 21–37. [20]

Campbell, N. A. (1982). Robust procedures in multivariate analysis II. Robust canonical variate analysis. *Appl. Statist.* **31**, 1–8.

Campbell, N. A. (1983). Some aspects of allocation and discrimination. "Multivariate Statistical Methods in Physical Anthropology" Eds van Vark, G. N. and Howells, W. W. D. Reidel, Dordrecht, Holland.

Campbell, N. A. and Atchley, W. R. (1981). The geometry of canonical variate analysis. *Syst. Zool.* **30**, 268–280. [53, 154]

Campbell, N. A. and Dearn, J. M. (1980). Altitudinal variation in, and morphological divergence between, three related species of Grasshopper, *Praxibulus* sp., *Kosciuscola cognatus* and *K. usitatus* (Orthoptera: Acrididae). *Aust. J. Zool.* **28**, 103–118. [57, 129, 185]

Campbell, N. A. and Kitchener, D. J. (1980). Morphological divergence in the genus *Eptesicus* (Microchiroptera: Vespertilionidae) in Western Australia: a multivariate approach. *Aust. J. Zool.* **28**, 457–474.

Campbell, N. A. and Mahon, R. J. (1974). A multivariate study of variation in two species of rock crab of the genus *Leptograpsus*. *Aust. J. Zool.* **22**, 417–425. [13, 187]

Campbell, N. A. and Reyment, R. A. (1978). Discriminant analysis of a Cretaceous foraminifer using shrunken estimators. *Math. Geol.* **10**, 347–359. [20, 56, 76, 154, 188]

Campbell, N. A. and Reyment, R. A. (1980). Robust multivariate procedures applied to the interpretation of atypical individuals of a Cretaceous foraminifer. *Cretaceous Res.* **I**. 207–221. [46, 175, 188]

Campbell, N. A. and Saunders, D. A. (1976). Morphological variation in the white-tailed Black Cockatoo, *Calyptorhynchus baudinii* in Western Australia: "a multivariate approach". *Aust. J. Zool.* **24**, 589–595. [183]

Campbell, N. A. and Tomenson, J. A. (1983). Canonical variate analysis for several sets of data. *Biometrics* **39**, 425–435. [53]

Carmelli, D. and Cavalli-Sforza, L. L. (1979). The genetic origin of the Jews: a multivariate approach. *Hum. Biol.* **51**, 41–61. [165]

Carson, H. L. (1975). The genetics of speciation at the diploid level. *Am. Nat.* **109**, 83–92.

Casetti, E. (1964). Multiple discriminant functions. Tech. Rept. No. II, O.N.R. Geography Branch, Evanston, Ill. 63 pp. [94]

Cassagnau, P. and Matsakis, J. T. (1966). Sur l'utilisation des comparisons multiples en biocénotique édaphique. *Rev. Ecol. Biol.* **2**, 463–474. [151]

Cassie, R. M. (1962). Multivariate analysis in the interpretation of numerical plankton data. *NZ J. Sci.* **6**, 36–59. [137]

Cassie, R. M. (1967). Principal component analysis of the zoo-plankton of Lake Maggiore. *Mem. Ist. Ital. Idrobiol.* **21**, 129–144. [141]

Cassie, R. M. and Michael, A. D. (1968). Fauna and sediments of an intertidal mud flat. *J. Exp. Mar. Biol. Ecol.* **2**, 1–23. [137]

Cassie, R. M. (1972). Fauna and sediments of an intertidal mud-flat: an alternative multivariate analysis. *J. Exp. Mar. Biol. Ecol.* **9**, 55–64. [178]

Castillo-Munoz, R. and Howarth, R. J. (1976). Application of the empirical discriminant function to regional geochemical data from the United Kingdom. *Geol. Soc. Am. Bull.* **87**, 1567–1581. [192]

Castro, A. (1965). La Realidad Histórica de España. Editorial Porrua SA, México. [161]

Cattell, R. B. (1965a). Factor analysis: an introduction to essentials. I. The purpose and underlying models. *Biometrics* **21**, 190–215. [103]

Cattell, R. B. (1965b). Factor analysis: an introduction to essentials. II. The role of factor analysis in research. *Biometrics* **21**, 405–435.   [103]

Cavalli-Sforza, L. L. (1966). Population structure and human evolution. *Proc. R. Soc. Lond. B* **164**, 362–379.   [114]

Cheetham, A. H. (1968). Morphology and systematics of the bryozoan genus *Metrarhabdotos*. *Smithson Misc. Colln.* **153**, 1–121.   [89]

Cheetham, A. H. and Lorenz, D. M. (1976). A vector approach to size and shape comparisons among zooids in cheilostome Bryozoans. *Smithsonian Contribs. Paleobiol.* No. 29, 55 pp.   [120, 189]

Cherry, L. M., Case, S. M., Kunkel, J. G., Wyles, J. S. and Wilson, A. C. (1982). Body shape metrics and organismal evolution. *Evolution* **36**, 914–933.   [126]

Cheverud, J. M. (1982a). Variation in highly and lowly heritable morphological traits among social groups of Rhesus macaques *(Macaca mulatta)* on Cayo Santiago. *Evolution* **35**, 75–83.   [119]

Cheverud, J. M. (1982b). Phenotypic, genetic, and environmental morphological integration in the cranium. *Evolution* **36**, 499–516.   [119]

Cheverud, J. M. (1982c). Relationships among ontogenetic, static, and evolutionary allometry. *Am. J. Phys. Anthropol.* **58**, 1–11.   [116]

Christensen, K. (1954). Ratios as a means of specific differentiation in Collembola. *Ent. News* **65**, 176–177.   [30]

Christensen, W. K. (1974). Morphometric analysis of *Actinocamax plenus* from England. *Bull. Geol. Soc. Denmark* **23**, 1–26.   [156]

Claringbold, P. J. (1958). Multivariate quantal analysis. *J. R. Statist. Soc. B* **20**, 113–121.   [61]

Clark, D., Davies, W. K. D. and Johnston, R. J. (1974). The application of factor analysis in human geography. *Statistician* **23**, 259–281.   [176]

Clark, P. J. and Spuhler, J. N. (1959). Differential fertility in relation to body dimensions. *Hum. Biol.* **31**, 121–137.   [113]

Clunies-Ross, C. W. (1969). Profiles and association: two exploratory techniques in the multivariate analysis of surveys. *Statistician* **19**, 49–60.   [149]

Cochran, W. G. (1962). On the performance of the linear discriminant function. *Bull. Inst. Int. Statist.* **39**, 435–447.   [37]

Cochran, W. G. and Hopkins, C. E. (1961). Some classification problems with multivariate qualitative data. *Biometrics* **17**, 10–32.   [44]

Cock, A. G. (1966). Genetical aspects of metrical growth and form in animals. *Q. Rev. Biol.* **41**, 131–190.   [109]

Colgan, P. W. (Ed.) (1978). "Quantitative Ethology", Wiley, New York.   [183]

Colless, D. H. (1967). An examination of certain concepts in phenetic taxonomy. *Syst. Zool.* **16**, 6–27.   [7, 116]

Cook, P. L. (1977). The genus *Tremogasterina* Canu (Bryozoa, Cheilostomata). *Bull. Br. Mus. Nat. Hist.* **32**, 103–165.   [190]

Cooley, W. W. and Lohnes, P. R. (1971). "Multivariate Data Analysis", 364 pp. Wiley, New York.   [44, 78, 80]

Cooper, P. W. (1963). Statistical classification with quadratic forms. *Biometrika* **50**, 439–448.   [33, 50]

Cooper, P. W. (1965). Quadratic discriminant functions in pattern recognition. *IEEE Trans. Inform. Theory* **11**, 313–315.   [50]

Corrucini, R. (1978). Multivariate analysis in biological anthropology: some considerations. *J. Human Evolution* **4**, 1–19.   [180, 182]

[30, 182] Corrucini, R. S. (1977). Correlation properties of morphometric ratios. *Syst. Zool.*, **26**, 211–214.

Corrucini, R. S. (1978). Primate skeletal allometry and hominoid evolution. *Evolution* **32**, 752–758.

[6, 109] Cousin, G. (1961). Analyse des équilibres morphogénétiques des types structureaux spécifiques et hybrides chez quelques gryllides. *Bull. Soc. Ent. Fr.* **86**, 500–521.

Cox, G. M. and Martin, W. P. (1937). The discriminant function applied to the differentiation of soil types. *Iowa State Coll. J. Sci.* **11**, 323–331.

[192] Craddock, J. M. (1965). A meteorological application of principal components analysis. *Statistician* **15**, 143–156.

[105] Crovello, T. J. (1968). Key communality cluster analysis as a taxonomic tool. *Taxon* **17**, 241–258.

[143, 177] Cugny, P. and Rey, J. (1975). Un exemple d'utilisation de l'analyse factorielle des correspondances en paléo-écologie: la répartition des microfossiles dans le Bédoulien D'Estremadura (Portugal). BSGF **17**, 787–796.

[120, 186] Cuzin-Roudy, J. and Laval, P. (1975). A canonical discriminant analysis of post-embryonic development in *Notonecta maculata* Fabricius (Insecta: Heteroptera). *Growth* **39**, 251–280.

[145] Dagnelie, P. (1965a). L'étude des communautés végétales par l'analyse statistique des liaisons entre les espèces et les variables écologiques: Principes fondamentaux. *Biometrics* **21**, 345–361.

[145] Dagnelie, P. (1965b). L'étude des communautés végétales par l'analyse statistique des liaisons entre les espèces et les variables écologiques: un exemple. *Biometrics* **21**, 890–907.

[153] Dale, M. B. (1979). On linguistic approaches to ecosystems and their classification. *In* "Multivariate Methods in Ecological Work", Eds Orlóci, L. Rao, C. R. and Stiteler, W. M., pp. 11–20. International Co-operative Publishing House, Fairland, Maryland.

[104] Darton, R. A. (1980). Rotation in factor analysis. *Statistician* **29**, 167–194.

[175] David, M., Campaglio, C. and Darling, R. (1974). Progress in *R*- and *Q*-mode analysis: correspondence analysis and its application to the study of geological processes. *Can. J. Earth Sci.* **11**, 131–146.

[185] Davies, R. G. and Boratynski, K. L. (1979). Character selection in relation to the numerical taxonomy of some male Diaspididae (Homoptera: Coccoidea). *Biol. J. Linn. Soc.* **12**, 95–165.

[186] Davies, R. G. and Brown, V. (1972). A multivariate analysis of post-embryonic growth in two species of *Ectobius* (Dictyoptera: Blattidae). *J. Zool. Lond.* **168**, 51–79.

[86] Davies, P. M. and Coxon, A. P. M. (Eds) (1982) "Key Texts in Multidimensional Scaling", Heinemann Educational Books, London.

[115] Davis, B. L. and Baker, R. J. (1974). Morphometrics, evolution and cytotaxonomy of mainland bats of the genus *Macrotus* (Chiroptera: Phyllostomatidae). *Syst. Zool.* **23**, 26–39.

[190] Décamps, O. and Bonnet, L. (1976). Application de l'analyse des données multidimensionnelles à l'étude morphologique des plantules de Ranunculacées. *CR Acad. Sc. Paris (D)* **282**, 597–600.

[73] Defrise-Gussenhoven, E. (1957). Mesure de divergence entre quelques fémurs fossiles et un ensemble de fémurs récents: étude biométrique. *Bull. Inst. Sci. Nat. Belg.* **33**, 13 pp.

de Gregorio, R. and Lauga, J. (1981). Analyse biométrique du polymorphisme saisonnier chez les imagos de *Zonocerus variegatus* (L.) (Orthoptera: Pyrgomorphidae). *Acrida* **10**, 15–24.   [184]

de Groot, M. H. and Li, C. C. (1966). Correlations between similar sets of measurements. *Biometrics* **22**, 781–790.

Delany, M. J. and Healy, M. J. R. (1964). Variation in the long-tailed field-mouse (*Apodemus sylvaticus* L.) in north-west Scotland. II. Simultaneous examination of all the characters. *Proc. R. Soc. Lond.* B **161**, 200–207.   [133]

Delany, M. J. and Whittaker, H. M. (1969). Variation in the skull of the long-tailed field-mouse *Apodemus sylvaticus* in mainland Britain. *J. Zool.* **157**, 147–157.   [133]

Dempster, A. P. (1963). Stepwise multivariate analysis of variance based on principal variables. *Biometrics* **19**, 478–490.   [50]

Dempster, A. P. (1964). Tests for the equality of two covariance matrices in relation to a best linear discriminator analysis. *Ann. Math. Statist.* **35**, 355–374.   [50]

de Virieu, F. -H. (1975). Le prix d'un Francais: l'inégalité des malchances. *Le Nouvel Observateur,* 22/9/1975, pp. 47–66.   [192]

Diday, E. and Lebart, L. (1977). L'analyse des données. *La Recherche* **8**, 15–25.   [192]

Digby, P. G. N. and Gower, J. C. (1981). Ordination between- and within-groups applied to soil classification. Down-To-Earth Statistics: Solutions Looking for Geological Problems. Syracuse University Geology Contributions No. 8 (Ed. Merriam, D. F.), pp. 63–75.   [154]

Dillon, R. T. (1980). Multivariate analysis of Desert snail distribution in an Arizona canyon. *Malacologia* **19**, 201–207.   [142]

Dodson, P. (1975). Functional and ecological significance of relative growth in *Alligator. J. Zool.* **175**, 315–355.   [189]

Dodson, P. (1976). Quantitative aspects of relative growth and sexual dimorphism in *Protoceratops. J. Paleontology* **50**, 929–940.   [189]

Dürer, A. (1613). Les quatres livres d'Albrecht Dürer de la proportion des parties et pourtraicts des corps humains. Arnheim.   [6]

Eckart, C. and Young, G. (1936). The approximation of one matrix by another of lower rank. *Psychometrika* **1**, 211–218.   [35]

Eger, J. L. and Peterson, R. L. (1979). Distribution and systematic relationship of *Tadarida bivittata* and *Tadarida ansorgei* (Chiroptera: Molossidae). *Can. J. Zool.* **57**, 1887–1895.   [189]

Ehrenberg, A. S. C. (1962). Some questions about factor analysis. *Statistician* **12**, 191–209.   [102]

Ehrenberg, A. S. C. (1963). Some queries to factor analysts. *Statistician* **13**, 257–262.   [102]

Ehrenberg, A. S. C. (1964). Replies to Dr Warburton, Mr Jeffers, and Dr Lawley. *Statistician* **14**, 51–61.   [102]

Eickwort, K. (1969). Differential variation of males and females in *Polistes exclamans. Evolution* **23**, 391–405.   [110]

Elder, W. H. and Hayden, C. M. (1977). Use of discriminant function in taxonomic determination of canids from Missouri. *J. Mammalogy* **58**, 17–24.   [189]

Eldredge, N. and Gould, S. J. (1972). Punctuated equilibria: an alternative to phyletic gradualism. *In* "Models in Paleobiology", Ed. Schopf, T., pp. 82–115. Freeman, Cooper and Co., San Francisco.   [76]

Ellis, W. M., Lee, B. T. O. and Calder, D. M. (1971). A biometric analysis of populations of *Poa annua. Evolution* **25**, 29–37.   [190]

Erez, J. and Gill, D. (1977). Multivariate analysis of biogenic constituents in recent sediments off Ras Burka, Gulf of Elat, Red Sea. *Math. Geol.* **9**, 77–98.   [142, 177]

Everitt, B. (1978). "Graphical techniques for multivariate data", Heinemann,
[126, 173]   London.
Everitt, B. S. and Nicholls, P. (1975). Visual techniques for representing multivariate
[173]   data. *Statistician* **24**, 37–49.
Eyles, A. C. and Blackith, R. E. (1965). Studies on hybridization in *Scolopostethus*
[57]   Fieber (Heteroptera Lygeidae). *Evolution* **19**, 465–479.
Fasham, M. J. R. (1977). A comparison of non-metric multidimensional scaling,
     principal component analysis, and reciprocal averaging for the ordination of
     simulated coenoclines and coenoplanes. *Ecology* **58**, 558–561.
Ferrari, T. J., Pijl, H. and Venekamp, J. T. N. (1957). Factor analysis in agricultural
[149]   research. *Neth. J. Agric. Sci.* **5**, 211–221.
Field, J. G. (1969). The use of the information statistic in the numerical classification
[151]   of heterogeneous systems. *J. Ecol.* **57**, 565–569.
Fisher, D. R. (1968). A study of faunal resemblance using numerical taxonomy and
     factor analysis. *Syst. Zool.* **17**, 48–63.
Fisher, R. A. (1936). The use of multiple measurements in taxonomic problems.
[15, 43, 46]   *Ann. Eugen. Lond.* **7**, 179–188.
Fisher, R. A. (1940). The precision of discriminant functions. *Ann. Eugenics* **10**,
[86]   422–429.
Flessa, K. W. and Bray, R. G. (1977). On the measurement of size-independent
     morphological variability: an example using successive populations of a Devonian
     spiriferid brachiopod. *Paleobiology* **3**, 350–359.
Fourt, D. F., Donald, D. G. M., Jeffers, J. N. R. and Binns, W. O. (1971). Corsican
     Pine (*Pinus nigra* var. *maritima* (Ait.) Melville) in southern Britain: a study of
[191]   growth and site factors. *Forestry* **44**, 189–207.
Fraisse, R. and Arnoux, J. (1954). Les caractères biométriques du cocon chez
     Bombyx mori L. et leurs variations sous l'influence de l'alimentation. *Rev. Ver*
[49]   *Soie* **6**, 43–62.
Fries, J. and Matérn, B. (1966). On the use of multivariate methods for the
     construction of tree-taper curves. *Res. Notes Dep. Forestry Biometry, Skogshögs-*
[88]   *kolan, Stockholm,* **9**, 85–117.
French, D. D. (1981). Multivariate comparisons of IBP tundra biome site charac-
     teristics. *In* "Tundra Ecosystems: a Comparative Synthesis", Eds Bliss, L. C.,
     Heal, O. W. and Moore, J. J., (The International Biological Programme 25, pp.
[190]   47–75). Cambridge University Press.
Fujii, K. (1969). Numerical taxonomy of ecological characteristics and the niche
[133]   concept. *Syst. Zool.* **18**, 151–153.
Gabriel, K. R. (1968). The biplot graphical display of matrices with application to
[86]   principal components analysis. *Biometrika* **58**, 453–467.
Gabriel, K. R. and Sokal, R. R. (1969). A new statistical approach to geographic
[132]   variation analysis. *Syst. Zool.* **18**, 259–278.
Garten, C. T. Jr (1978). Multivariate perspectives on the ecology of plant mineral
[190]   element composition. *Am. Nat.* **112**, 533–544.
Gates, M. A. and Berger, J. (1974). A biometric study of three strains of *Tet-*
[188]   *rahymena pyriformis* (Cileata: Hymenostomatidae). *Can. J. Zool.* **52**, 1167–1183.
Gates, M. A., Powelson, E. E. and Berger, J. (1974). Syngenic ascertainment in
[188]   *Paramecium aurelia. Syst. Zool.* **23**, 482–489.
Gatty, R. (1966). Multivariate analysis for marketing research: an evaluation. *Appl.*
     Statist. **15**, 157–172.

Geisser, S. (1977). Discrimination, allocation and separatory, linear aspects. "Classification and Clustering" Ed. van Ryzin, J. Academic Press, New York.

Gilchrist, B. M. (1960). Growth and form in the brine shrimp *Artemia salina* (L.) *Proc. R. Soc. Lond.* **134**, 221–235. [129]

Giles, E. and Bleibtrau, H. K. (1961). Cranial evidence in archaeological reconstructions: a trial of multivariate techniques for the South-West. *Am. Anthrop.* **63**, 48–61. [73]

Giles, E. and Elliot, O. (1962). Race identification from cranial measurements. *J. Forensic Sci.* **7**, 147–157. [73]

Giles, E. and Elliot, O. (1963). Sex determination by discriminant function. *Am. J. Phys. Anthrop.* **21**, 53–68. [73]

Gillett, S. D. (1978). Environmental determinants of phase polymorphism of the Desert Locust, *Schistocerca gregaria* (Forsk.) reared crowded. *Acrida* **7**, 267–288. [184]

Gittins, R. (1965a). Multivariate approaches to a limestone grassland community. II. A direct species ordination. *J. Ecol.* **53**, 403–409. [146]

Gittins, R. (1965b). Multivariate approaches to a limestone grassland community. III. A comparative study of ordination and association analysis. *J. Ecol.* **53**, 411–425. [146]

Gittins, R. (1969). The application of ordination techniques. *In* "Ecological Aspects of the Mineral Nutrition of Plants", Ed Rorison, I. H., pp. 37–66. Brit. Ecol. Soc. Symp. No. 9. [146, 149]

Gittins, R. (1979). Ecological applications of canonical analysis. *In* "Multivariate Methods in Ecological Work", pp. 309–535, Eds Orlóci, L., Rao, C. R. and Stiteler, W. M., International Co-operative Publishing House, Fairland, Maryland. [142, 146, 191]

Glahn, H. R. (1968). Canonical correlation and its relationship to discriminant analysis and multiple regression. *J. Met.* **25**, 23–31. [78]

Gnanadesikan, R. (1977). "Methods for Statistical Data Analysis of Multivariate Observations", Wiley, New York. [20, 45, 173]

Gökçen, S. L. and Ozkaya, I. (1981). Olistostrom ve turbidit fasiyeslerinin diskriminant analizi ile ayirimi. *Yerbilimleri* **8**, 53–60. [144]

Goldstein, M. and Dillon, W. R. (1978). "Discrete Discriminant Analysis", 186 pp. Wiley, New York. [44]

Goodall, D. W. (1954a). Objective methods for the classification of vegetation. III. An essay in the use of factor analysis. *Aust. J. Bot.* **2**, 304–324. [146]

Goodall, D. W. (1954b). Vegetational classification and vegetational continua. *Angew. Pfl. Soziol.* **1**, 168–182. [146]

Goodman, M. M. (1969). Measuring evolutionary divergence. *Jap. J. Genet.* **44** (suppl. 1), 310–316.

Goodman, M. M. (1972). Distance analysis in biology. *Syst. Zool.* **21**, 174–186. [114]

Gordon, A. D. (1981). "Classification", Chapman and Hall, London. [151]

Gordon, A. D. and Birks, H. J. B. (1974). Numerical methods in Quaternary palaeoecology. II. Comparison of pollen diagrams. *New Phytol.* **73**, 221–249. [192]

Gordon, A. D. and Reyment, R. A. (1979). Slotting of borehole sequences. *Math. Geol.* **11**, 309–327. [175]

Gould, S. J. (1966). Allometry and size in ontogeny and phylogeny. *Biol. Rev.* **41**, 587–640. [30, 124]

Gould, S. J. (1967). Evolutionary patterns in pelycosaurian reptiles; a factor analytical study. *Evolution* **21**, 385–401. [84, 103]

Gould, S. J. (1975). Allometry in primates, with emphasis on scaling and the evolution of the brain. *In* "Approaches to Primate Paleobiology", Ed. Szalay [182] (*Contrib. Primat.* **5**, 244–292).

Gould, S. J. (1977). "Ontogeny and Phylogeny", Belknap Press, Harvard.

[103, 104] Gould, S. J. (1981). "The Mismeasure of Man", 352 pp. Norton, New York.

Gould, S. J. and Johnson, R. F. (1972). Geographic variation. *Ann. Rev. Ecol. Systematics* **3**, 457–498.

Gould, S. J. and Paull, C. (1977). Natural history of *Cerion*. VII Geographic variation of *Cerion* (Mollusca: Pulmonata) from the eastern end of its range (Hispaniola to the Virgin Islands): coherent patterns and taxonomic simplification. [115, 187] *Breviora* **445**, 1–24.

Gould, S. J., Woodruff, D. S. and Martin, J. P. (1974). Genetics and morphometrics of *Cerion* at Pongo Carpet: a new systematic approach to this enigmatic land snail. [115] *Syst. Zool.* **23**, 518–535.

Gower, J. C. (1966). Some distance properties of latent root and vector methods [25, 86, 103] used in multivariate analysis. *Biometrika* **53**, 13–28.

Gower, J. C. (1967). Multivariate analysis and multidimensional geometry. *Statisti-* [95, 108, 149] *cian* **17**, 13–28.

Gower, J. C. (1968). Adding a point to vector diagrams in multivariate analysis. *Biometrika* **55**, 582–585.

Gower, J. C. (1972). Measurements of taxonomic distance and their analysis. *In* "The Assessment of Population Affinities in Man", Eds Weiner, J. S. and [154] Huizinga, J., pp. 1–23, Clarendon Press, Oxford.

Gower, J. C. (1975a). Statistical methods of comparing different multivariate analysis of the same data. *In* "Mathematics in the Archaeological and Historical sciences", Eds Hodson, F. R., Kendall, D. G. and Tautu, P., pp. 138–149, [154] Edinburgh University Press.

[176] Gower, J. C. (1975b). Generalised procrustes analysis. *Psychometrika* **40**, 33–51.

Gower, J. C. (1976). Growth-free canonical variates and generalised inverses. *Bull.* [31, 135, 176] *Geol. Inst. Univ. Uppsala, n.s.* **7**, 1–10.

Grafius, J. E. (1965). A geometry of plant breeding. *Mich. St. Univ. Res. Bull.* **7**, 59 [5, 111] pp.

Green, R. G. (1980). Multivariate approaches in ecology: the assessment of ecologic [178] similarity. *Ann. Rev. Ecol. Syst.* **II**, 1–14.

Greenacre, M. (1978). Some objective methods of graphical display of a data matrix, Special Report, University of South Africa.

Greenacre, M. (1981). Practical Correspondence Analysis "Interpreting Multivariate Data" Ed. Barnett, V. Wiley, Chichester.

Greig-Smith, P. (1964) "Quantitative plant ecology", 2nd. Edn., 256 pp., Butter- [149] worths, London.

Greig-Smith, P., Austin, M. P. and Whitmore, T. C. (1967). The application of quantitative methods to vegetation survey. I. Association analysis and principal [149] component analysis. *J. Ecol.* **55**, 483–503.

Grewal, M. S. (1962). The rate of genetic divergence in the C57BL strain of mice. [114] *Genet. Res.* **3**, 226–237.

Griffiths, J. C. (1966). Applications of discriminant functions as a classificatory tool in the geosciences. *In* "Computer Applications in the Earth Sciences", pp. 48–52, Lawrence, Kansas, State Geol. Survey.

Groves, C. P. (1963). Results of a multivariate analysis on the skulls of Asiatic Wild

Asses: with a note on the status of *Microhippus hemionus blandfordi* Pocock. *Ann. Mag. Nat. Hist. ser. 13* **6**, 329–336. [47]

Guillaumin, M. and Lefèbvre, J. (1974). Etude biométrique des populations de *Pyrgus carlinae* Rbr. et de *Pyrgus cirsh* Rbr. (Lépid. Hesperidae). 11. Utilisation du D^2 de Mahalanobis dans l'analyse et la classification des populations naturelles. *Archs Zool. Exp. Gén.* **115**, 505–548. [186]

Guthrie, W. K. C. (1962). "A history of Greek Philosophy", 539 pp. University Press, Cambridge. [2]

Habbema, J. D. F. (1976). A discriminant analysis approach to the identification of human chromosomes. *Biometrics* **32**, 919–928. [181]

Hansell, R. I. C., Bookstein, F. L. and Rowell, H. J. (1980). Operational point homology by Cartesian transformations to standard shape: examples from setal positions in phytoseiid mites. *Syst. Zool.* **29**, 43–49. [189]

Harberd, D. J. (1962). Application of a multivariate technique to an ecological survey. *J. Ecol.* **50**, 1–17. [146]

Harman, H. H. (1960). "Modern Factor Analysis", 474 pp. Chicago University Press, Chicago, Ill. [105]

Hashiguchi, S. and Morishima, H. (1969). Estimation of genetic contribution of principal components to individual variates concerned. *Biometrics* **25**, 9–15. [111]

Hazel, J. E. (1977). Use of certain multivariate and other techniques in assemblage zonal biostratigraphy: examples utilizing Cambrian, Cretaceous and Tertiary benthic assemblages. *In* "Concepts and Methods of Biostratigraphy", Eds Kauffman, E. G. and Hazel, J. E., pp. 187–212, Dowden, Hutchinson and Ross, Stroudsburg. [112, 167, 177]

Hazel, L. N. (1943). The genetic basis for constructing selection indices. *Genetics NY* **28**, 476–490.

Heal, O. W., Howson, G., French, D. D. and Jeffers, J. N. R. (1974). Decomposition of cotton strips in tundra. *In* "Soil Organisms and Decomposition in Tundra", Eds Holding, A. J., Heal, O. W., MacLean, S. F. Jr and Flanagan, P. W., pp. 341–362, Stockholm, Tundra Biome Steering Committee. [190]

Healy, M. J. R. (1968). Multivariate normal plotting. *Appl. Statist.* **17**, 157–161. [20]

Healy, M. J. R. and Tanner, J. M. (1981). Size and shape in relation to growth and form. *In* "Perspectives in Primate Biology", Eds Ashton, E.H. and Holmes, R. L., pp. 19–35, Academic Press, London. [181]

Hendrickson, J. A. Jr (1979). Examples of discrete multivariate methods in ecology. *In* "Multivariate Methods in Ecological Work", Eds Orlóci, L., Rao, C. R. and Stiteler, W. M., pp. 55–63. International Co-operative Publishing House, Fairland, Maryland. [141]

Hengeveld, R. and Hogeweg, P. (1979). Cluster analysis of the distribution patterns of Dutch Carabid species (Col.). *In* "Multivariate Analysis in Ecological Work", Eds Orlóci, L., Rao, C. R. and Stiteler, W. M., pp. 65–86. [141]

Hill, M. O. (1974). Correspondence analysis: a neglected multivariate method. *J. Roy. Stat. Soc. (Applied Statistics)* **23**, 340–354. [86, 192]

Hill, M. O. and Smith, A. J. E. (1976). Principal component analysis of taxonomic data with multistate characters. *Taxon* **25**, 249–255.

Hill, M. O. and Gauch, H. G. Jr (1980). Detrended correspondence analysis, an improved ordination technique. *Vegetatio*, **42**, 47–58.

Hills, M. (1978). On ratios—a response to Atchley, Gaskins and Anderson. *Syst. Zool.* **27**, 61–62.

Hocker, H. W. (1956). Certain aspects of climate as related to the distribution of
[141]   Loblolly pine. *Ecology* **37**, 824–834.

Hodson, F. R., Sneath, P. H. A. and Doran, E. J. (1966). Some experiments on the
[192]   numerical analysis of archaeological data. *Biometrika* **53**, 311–324.

Hohn, M. E. (1978). Stratigraphical correlation by principal components: the effects
of missing data. *J. Geol.* **86**, 524–532.

Holland, D. A. (1968a). Component analysis: an aid to the interpretation of data.
[91, 109]   *Expl. Agric.* **5**, 151–164.

Holland, D. A. (1968b). Component analysis—an approach to the interpretation of
[91, 109]   soil data. *J. Sci. Fd Agric.* **20**, 26–31.

Holligan, P. M., Maddock, L. and Dodge, J. D. (1980). The distribution of
dinoflagellates around the British Isles in July 1977: a multivariate analysis. *J. Mar.*
[153]   *Biol. Ass. UK* **60**, 851–867.

Holmes, J. M. C. (1975). A comparison of numerical taxonomic techniques using
measurements on the genera *Gammarus* and *Marinogammarus* (Amphipoda).
[187]   *Biol. J. Linn. Soc.* **7**, 183–214.

Hopkins, J. W. (1966). Some considerations in multivariate allometry. *Biometrics*
[31, 122]   **22**, 747–760.

Hopper, S. D. (1978). Speciation in the Kangaroo Paws of South-Western Australia
(*Anigozanthos* and *Macropidia*: Haemodoraceae). PhD Thesis, University of
Western Australia.

Hopper, S. D. and Campbell, N. A. (1977). A multivariate morphometric study of
species relationships in Kangaroo Paws (*Anigozanthos* Labill. and *Macropidia*
[129, 191] Drumm. ex. Harv.: Haemodoraceae) *Aust. J. Bot.* **25**, 523–524.

Hopper, S. D., Campbell, N. A. and Moran, G. F. (1982). *Eucalyptus caesia*, a rare
mallee of granite rocks from South-Western Australia "Species at Risk" Eds
Groves, R. H. and Ride, W. D. L.

Hotelling, H. (1933). Analysis of a complex of statistical variables into principal
components. *J. Educ. Psychol.* **24**, 417–441.

[78]   Hotelling, H. (1935). The most predictable criterion. *J. Educ. Psychol.* **26**, 139–142.

Hotelling, H. (1936). Relations between two sets of variables. *Biometrika*, **28**,
[28, 78]   321–377.

Hotelling, H. O. (1957). The relations of the newer multivariate statistical methods
[104]   to factor analysis. *Br. J. Statist. Psychol.* **10**, 69–79.

Howarth, R. J. (1973). The pattern recognition problem in applied geochemistry. *In*
"Exploration Geochemistry 1972", Ed. Jones, M. B., pp. 259–273. Inst. Mining
Metallurgy, London.

Howarth, R. J. and Murray, J. W. (1969). The Foraminiferida of Christchurch
[137, 179] harbour, England. *J. Paleont.* **43**, 660–675.

Humphries, J. M., Bookstein, F. L., Chernoff, B., Smith, G. R., Elder, R. L. and
[20, 22, 31, 122, Poss, S. G. (1981). Multivariate discrimination by shape in relation to size. *Syst.*
124, 126, 135] *Zool.* **30**, 291–308.

Huxley, J. S. and Teissier, G. (1936). Terminologie et notation dans la description
[6]   de la croissance relative. *CR Séance Soc. Biol. Paris* **150**, 934–936.

Imbrie, J. and Kipp, N. G. (1971). A new micropaleontological method for
quantitative paleoclimatology: application to a late Pleistocene Caribbean core. *In*
"The late Cenozoic Glacial Ages", Ed. Turekian, K., pp. 71–181, Yale University
Press, New Haven.

Ivert, H. (1980). Relationship between stratigraphical variation in the morphology
of *Gabonella elongata* and geochemical composition of the host sediment. *Cretace-*
[81, 176]   *ous Research* **1**, 223–233.

Ivimey-Cook, R. B. and Proctor, M. C. F. (1967). Factor analysis of data from an East Devon heath: a comparison of principal component and rotated solutions. *J. Ecol.* **55**, 405–413. [147]

Jackson, J. F., Ingram, W. and Campbell, H. W. (1976). The dorsal pigmentation pattern of snakes as an antipredator strategy: a multivariate approach. *Am. Nat.* **110**, 1029–1053. [184]

Jacobshagen, B. (1979). Morphometric studies in the taxonomy of the orang-utan (*Pongo pygmaeus* L. 1760). *Folia Primatologica* **32**, 29–34. [182]

Jameson, D. L. and Richmond, R. C. (1971). Parallelism and convergence in the evolution of size and shape in holarctic *Hyla*. *Evolution* **25**, 497–508. [189]

Jardine, N. (1969). The observational and theoretical components of homology: a study based on the morphology of the dermal skull-roofs of rhipidistian fishes. *Biol. J. Linn. Soc. Lond.* **1**, 327–361. [35]

Jayakar, S. D., Sgaramella, L. Z., Galante, A. and Pennetti, V. (1978). The use of discriminant functions to define obesity in children aged 7–11 years. *Ann. Hum. Biol.* **5**, 519–528. [182]

Jeannel, R. (1943). "La Genèse des Faunes Terrestres", Presses Universitaires de France, Paris. [153]

Jeffers, J. N. R. (1962). Principal component analysis of designed experiment. *Statistician* **12**, 230–242. [103]

Jeffers, J. N. R. (1964). A reply to Mr Ehrenberg's questions. *Statistician* **14**, 54–55. [103]

Jeffers, J. N. R. (1965). Correspondence. *Statistician* **15**, 207–208. [103]

Jeffers, J. N. R. (1967a). The study of variation in taxonomic research. *Statistician* **17**, 29–43. [30]

Jeffers, J. N. R. (1967b). Two case studies in the application of principal component analysis. *Appl. Statist.* **16**, 225–236. [88, 147]

Jenkins, C. L., Kuhn, D. A. and Daly, K. R. (1977). Fatty acid composition of *Simonsiella* strains. *Archs Microbiol.* **113**, 209–213. [190]

Johansson, C. (1982a). The ecological characteristics of 314 algal taxa found in Jämtland streams, Sweden. *Medd. Växtbiolä Inst.* **2**, 170.

Johansson, C. (1982b). Attached algal vegetation in running waters of Jämtland, Sweden. *Acta Phytographica Suecica* **71**, 1–80.

Johnstone, R. F. (1973). Evolution in the house sparrow. IV. Replicate studies in phenetic variation. *Syst. Zool.* **22**, 219–226. [183]

Johnstone, R. F., Niles, D. M. and Rohwer, S. A. (1972). Hermon Bumpus and natural selection in the sparrow, *Passer domesticus. Evolution* **26**, 20–31. [116, 183]

Jolicoeur, P. (1959). Multivariate geographical variation in the wolf, *Canis lupus* L. *Evolution* **13**, 283–299. [132]

Jolicoeur, P. (1963a). Les combinaisons multidimensionelles de caractères anatomiques quantitatifs. *Proc. XVI. Int. Congr. Zool.* **1**, 183. [123, 126]

Jolicoeur, P. (1963b). The multivariate generalisation of the allometry equation. *Biometrics* **19**, 497–499. [30, 122]

Jolicoeur, P. (1963c). Bilateral symmetry and asymmetry in limb bones of *Martes americana* and man. *Rev. Can. Biol.* **22**, 409–432. [132]

Jolicoeur, P. (1963d). The degree of generality of robustness in *Martes americana*. *Growth* **27**, 1–27. [6, 11, 123]

Jolicoeur, P. (1968). Interval estimation of the slope of the major axis of a bivariate normal distribution in the case of a small sample. *Biometrics* **24**, 679–682. [30]

Jolicoeur, P. and Mosimann, J. E. (1960). Size and shape variation in the Painted Turtle, a principal component analysis. *Growth* **24**, 339–354. [30, 87, 156]

Jolicoeur, P. and Mosimann, J. E. (1968). Intervalles de confiance pour la pente de l'axe majeur d'une distribution normale bidimensionelle. *Biométr. Praxim.* **9**, 121–140.

Jöreskog, K. G. (1963). "Statistical Estimation in Factor Analysis", 145 pp. Almqvist [103] and Wiksell, Stockholm.

Jöreskog, K. G. (1967). Some contributions to maximum-likelihood factor analysis. [103] *Psychometrika* **32**, 443–482.

[82, 86, 103, Jöreskog, K. G., Klovan, J. E. and Reyment, R. A. (1976). "Geological Factor 104, 105] Analysis". 178 pp. Elsevier, Amsterdam.

Jouffroy, F. K. and Lessertisseur, J. (1978). Etude écomorphologique des proportions des membres des primates et spécialement des prosimiens. *Ann. Sci. Nat.* [181] *(Zoologie)* **20**, 99–128.

Jux, U. and Strauch, F. (1966). Die mitteldevonische Brachiopodengattung *Uncites* [69] de France 1825. *Palaeontographica Abt. A* **125**, 176–227.

Kaesler, R. L., Mulvany, P. S. and Kornicker, L. S. (1977). Delimitation of the Antarctic convergence by cluster analysis and ordination of benthic Myodocopid [178] ostracods. Proc. 6th Int. Ostracod Symp., Saalfelden, pp. 235–244.

Karr, J. R. and James, F. C. (1975). Ecomorphological configurations and convergent evolution in species communities. *In* "Ecology and Evolution of Communities", Eds Cody, M. L. and Diamond, J. M., pp. 258–291, The Belknap Press, [177, 179] Cambridge, Mass.

Kendall, D. G. (1971). Seriation from abundance matrices. *In* "Mathematics in Archaeological and Historical Sciences", Eds Hodson, F. R., Kendall, D. G. and [192] Tautu, P., pp. 215–252, Edinburgh University Press.

Kendall, M. G. (1957). "A Course in Multivariate Analysis", 185 pp., Griffin, London.

Kendall, M. G. and Stuart, A. (1966). "The Advanced Theory of Statistics", Griffin, [45] London.

Kendrick, W. B. and Weresub, L. K. (1966). Attempting neo-Adansonian computer [35] taxonomy at the ordinal level in the basidiomycetes. *Syst. Zool.* **15**, 307–329.

Kennedy, M. L. and Schnell, G. D. (1978). Geographic variation and sexual [176] dimorphism in Ord's kangaroo rat, *Dipodomys ordii*. *J. Mammalogy* **59**, 45–59.

Kermack, K. A. (1954). A biometrical study of *Micraster coranguinum* and *M.* [160] *(Isomicraster) senonensis*. *Phil. Trans. R. Soc. B.* **237**, 375–428.

Kermack, K. A. and Haldane, J. B. S. (1950). Organic correlation and allometry. *Biometrika* **37**, 30–41.

[35] Key, K. H. L. (1967). Operational homology. *Syst. Zool.* **16**, 275–276.

Key, K. H. L. (1982). The genus *Geckomima* (Orthoptera: Eumastacidae: [185] Morabinae). *Aust. J. Zool.* **30**, 931–1026.

Kokawa, S. (1958). Some tentative methods for the age-estimation by means of [30] morphometry of *Menyanthes* remains. *J. Inst. Poly. Osaka* **D9**, 111–118.

Kraus, B. S. and Choi, S. C. (1958). A factorial analysis of the prenatal growth of the [109] human skeleton. *Growth* **22**, 231–242.

Kres, H. (1975). "Statistische Tafeln zur multivariaten Analysis", 431 pp. Springer-Verlag, Berlin.

Krumbein, W. C. (1959). Trend-surface analysis of contour-type maps with irregular control-point spacing. *J. Geophys. Res.* **64**, 823–824.

Krumbein, W. C. and Graybill, F. A. (1965). "An Introduction to Statistical Models in Geology", 465 pp. McGraw-Hill, New York.

Kruskal, J. B. and Wish, M. (1978). Multidimensional scaling. *In* "Quantitative Applications in the Social Sciences", No. 11, Sage, Beverley Hills.

Kshirsagar, A. M. (1972). "Multivariate Analysis", Marcel Dekker, New York.   [41, 44]

Kullback, S. (1959). "Information Theory and Statistics", 395 pp. John Wiley, New York.

Kurczynski, T. W. (1970). Generalized distance and discrete variables. *Biometrics* **26**, 525–534.   [20]

Lachenbruch, H. P. A. (1968). On expected probabilities of misclassification in discriminant analysis, necessary sample size, and a relation with the multiple correlation coefficient. *Biometrics* **24**, 823–834.   [50]

Lachenbruch, P. A. and Goldstein, M. (1979). Discriminant analysis. *Biometrics* **35**, 69–85.

Laird, A. K., Tyler, S. A. and Darton, A. D. (1965). Dynamics of normal growth. *Growth* **29**, 233–248.   [30]

Lambert, J. M. and Williams, W. T. (1962). Multivariate methods in plant ecology. IV. Nodal analysis. *J. Ecol.* **50**, 775–802.   [146]

Lamont, B. B. and Grant, K. J. (1979). A comparison of 21 measures of site dissimilarity. *In* "Multivariate Analysis in Ecological Work", Eds Orlóci, L., Rao, C. R. and Stiteler, W. M., pp. 101–126, International Co-operative Publishing House, Fairland, Maryland.   [16, 150, 151]

Lande, R. (1976). Natural selection and random genetic drift in phenotype evolution. *Evolution* **30**, 314–334.   [135]

Lande, R. (1979). Quantitative genetic analysis of multivariate evolution, applied to the brain: body size allometry. *Evolution* **33**, 402–416.   [34, 76, 116, 118, 119, 121, 122, 182]

Lane, R. P. (1981). A quantitative analysis of wing pattern in the *Culicoides pulicaris* species group (Diptera: Ceratopogonidae). *Zool. J. Linn. Soc.* **72**, 21–41.   [186]

Largo, P. H., Gasser, T., Prader, A., Stuetzle, W. and Huber, P. J. (1978). Analysis of the adolescent growth spurt using smoothing spline functions. *Ann. Hum. Biol.* **5**, 421–434.   [180, 183]

Lauga, J. (1974). Analyse biométrique du polymorphisme des nouveau-nées chez *Locusta migratoria. Acrida* **3**, 277–284.   [185]

Lauga, J. (1977). Nature et détermination du polymorphisme phasaire morphologique des larves nouveau-nées de *Locusta migratoria migratorioides* (R. & F.) en milieu humide et sec. *Acrida* **6**, 239–248.   [184]

Laurec, A., Chardy, P., de la Salle, P. and Rockaert, M. (1979). Use of dual structures in inertia analysis: ecological implications. *In* "Multivariate Analysis in Ecological Work", Eds Orlóci, L., Rao, C. R. and Stiteler, W. M., pp. 124–174. International Co-operative Publishing House, Fairland, Maryland.   [142]

Laval, P. (1975). Une analyse multivariable du développement au laboratoire de *Phronima sedentaria* (Forsk.), Amphipode Hypéride, étude de l'influence de la température et de la quantité de nourriture. *Ann. Inst. Océanog. n.s.* **51**, 5–41.   [187]

Lavelle, L. B. (1977). Relationship between tooth and long-bone size. *Am. J. Phys. Anthropol.* **46**, 423–425.   [182]

Lawley, D. N. (1940). The estimation of factor loadings by the method of maximum likelihood. *Proc. R. Soc. Edin.* **60**, 64–82.   [103]

Lawley, D. N. (1958). Estimation in factor analysis under various initial assumptions. *Br. J. Statist. Psychol.* **11**, 1–12.   [103]

Lawley, D. N. (1960). Approximate methods in factor analysis. *Br. J. Statist. Psychol.* **13**, 11–17.   [103]

Lawley, D. N. and Maxwell, A. E. (1963). "Factor Analysis as a Statistical Method", 117 pp. Butterworths, London. Second edition (1971) by American Elsevier, 153 pp.   [103]

[116] Leamy, L. (1977). Genetic and environmental correlations of morphometric traits in randombred house mice. *Evolution* **31**, 357–369.

[113] Lecher, P. (1967). Cytogénétique de l'hybridisation expérimentale et naturelle chez l'Isopode *Jaera (albifrons) syei* Bocquet. *Archs Zool. éxp. gén.* **108**, 633–698.

[78] Lee, P. J. (1969). The theory and application of canonical trend surfaces. *J. Geol.* **77**, 303–318.

[109, 112] Lefèbvre, J. (1966). Etude, á l'aide de mensuration, de la conformation et de la croissance des bovins normands, 124 pp. Thése, Fac. des Sciences, Univ. Caen.

[88] Lefèbvre, J. (1980). "Introduction aux Analyses Statistiques Multidimensionelles", 2nd edn, 259 pp. Masson, Paris.

Lefèbvre, J. Boitard, M. and Frey, J.-F. (1981). "Logiciels d'analyse Multidimensionelles sur Micro-ordinateurs", 152 pp. Masson, Paris.

[178] Lefèbvre, J., Laurent, P. and Louis, J. (1972). Application de l'analyse des correspondances á l'étude de l'évolution du phytoplancton du lac d'Annecy. *Investigaçion pesquera* **36**, 119–126.

[116] Legendre, P. (1970). The bearing of *Phoxinus* (Cyprinidae) hybridity on the classification of its North American species. *Can. J. Zool.* **48**, 1167–1177.

Legendre, L. and Legendre, P. (1979). "Ecologie Numérique" (2 vols), 197 + 254 pp. Masson, Paris.

[180] Leguebe, A. and Vrydagh, S. (1981). Geographic variability of digital ridge-counts: principal component analysis of male and female world samples. *Ann. Hum. Biol.* **8**, 519–528.

[103, 104, 155] Le Maitre, R. W. (1982). "Numerical Petrology". Developments in Petrology, No. 8, 281 pp. Elsevier, Amsterdam.

[50, 113] Lerman, A. (1965). On rates of evolution of unit characters and character complexes. *Evolution* **19**, 16–25.

[117] Lessios, H.A. (1980). Divergence in allopatry: molecular and morphological differentiation between sea urchins separated by the isthmus of Panama. *Evolution* **35**, 618–634.

[2] Levy, I. (1926). "Recherches sur les Sources de la Légende de Pythagore", 149 pp. Laroux, Paris.

[116] Lin, C. Y. (1978). Index selection for genetic improvement of quantitative characters. *Theoret. Appl. Genet.* **52**, 49–56.

[103] Lindley, D. V. (1962). Factor analysis. *Statistician* **12**, 169–171.

[103] Lindley, D. V. (1964). Factor analysis, a summary of discussion. *Statistician* **14**, 47–49.

[186] Louis, J. and Lefèbvre, J. (1971). Les races d'abeilles (*A. mellifica* L.) I. Détermination par l'analyse canonique (étude préliminaire). *Biométrie-Praximétrie* **12**, 1–4.

Lu, K. H. (1965). Harmonic analysis of the human face. *Biometrics* **21**, 491–505.

[47] Lyubischev, A. A. (1959). The application of biometrics in taxonomy. *Vestnik Leningradskogo Univ.* **14**, 128–136.

Maarel, E. van der (1979). Multivariate methods in phytosociology, with reference to the Netherlands. *In* "The Study of Vegetation", Werger, M. J. A. Ed. Junk, The Hague.

[128] McCammon, H. M. (1970). Variation in recent brachiopod populations (see Reyment (1970c) (Symposium on Biometry in Palaeontology.) *Bull. Geol. Inst. Univ. Uppsala ns* **3**, 1–18.

[137] McCammon, R. B. (1966). Principal component analysis and its application in large-scale correlation studies. *J. Geol.* **74**, 721–733.

McCammon, R. B. (1968). Multiple component analysis and its application in classification of environments. *Bull. Am. Ass. Petrol. Geol.* **52**, 2178–2196. [137]

McHenry, H. M. and Corrucini, R. S. (1975). Multivariate analysis of early hominid pelvic bones. *Am. J. Phys. Anthropol.* **43**, 263–270. [180]

McHenry, H. M. and Corrucini, R. S. (1978). The femur in early human evolution. *Am. J. Phys. Anthropol.* **49**, 473–487. [180]

McMahon, T. (1975). Allometry and biomechanics: limb bones in adult ungulates. *Am. Nat.* **108**, 547–563. [189]

Madsen, K. S. (1977). A growth curve model for studies in morphometrics. *Biometrics* **33**, 659–669. [183]

Mahalanobis, P. C. (1928). A statistical study of the Chinese head. *Man in India* **8**, 107–122. [73]

Mahalanobis, P. (1936). On the generalized distance in statistics, *Proc. Natn. Inst. Sci. India* **2**, 49–55. [16]

Mahé, J. (1974). L'analyse factorielle des correspondances et son usage en paléontologie et dans l'étude de l'évolution. *Bull. Soc. Geol. France* **16**, 336–340. [176, 186, 192]

Mahon, R. J. (1974). A study of Rock Crabs of the genus *Leptograpsus*. PhD Thesis, University of Western Australia.

Malmgren, B. (1970). Morphometric analysis of two species of *Floridina* (Cheilostomata: Bryozoa). *Stockh. Contr. Geol.* **23**, 73–89. [132]

Malmgren, B. A. (1972). Morphometric studies of "*Globorotalia*" *pseudobulloides*. *Stockh. Contr. Geol.* **24**, 33–49. [174]

Malmgren, B. (1974). Morphometric studies of planktonic foraminifers from the type Danian of southern Scandinavia. *Stockh. Contr. Geol.* **29**, 1–126. [135]

Malmgren, B. and Kennett, J. P. (1972). Biometric analysis of phenotypic variation: *Globigerina pachyderma* (Ehrenberg) in the South Pacific Ocean. *Micropalaeontology* **18**, 241–248. [189]

Malmgren, B. A. and Kennett, J. P. (1973). Recent planktonic foraminiferal distribution in high latitudes of the South Pacific: a multivariate statistical study. *Palaeogeog. Palaeoclim. Palaeoecol.* **14**, 127–136. [178]

Malmgren, B. and Kennett, J. P. (1978). Test variation in *Globigerina bulloides* in response to Quaternary paleooceanographic changes. *Nature*, **275**, 123–124.

Manaster, B. J. (1979). Locomotor adaptations within the *Cercopithecus* genus: a multivariate approach. *Am. J Phys. Anthropol.* **50**, 169–182. [182]

Marcellino, A. J., Da Rocha, F. J. and Salzano, F. M. (1978). Size and shape differences among six South American tribes. *Ann. Hum. Biol.* **5**, 69–74. [180]

Marcus, L. F. (1969). Measurement of selection using distance statistics in the prehistoric orang-utan *Pongo pygmaeus palaeosumatrensis*. *Evolution* **23**, 301–307. [113]

Mardia, K. V., Kent, J. T. and Bibby, J. M. (1979). "Multivariate Analysis", Academic Press, London. [16, 41]

Margetts, B. M., Campbell, N. A. and Armstrong, B. K. (1981). Summarizing dietary patterns using multivariate analysis. *J. Hum. Nutr.* **35**, 281–286. [181]

Markham, J. and O'Farrell, P. N. (1974). Choice of mode for the journey-to-work in Dublin: some multivariate aspects. *J. Environ. Management* **2**, 123–148. [192]

Marriott, F. H. C. (1974). "The Interpretation of Multiple Observations", 177 pp. Academic Press, London.

Martin, E. S. (1936). A study of an Egyptian series of mandibles with special reference to mathematical methods of sexing. *Biometrika* **28**, 148–178. [73]

Martin, L. (1960). Homométrie, allométrie et cogradation en biométrie générale. *Biometrische Zeitschr.* **2**, 73–97. [31, 73]

[191] Mastrogiuseppe, T. D., Cridland, A. A. and Bogyo, T. P. (1970). Multivariate comparison of fossil and recent Gingko wood. *Lethaia* **3**, 271–277.

[49] Mather, K. (1949). "Biometrical Genetics", 162 pp. Methuen, London.

[176] Mather, P. and Openshaw, S. (1974). Multivariate methods and geographic data. *Statistician* **23**, 283–308.

[151] Matsakis, J. T. (1964). "Comparaisons Multiples en Biologie", 201 pp. Imp. du Sud-Ouest, Toulouse.

[61] Maxwell, A. E. (1961). Canonical variate analysis when the variables are dichotomous. *Educ. Psychol. Measur.* **21**, 259–271.

[161] Medina Molera, A. (1980–1981). "Historia de Andalucia", 2 volumes. Biblioteca de Ediciones Andaluzas, Sevilla.

[175] Melguen, M. (1972). Exemple de traitement statistique de données sedimentologiques: differenciation de facies dans un sediment d'apparence homogène. *Bull. Union Océanographes de France*, **8**, 3–23.

[175] Melguen, M. (1973). Correspondance analysis for recognition of facies in homogeneous sediments off an Iranian river mouth. *In* "The Persian Gulf", Ed. Purser, B. H., pp. 99–113. Springer-Verlag, Berlin.

Merriam, D. F. and Harbaugh, J. W. (1963). Computer helps map oil structures. *Oil Gas J.* **61**, 158–159.

[133] Michaelis, J. (1972). Zur Anwendung der Diskriminanzanalyse für die medizinische Diagnostik. Habilitationsschrift der Medizinischen Fakultät zu Mainz, 121 pp.

Michaelis, J. (1973). "Simulation Experiments with Multiple Group Linear and Quadratic Discriminant Analysis", pp. 225–238. See Cacoullos, T. (1973).

[110] Michener, C. D. and Sokal, R. R. (1966). Two tests of the hypothesis of non-specificity in the *Hoplitis* complex (Hymenoptera: Megachilidae). *Ann. Ent. Soc. Am.* **59**, 1211–1217.

[32] Middleton, G. V. (1962). A multivariate statistical technique applied to the study of sandstone composition. *Trans. R. Soc. Can. Ser. III* **56**, 119–126.

[105] Middleton, G. V. (1964). Statistical studies on scapolites. *Can. J. Earth Sci.* **1**, 23–34.

Miesch, A. T. (1973). Q-mode factor analysis of compositional data. *Computers and Geosciences* **1**, 147–159.

Miesch, A. T. and Morton, D. M. (1977). Chemical variability in the Lakeview Mountains pluton, southern California batholith—a comparison of the methods of correspondence analysis and extended Q-mode factor analysis. *J. Res. US Geol. Survey* **5**, 103–116.

Miller, R. L. and Kahn, J. S. (1962). "Statistical Analysis in the Geological Sciences", 483 pp. Wiley, New York.

[156, 157] Misra, R. K. (1966). Vectorial analysis for genetic clines in body dimensions in populations of *Drosophila subobscura* Coll. and a comparison with those of *D. robusta* Sturt. *Biometrics* **22**, 469–487.

[109] Moore, C. S. (1968). Response of growth and cropping of apples to treatments studied by the method of component analysis. *J. Hort. Sci.* **43**, 249–262.

[190] Moore, C. S. (1975). Relative importance of rootstock and scion in determining growth and fruiting in young apple trees. *Ann. Bot.* **39**, 113–123.

[149, 150] Moore, J. J., Fitzsimons, P., Lambe, E. and White, J. (1970). A comparison and evaluation of some phytosociological techniques. *Vegetatio* **20**, 1–20.

[104] Moran, P. A. P. (1975). The use of statistics in geography. *Statistician* **24**, 161–162.

[180] Morgan, B. J. T. (1981). Three applications of methods of cluster analysis. *Statistician* **30**, 205–223.

Morrison, D. F. (1976). "Multivariate Statistical Methods", 2nd edn, McGrawhill, New York. [44]

Moser, C. A. and Scott, W. (1961). "British Towns: a Statistical Study of Their Social and Economic Differences", 169 pp. Oliver and Boyd, London. [92]

Mosimann, J. E. (1970). Size allometry: size and shape variables with characterizations of the lognormal and generalized gamma distributions. *J. Am. Statist. Ass.* **65**, 930–945. [5, 22, 31, 122, 123, 124, 157, 183]

Mosimann, J. E. and James, F. C. (1979). New statistical methods for allometry with application to the Florida Red-winged Blackbirds. *Evolution* **33**, 444-459. [30, 123, 183]

Mosimann, J. E. and Malley, J. D. (1979). Size and shape variables. *In* "Multivariate Analysis in Ecological Work", Eds Orlóci, L., Rao, C. R. and Stiteler, W. M., pp. 175–189, International Co-operative Publishing House, Fairland, Maryland. [124]

Moss, W. W. and Webster, W. A. (1970). Phenetics and numerical taxonomy applied to systematic nematology. *J. Nemat.* **2**, 16–25. [188]

Moss, W. W., Peterson, P. C. and Atyeo, W. T. (1977). A multivariate assessment of phenetic relationships within the feather mite family Eustathiidae (Acari). *Syst. Zool.* **26**, 386–409. [189]

Mound, L. A. (1963). Host correlated variation in *Bemisia tabaci* (Gennadius) (Homoptera: Aleyrodidae) *Proc. R. Ent. Soc. Lond. A* **38**, 171–180. [48]

Mourant, A. E., Kopeč, A. C. and Domaniewska-Sobczak, K. (1976). "The Distribution of Human Blood Groups and other Polymorphisms", Oxford University Press. [161]

Mourant, A. E., Kopeč, A. C. and Domaniewska-Sobczak, K. (1978). "The Genetics of the Jews", Clarendon Press, Oxford. [161]

Mukherjee, R., Rao, C. R. and Trevor, J. C. (1955). "The Ancient Inhabitants of the Jebel Moya", 123 pp., Cambridge University Press, Cambridge. [39, 74]

Murdie, G. (1969). Some causes of size variation in the pea aphid, *Acyrtosiphon pisum* Harris. *Trans. R. Ent. Soc. Lond.* **121**, 423–442. [88]

Needham, R. M. (1967). Automatic classification in linguistics. *Statistician* **17**, 45–54.

Neff, N. A. and Smith, G. R. (1979). Multivariate analysis of hybrid fishes. *Syst. Zool.* **28**, 176–196. [116]

Neff, N. A. and Marcus, L. F. (1980). A survey of multivariate methods for systematics. (Workshop on Numerical methods in systematic mammalogy: Annual meeting, 9th June 1980) American Society of Mammalogists, 243 pp. [172]

Nichols, F. (1959). Changes in the chalk heart-urchin *Micraster* interpreted in relation to living forms. *Phil. Trans. R. Soc. Lond. B.* **242**, 347–437. [158, 159, 160]

Niles, D. M. (1973). Adaptive variation in body size and skeletal proportions in horned larks of the South-West United States. *Evolution* **27**, 405–426. [176]

Nishisato, S. (1980). "Analysis of Categorical Data: Dual Scaling and its Applications" (Mathematical expositions No. 24), 276 pp. University Press, Toronto.

Noy-Meir, I. and Austin, M. P. (1970). Principal coordinate ordination and simulated vegetational data. *Ecology* **51**, 551–552. [149]

O'Brien, F. I. and Rock, N. J. (1978). An interpretation of the chaetognatha in Galway Bay during October 1973 using multivariate techniques. *Proc. R. Irish Acad.* **78** (B), 213–232. [179]

Orlóci, L. (1966). Geometric methods in ecology. I. The theory and application of some ordination methods. *J. Ecol.* **54**, 193–215. [147, 148]

Orlóci, L. (1967). Data centering: a review and evaluation with reference to component analysis. *Syst. Zool.* **16**, 208–212. [149]

Orlóci, L. (1968a). Definitions of structure in multivariate phytosociological samples. *Vegetatio* **15**, 281–291.

Orlóci, L. (1968b). Information analysis in phytosociology. *J. Theoret. Biol.* **20**,
[146]   271–284.

Orlóci, L. (1978). "Multivariate Analysis in Vegetation Research", 2nd edn, Junk,
[146]   The Hague.

Orlóci, L. (1979). Non-linear data structures and their description. *In* "Multivariate Methods in Ecological Work", Eds Orlóci, L., Rao, C. R. and Stiteler, W. M., pp.
[146, 147]   191–202.

Osborne, R. H. (1969). The American Upper Ordovician standard XI, Multivariate classification of typical Cincinnatian calcarenites. *J. Sedim. Petrol* **39**, 769–776.

Ottestad, P. (1975). Component analysis: an alternative system. *Int. Stat. Rev.* **43**,
[158]   83–108.

Oxnard, C. E. (1967). The functional morphology of the primate shoulder as revealed by comparative anatomical osteometric, and discriminant function
[62]   techniques. *Am. J. Phys. Anthrop.* **26**, 219–240.

Oxnard, C. E. (1968). The architecture of the shoulder in some mammals. *J. Morph.*
[62, 64, 65]   **126**, 249–290.

Oxnard, C. E. (1969). Mathematics, shape and function: a study in primate anatomy.
[114]   *Am. Scient.* **57**, 75–96.

Oxnard, C. E. (1973). Functional inferences from morphometrics: problems posed
[5, 182]   by uniqueness and diversity among the primates. *Syst. Zool.* **22**, 409–424.

Oxnard, C. E. and Yang, H. C. (1981). Beyond biometrics: studies of complex biological patterns. *In* "Perspectives in Primate Biology". Eds Ashton, E. H.,
[182]   Holmes, R. L., pp. 127–167. Academic Press, London.

Parks, J. M. (1969). Multivariate facies maps. *In* "Symp. Computer Applications in Petroleum Exploration", pp. 6–18, State Geol. Survey, Lawrence, Kansas.

Parsons, P. A. and White, N. G. (1973). Genetic differences among Australian aborigines with special reference to dermatoglyphics and other anthropometric traits. *In* "The Human Biology of Aborigines in Cape York", Ed. Kirk, R. L., pp.
[181]   81–94. Canberra, Australian Inst. Aboriginal Studies.

Pauken, R. J. and Metter, D. E. (1971). Geographic representation of morphologic
[176]   variation among populations of *Ascaphus truei* Stejneger. *Syst. Zool.* **20**, 434–441.

Pearce, S. C. (1959). Some recent applications of multivariate analysis to data from
[21, 107]   fruit trees. *Ann. Rep. E. Malling. Res. Stat.* (1958), 73–76.

Pearce, S. C. and Holland, D. A. (1960). Some applications of multivariate analysis in botany. *Appl. Statist.* **9**, 1–7.

Pearce, S. C. and Holland, D. A. (1961). Analyse des composantes, outil en
[91]   recherche biométrique. *Biométre.-Praxim* **2**, 159–177.

Pearson, K. (1901) On lines and planes of closest fit to systems of points in space.
[91]   *Phil. Mag. ser.* **6**, **2**, 559–572.

[137] Pelto, C. R. (1954). Mapping of multicomponent systems. *J. Geol.* **62**, 501–511.

Pennington, W. and Sackin, M. J. (1975). An application of principal components
[192]   analysis to the zonation of two late-Devensian profiles. *New Phytol.* **75**, 419–453.

[31] Penrose, L. S. (1954). Distance, size and shape. *Ann. Eugen. Lond.* **18**, 337–343.

Petit-Maire, N. and Ponge, J. F. (1979). Primate cranium morphology through ontogenesis and phylogenesis: factorial analysis of global variation. *J. Hum. Evol.*
[180]   **8**, 233–234.

Phillips, B. F., Campbell, N. A. and Wilson, B. R. (1973). A multivariate study of
[187]   geographic variation in the whelk *Dicathais*. *J. Exp. Mar. Biol. Ecol.* **11**, 27–69.

Piazza, A., Menozzi, P. and Cavalli-Sforza, L. L. (1981). Synthetic gene-frequency maps of man and selective effects of climate. *Proc. Nat. Acad. Sci. USA* **78**, 2638–2642. [165]

Pilbeam, D. and Gould, S. J. (1974). Size and scaling in human evolution. *Science* **186**, 892–901. [182]

Pimentel, R. A. (1958). Taxonomic methods, their bearing on subspeciation. *Syst. Zool.* **7**, 139–156. [118, 131]

Pimentel, R. A. (1959). Mendelian infra-specific divergence levels and their analysis. *Syst. Zool.* **8**, 139–159. [131]

Pimentel, R. A. (1979). "Morphometrics: the Multivariate Analysis of Biological Data", 276 pp. Kendall/Hunt, Dubuque, Iowa.

Pimentel, R. A. (1981). A comparative study of data and ordination techniques based on a hybrid swarm of sand verbenas (*Abronia* Juss.). *Syst. Zool.* **30**, 250–267. [118]

Pirson, S. J. (1977). "Geologic Well Log Analysis", Gulf Publishing Co., Houston. [135]

Pitcher, M. (1966). "A factor-analytic scheme for grouping and separating types of fossils. Computer Applications in the Earth Sciences", pp. 30–41. State Geol. Survey, Lawrence, Kansas. [105, 106]

Pons, J. (1955). The sexual diagnosis of isolated bones of the skeleton. *Hum. Biol.* **27**, 12–21. [73]

Powelson, E. E., Gates, M. A. and Berger, J. (1975). A biometrical analysis of 22 stocks of four syngens of *Paramecium aurelia. Can. J. Zool.* **53**, 19–32. [188]

Prim, R. C. (1957). Shortest connection networks and some generalizations. *Bell Syst. Tech. J.* **36**, 1389–1401. [164]

Proctor, M. C. F. (1967). The distribution of British liverworts: a statistical analysis. *J. Ecol.* **55**, 119–136. [147]

Prunus, G. and Lefèbvre, J. (1971). L'analyse canonique appliquée à l'étude de la systématique évolutive chez l'Isopode *Jaera (albifrons) albifrons* Forsman. *Archs Zool. Éxp. Gén.* **112**, 793–804. [117]

Rao, C. R. (1952). "Advanced Statistical Methods in Biometric Research", 292 pp. Wiley, New York. [17, 44, 55, 98]

Rao, C. R. (1953). Discriminant functions for genetic differentiation and selection. *Sankhya* **12**, 229–246. [112]

Rao, C. R. (1961). Some observations on multivariate statistical methods in anthropological research. *Bull. Int. Statist. Inst.* **33**, 99–109. [73]

Rao, C. R. (1964). The use and interpretation of principal components analysis in applied research. *Sankhya* **26**, 329–358. [87]

Rao, C. R. (1966). Discriminant function between composite hypotheses and related problems. *Biometrika* **53**, 339–345. [135]

Rao, C. R. (1972). Recent trends of research work in multivariate analysis. *Biometrics* **28**, 3–22.

Rao, C. R. and Slater, P. (1949). Multivariate analysis applied to differences between neurotic groups. *Br. J. Psychol. Statist. Section* **2**, 17–29. [62]

Rasch, D. (1962). Die Factoranalyse und ihre Anwendung in der Tierzucht. *Biometr. Z.* **4**, 15–39. [103]

Rayner, J. H. (1966). Classification of soils by numerical methods. *J. Soil Sci.* **17**, 79–82. [143]

Rayner, J. H. (1969). The numerical approach to soil systematics. *In* "The Soil Ecosystem", Ed. Sheals, J. G., pp. 33–39. Syst. Assn. Publ. No. 8, London. [143]

Reed, T. and Young, R. S. (1979). Genetic analysis of multivariate fingertip dermatoglyphics factors and comparison with corresponding individual variables. *Ann. Hum. Biol.* **6**, 357–362. [180]

Rees, J. W. (1970). A multivariate morphometric analysis of divergence in skull

morphology among geographically contiguous groups of Whitetailed deer
[118] *(Odocoileus virginianus)* in Michigan. *Evolution* **24**, 220–229.

Reeve, E. C. R. (1941). A statistical analysis of taxonomic differences within the
[66, 68] genus *Tamandua* Gray *(Xenarthra)*. *Proc. R. Soc. Lond.* A **111**, 279–302.

Relethford, J. H., Lees, F. C. and Crawford, M. H. (1980). Population structure and
anthropometric variation in rural western Ireland: migration and biological dif-
[180] ferentiation. *Ann. Hum. Biol.* **7**, 411–428.

Rempe, U. (1965). Lassen sich bei Säugertieren Introgression mit multivariaten
[112] Verfahren nachweisen? *Z. zool. Syst. Evol.* **3**, 388–412.

Rempe, U. and Bühler, P. (1969). Zum Einfluss der geographischen und
altersbedingten Variabilität bei der Bestimmung von *Neomys*—Mandibeln mit
[134] Hilfe der Diskriminanzanalyse. *Ztg. Säugetierk.* **34**, 148–164.

Reyment, R. A. (1962). Observations on homogeneity of covariance matrices in
paleontologic biometry. *Biometrics* **18**, 1–11.

Reyment, R. A. (1963a). Multivariate analytical treatment of quantitative species
[75, 134, 177] associations: an example from Palaeoecology. *J. Anim. Ecol.* **32**, 535–547.

Reyment, R. A. (1963b). Studies on Nigerian Upper Cretaceous and Lower Tertiary
Ostracoda. II. Danian, Paleocene, and Eocene Ostracoda. *Stockh. Contr. Geol.*
10, 286.

Reyment, R. A. (1966a). Studies on Nigerian Upper Cretaceous and Lower Tertiary
[50, 129, 130, Ostracoda. III. Stratigraphical, Palaeoecological, and Biometrical conclusions.
130, 156] *Stockh. Contr. Geol.* **14**, 151 pp.

Reyment, R. A. (1966b). *Afrobolivina africana* (Graham, de Klasz, Rérat) quantitative Untersuchung der Variabilität einer palaeozänen Foraminifera. *Ec. Geol.*
[87] *Helv.* **59**, 319–338.

Reyment, R. A. (1968). Multivariate statistical analysis in geology. *Pub. Palaeontol.*
[79] *Inst. Univ. Uppsala* **74**, 18 pp.

Reyment, R. A. (1969). Biometrical techniques in systematics. *In* "Systematic
Biology", pp. 542–587, National Academy of Science, Washington, DC.

Reyment, R. A. (1969a). A multivariate paleontological growth problem. *Biometrics* **25**, 1–8.

Reyment, R. A. (1969b). *Textilina mexicana* (Cushman) from the western Niger
[88] delta. *Bull. Geol. Inst. Univ. Uppsala* n. *Ser.* **1**, 75–81.

Reyment, R. A. (1969c). Upper Sinemurian (Lias) at Gantofta, Skåne. *Geol. För.*
[60, 88] *Stockh. Förh.* **91**, 208–216.

Reyment, R. A. (1969d). Some case studies of the statistical analysis of sexual
[51] dimorphism. *Bull. Geol. Inst. Univ. Uppsala* n. *Ser.* **1**, 97–119.

Reyment, R. A. (1969e). Interstitial ecology of the Niger delta; an actuopaleoecolog-
ical study. *Bull. Geol. Inst. Univ. Uppsala* n. *Ser.* **1**, 121–159.

Reyment, R. A. (1969f). Covariance structure and morphometric analysis. *Math.*
[50] *Geol.* **I**, 185–197.

Reyment, R. A. (1970). Eigen-theory in numerical taxonomy. *Bull. Geol. Inst.
Univ. Uppsala* n. *Ser.* **2**, 67–72.

Reyment, R. A. (1971a). Vermuteter Dimorphismus bei der Ammonitengattung
[175, 190] *Benueites. Bull. Geol. Inst. Univ. Uppsala* n. *Ser.* **3**, 1–18.

Reyment, R. A. (1971b). Multivariate normality in morphometric analysis. *Math.*
[20, 134] *Geol.* **3**, 357–368.

Reyment, R. A. (1972). Models for studying the occurrence of lead and zinc in a
deltaic environment. *In* "Mathematical Models of Sedimentary Processes", Ed.
[80] Merriam, D. F., pp. 233–245. Plenum Press, New York.

Reyment, R. A. (1973). Factors in the distribution of fossil cephalopods. 3. Experiments with exact models of certain shell types. *Bull. Geol. Inst. Univ. Uppsala n. Ser.* **4**, 7–41. [126, 175, 187, 189]

Reyment, R. A. (1974). Multivariate analysis of populations split by continental drift. *Math. Geol.* **6**, 173–181. [187]

Reyment, R. A. (1975a). Analysis of a generic level transition in Cretaceous ammonites. *Evolution* **28**, 665–676. [80]

Reyment, R. A. (1975b). Canonical correlation analysis of hemicytherinid and trachyleberinid ostracodes in the Niger delta. *Bull. Am. Paleontol.* **65**, 141–145. [80]

Reyment, R. A. (1976). Chemical components of the environment and late Campanian microfossil frequencies. *Geol. För. Stockh. Förh.* **98**, 322–328. [80, 176]

Reyment, R. A. (1978). Biostratigraphical logging methods. *Computers Geosci.* **4**, 261–268. [175]

Reyment, R. A. (1978a). Analyse quantitative des vascoceratides à carènes. *Cahiers Micropaléont.* **4**, 57–64. [175, 190]

Reyment, R. A. (1978b). Graphical display of growth-free variation in the Cretaceous benthonic foraminifer *Afrobolivina afra. Paleogeog. Paleoclimat. Paleoecol.* **25**, 267–276. [76, 173, 175, 188]

Reyment, R. A. (1979a). Variation and ontogeny in *Bauchioceras* and *Gombeoceras. Bull. Geol. Inst. Univ. Uppsala n. Ser.* **8**, 89–111. [169, 172, 175]

Reyment, R. A. (1979b). Multivariate analysis in statistical paleoecology. *In* "Multivariate Methods in Ecological Work", Eds. Orlóci, L., Rao, C. R. and Stiteler, W. M., pp. 211–235. International Publishing House, Fairland, Maryland.

Reyment, R. A. (1980a). On the interpretation of the smallest principal component. *Bull. Geol. Inst. Univ. Uppsala n. Ser.* **8**, 1–4. Translation from the Russian article in *Issledovaniya po Matematicheskoi Geologii* (1978), Akad. Nauk USSR. [75, 76, 85]

Reyment, R. A. (1980b). "Morphometric Methods in Biostratigraphy", Academic Press, London. [134, 167, 175, 188]

Reyment, R. A. (1981a). The Ostracoda of the Kalambaina Formation (Paleocene), northwestern Nigeria. *Bull. Geol. Inst. Univ. Uppsala, n. Ser.* **9**, 51–65.

Reyment, R. A. (1981b). Les voyageurs suédois: aspects physiques et linguistiques. *Etudes Tsiganes* **27**, 1–14. [181]

Reyment, R. A. (1982a). Application of quantitative genetics to evolutionary series of microfossils. *In* "Statistik og Genetik", pp. 307–385. Forlag NEUCC, Copenhagen. [76, 122]

Reyment, R. A. (1982b). Analysis of secular changes in foraminifers and ostracods in the light of quantitative-genetic principles. *In* "Modes, Tempi and Mechanisms of Evolution: Phyletic Gradualism or Punctuated Equilibria", 22 pp. Coll. CNRS, Dijon. [76, 122, 187, 188]

Reyment, R. A. (1982c). Quantitative-genetic analysis of evolution in two late Cretaceous species of ostracods, 7 pp. Symposium paper, Proc. N. American Paleontological Convention III, Montreal, August 1982. [118, 122, 187]

Reyment, R. A. (1982d). Size and shape variation in some Japanese Upper Turonian (Cretaceous) ammonites. *Stockh. Contr. Geol.* **37**, 201–214. [124]

Reyment, R. A. (1982e). Phenotypic evolution in a Cretaceous foraminifer. *Evolution* **36**, 1182–1199. [88, 135, 189]

Reyment, R. A. (1982f). Analysis of trans-specific evolution in Cretaceous ostracods. *Paleobiology* **8**, 293–306. [33, 187]

Reyment, R. A. (1983). Moors and Christians: an example of multivariate analysis applied to human blood-groups. *Annal. Hum. Biol.* **10**, 505–522.

Reyment, R. A. and Banfield, C. F. (1976). Growth-free canonical variates applied to fossil foraminifers. *Bull. Geol. Inst. Univ. Uppsala n. Ser.* **7**, 11–21. [31, 135]

Reyment, R. A. and Banfield, C. F. (1981). Analysis of asymmetric relationships in geological data. *In* "Future Trends in Geomathematics", Eds Craig, R. C. and Labovitz, M. C., pp. 236–253. Pion Ltd, London. [135, 176, 188]

Reyment, R. A., Berthou, P. Y. and Moberg, B. Å. (1976). Statistical recognition of terrestrial and marine sediments in the Lower Cretaceous. *In* "Quantitative Techniques for the Analysis of Sediments", pp. 53–59. Ed. Merriam, D. F., Pergamon Press, London. [176]

Reyment, R. A. and Brännström (1962). Certain aspects of the physiology of *Cypridopsis* (Ostracoda: Crustacea). *Stockh. Contr. Geol.* **9**, 207–242. [131, 132]

Reyment, R. A., Hayami, I. and Carbonnel, G. (1977). Variation of discrete morphological characters in *Cytheridea* (Crustacea: Ostracoda). *Bull. Geol. Inst. Univ. Uppsala n. Ser.* **7**, 23–36. [176]

Reyment, R. A. and Naidin, D. P. (1962). Biometric study of *Actinocamax verus* s.l. from the Upper Cretaceous of the Russian platform. *Stockh. Contr. Geol.* **9**, 147–206. [75, 154]

Reyment, R. A. and Neufville, E. M. H. (1974). Multivariate analysis of populations split by continental drift. *J. Int. Ass. Math. Geol.* **6**, 173–181. [126, 175]

Reyment, R. A., Ramdén, H.-Å. and Wahlstedt, W. J. (1969). Fortran IV program for the generalised statistical distance and analysis of covariance matrices for the CDC 3600 computer. Computer contr. 39, 42 pp. State Geol. Survey, Lawrence, Kansas.

Reyment, R. A. and Ramdén, H.-Å. (1970). Fortran IV program for canonical variates analysis for the CDC 3600 computer. Computer contr. 47, 39 pp. State. Geol Survey, Lawrence, Kansas. [68, 158]

Reyment, R. A. and Reyment, E. R. (1981). The Paleocene Trans-Saharan Transgression and its Ostracod Fauna. *Publ. Palaeontol. Inst. Univ. Uppsala*, No. 234, 245–254. *In* "The Geology of Libya", Vol. 1. Eds Salem, M. J. and Busrewil, M. T., Academic Press, London.

Reyment, R. A. and Sandberg, P. (1963). Biometric study of *Barramites subdifficilis* (Karakasch). *Palaeontology* **6**, 727–730. [87]

Reyment, R. A. and Van Valen, L. (1969). *Buntonia olokundudui* sp. nov. (Ostracoda, Crustacea): a study of meristic variation in Paleocene and recent ostracods. *Bull. Geol. Inst. Univ. Uppsala n. Ser.* **1**, 83–94. [98, 110]

Riggins, R., Pimentel, R. A. and Walters, D. (1978). Morphometrics of *Lupinus nanus* (Leguminosae). I. Variation in natural populations. *Syst. Bot.*, **2**, 317–326. [190]

Rightmire, G. P. (1976). Multidimensional scaling and the analysis of human biological diversity on subsaharan Africa. *Am. J. Phys. Anthropol.* **44**, 445–451. [181]

Ringo, J. M. and Hodosh, R. J. (1978). A multivariate analysis of behavioural divergence among closely related species of endemic Hawaiin *Drosophila*. *Evolution* **32**, 389–397. [186]

Rising, J. D. (1968). A multivariate assessment of interbreeding between the chickadees, *Parus atricapillis* and *P. caroliensis*. *Syst. Zool.* **17**, 160–169. [113]

Riska, B. (1981). Morphological variation in the horseshoe crab *Limulus polyphemus*. *Evolution* **35**, 647–658. [187]

Roberts, D. F., Mitchell, R. J., Creen, C. K. and Jorde, L. B. (1981). Genetic variation in Cumbrians. *Ann. Hum. Biol.* **8**, 135–144. [180]

Robinson, J. W. and Hoffmann, R. S. (1975). Geographical and interspecific cranial variation in Big-eared Ground Squirrels (*Spermophilus*): a multivariate study. *Syst. Zool.* **24**, 79–88. [115]

Rochford, J. R. (1982). PhD Thesis, Trinity College, Dublin, Ireland. [183]

Rohlf, F. J. (1967). Correlated characters in numerical taxonomy. *Syst. Zool.* **16**, 109–126. [37]

Rohlf, F. J. (1972). An empirical comparison of three ordination techniques in numerical taxonomy. *Syst. Zool.* **21**, 271–280. [186]

Rohlf, F. J. and Sokal, R. R. (1972). Comparative morphometrics by factor analysis in two species of Diptera. *Z. Morph. Tiere* **72**, 36–45. [186]

Rohwer, S. A. (1972). A multivariate assessment of interbreeding between the meadowlarks, *Sturnella. Syst. Zool.* **21**, 313–338. [117]

Rohwer, S. A. and Kilgore, D. L. (1973). Interbreeding in the arid-land foxes, *Vulpes velox* and *V. macrotis. Syst. Zool.* **22**, 157–165. [117]

Rouvier, R. R. (1966). L'analyse en composantes principales: son utilisation en génetique et ses rapports avec l'analyse discriminante. *Biometrics* **22**, 343–357. [154]

Rouvier, R. R. (1969). Pondération des valeurs génotypiques dans la sélection par index sur plusieurs caractères. *Biometrics* **25**, 295–307. [112]

Rouvier, R. and Ricard, F. H. (1965). Etude de conformation du poulet II. Recherche des composantes de la variabilité morphologique du poulet vivant. *Ann. Zootech.* **14**, 213–227. [109, 112]

Sakai, S. (1978). Dermapterorum Catalogus Praeliminaris XI Multivariate analysis of the zoogeographical distribution of the world Dermaptera with special reference to the Labiidae. *Bull. Daito Bunka Univ.* **16**, 1–53. [186]

Schnell, G. D. (1970). A phenetic study of the suborder Lari (Aves) I. Methods and results of principal component analysis. *Syst. Zool.* **19**, 35–57. [183]

Schreider, E. (1960). "La Biométrie" (Collection Que sais-je), 126 pp. Presses Universitaires, Paris. [111]

Schueler, F. W. and Rising, J. D. (1976). Phenetic evidence of natural hybridization. *Syst. Zool.* **25**, 283–289. [116]

Scopoli, G. A. (1777). "Introductio ad Historiam Naturalem", Praga. [7]

Scossiroli, R. E. (1959). Selezione artificiale per caratteri quantitativi e selezione naturale in *Drosophila melanogaster. Ric. Sci.* **29**, 1–10. [111]

Scott, J. F. (1975). Multivariate analysis in geography: Some comments. *Statistician* **24**, 211–216. [177]

Scott, J. H. (1957). Muscle growth and function in relation to skeletal morphology *Am. J. Phys. Anthropol. n. Ser.* **15**, 277–234. [112]

Seal, H. (1964). "Multivariate Statistical Analysis for Biologists", 207 pp. Methuen, London. Second edition 1967. [39, 66, 67, 103]

Sepkoski, J. J. Jr. (1981). A factor analytic description of the Phanerozoic marine fossils record. *Paleobiology* **7**, 36–53. [175]

Shaklee, J. B. and Tamaru, C. S. (1981). Biochemical and morphological evolution of Hawaiian bonefishes *(Albula). Syst. Zool.* **30**, 125–146. [116]

Sheldon, A. L. (1972). The classification of inland waters. *In* "Natural Environments", Ed. Krutkilla, J. V., Resources for the Future, Inc, Baltimore.

Siegel, A. F. and Benson, R. H. (1982). A robust comparison of biological shapes. *Biometrics* **38**, 341–350. [124, 157]

Sinha, R. N. and Shankarnarayan, D. (1955). Concepts of insect taxonomy in Ancient India. *Ent. News* **46**, 243–247. [3]

Skeel, M. A. and Carbyn, L. N. (1977). The morphological relationship of gray wolves *(Canis lupus)* in national parks of central Canada. *Can. J. Zool.* **55**, 737–747. [176, 189]

Smith, H. F. (1936). A discriminant function for plant selection. *Ann. Eugen. Lond.* **7**, 240–250. [112]

[49] Smith, J. E. K. and Klem, L. (1961). Vowel recognition using a multiple discriminant function. *J. Acoust. Soc. Am.* **33**, 358.

[192] Smith, R. F., Campbell, N. A. and Perdrix, J. L. (1982). Identification of some Western Australian gossans by multi-element geochemistry. *In* "Geochemical Exploration in Deeply Weathered Terrains", Ed. Smith, R. E., CSIRO, Perth.

Sneath, P. H. A. (1967). Trend-surface analysis of transformation grids. *J. Zool. Lond.* **151**, 65–122.

[152] Sneath, P. H. A. and McKenzie, K. G. (1973). Statistical methods for the study of biogeography. *In* "Organisms and Continents Through Time", Ed. Hughes, N. F. Spec. papers in Palaeontology, No. 12, 45–60.

[7, 37, 86] Sneath, P. H. A. and Sokal, R. R. (1973). "Numerical Taxonomy", Freeman, San Fransisco.

[39] Sokal, R. R. (1961). Distance as a measure of taxonomic similarity. *Syst. Zool.* **10**, 70–79.

[9, 111] Sokal, R. R. (1962). Variation and covariation of characters of alate *Pemphigus populi-transversus* in Eastern North America. *Evolution* **16**, 227–245.

Sokal, R. R., Bird, J., and Riska, B. (1980). Geographic variation in *Pemphigus populicaulis* (Insecta: Aphididae) in eastern North America. *Biol. J. Linn. Soc.* **14**, 163–200.

[103] Sokal, R. R., Daly, H. V. and Rohlf, F. J. (1961). Factor analytical procedures in a biological model. *Kans. Univ. Sci. Bull.* **42**, 1099–1121.

[131] Sokal, R. R. and Rinkel, R. C. (1963). Geographic variation of alate *Pemphigus populi-transversus* in Eastern North America. *Kans. Univ. Sci. Bull.* **44**, 467–507.

[106] Sokal, R. R., Rohlf, F. J., Zang, E. and Osness, W. (1980). Reification in factor analysis: a plasmode based on human physiology-of-exercise variables. *Multivariate Behav. Res.* **15**, 181–202.

[35, 89, 109, 137] Sokal, R. R. and Sneath (1963). "Principles of Numerical Taxonomy", 359 pp., Freeman, San Francisco.

[189] Spain, A. V., Heinsohn, G. E., Marsh, H. and Correll, R. L. (1976). Sexual dimorphism and other sources of variation in a sample of dugong skulls from North Queensland. *Aust. J. Zool.* **24**, 491–497.

[104] Spearman, C. (1904). General intelligence objectively determined and measured. *Am. J. Psychol.* **15**, 201–293.

[178] Spicer, R. A. and Hill, C. R. (1979). Principal components and correspondence analysis of quantitative data from a Jurassic plant bed. *Rev. Paleobot. Palynology* **28**, 273–299.

[124, 181] Spielman, R. S. (1973). Do the natives all look alike? Size and shape components of anthropometric differences among Yanomana Indian villagers. *Am. Nat.* **107**, 694–708.

[36] Sporne, K. R. (1960). The correlation of biological characters. *Proc. Linn. Soc. Lond.* **171**, 83–88.

[30, 122] Sprent, P. (1968). Linear relationships in growth and size studies. *Biometrics* **24**, 639–656.

[5] Sprent, P. (1972). The mathematics of size and shape. *Biometrics* **28**, 23–38.

[157] Stalker, H. D. and Carson, H. L. (1947). Morphological variation in natural populations of *Drosophila robusta* Sturtevant. *Evolution* **1**, 237–248.

[1] Stapleton, H. E. (1956). The hand, with its five fingers, as the primitive basis of geometry, arithmetic and algebra. *Actes du 8-me Cong. Int. Hist. Sci.* 1103.

[1] Stapleton, H. E. (1958). Ancient and modern aspects of Pythagoreanism. *Osiris* **30**, 12–53.

Steinhorst, R. K. (1979). Analysis of niche overlap. *In* "Multivariate Analysis in Ecological Work", Eds Orlóci, L., Rao, C. R. and Stiteler, W. M., pp. 263–278. International Co-operative Publishing House, Fairland, Maryland. [142]

Stiteler, W. M. (1979). Multivariate statistics with applications in statistical ecology. *In* "Multivariate Methods in Ecological Work", Eds Orlóci, L., Rao, C. R. and Stiteler, W. M., pp. 279–300. International Co-operative Publishing House, Fairland, Maryland.

Storms, L. H. (1958). Discrepancies between factorial and multivariate discrimination among groups as applied to personality theory. *J. Ment. Sci.* **104**, 713–721.

Stower, W. J., Davies, D. E. and Jones, I. B. (1960). Morphometric studies of the Desert Locust *Schistocerca gregaria* (Forsk.), *J. Anim. Ecol.* **29**, 309–339. [71]

Stroud, C. P. (1953). An application of factor analysis to the systematics of *Kalotermes. Syst. Zool.* **2**, 76–92.

Stuart, A. (1964). Review of Lawley and Maxwell's "Factor analysis as a statistical method". *Biometrika* **51**, 533. [108]

Stützle, W., Gasser, T., Molinari, L., Largo, R. H., Prader, A. and Huber, P. J. (1980). Shape invariant modelling of human growth. *Ann. Hum. Bio.* **7**, 507–528. [120, 183]

Susanne, C. and Sharma, P. D. (1978). Multivariate analysis of head measurements in Punjabi families. *Ann. Hum. Biol.* **5**, 179–183. [180]

Symmons, P. M. (1969). A morphometric measure of phase in the desert locust *Schistocerca gregaria* (Forsk.). *Bull. Ent. Res.* **58**, 803–809. [71]

Talbot, P. A. and Mulhall, H. (1962). "The Physical Anthropology of Southern Nigeria", 127 pp. Cambridge University Press, Cambridge. [40, 68]

Tantravahi, R. V. (1971). Multiple character analysis chromosome studies in the *Tripsacum lanceolatum* complex. *Evolution* **25**, 38–50. [118]

Teissier, G. (1938). Un essai d'analyse factorielle. Les variants sexuels de *Maia squinada. Biotypologie* **7**, 73–96. [6, 87, 157]

Teissier, G. (1960). Relative growth. *In* "The Physiology of the Crustacea", Ed. Waterman, T. H., Vol. I, pp. 537–560, Academic Press, New York and London. [156]

Temple, J. T. (1968). The Lower Llandovery (Silurian) brachiopods from Keisley, Westmorland. *Palaeontol. Soc. Publ.* No 521, 58 pp. [91, 93]

Thomas, B. A. M. (1961). Some industrial applications of multivariate analysis. *Appl. Statist.* **10**, 1–8.

Thomas, P. A. (1968). Geographic variation of the rabbit tick *Haemaphysalis leporispalustris* in North America. *Kans. Univ. Sci. Bull.* **47**, 787–828.

Thomas, E. (1980). Details of *Uvigerina* development in the Cretan Mio-Pliocene. *Utrecht Micropaleontol. Bull.* **23**, 1–167.

Thompson, D'Arcy W. (1942). "On Growth and Form", 1116 pp. 2nd Edn, University Press, Cambridge. [5, 124]

Thorpe, R. S. (1975). Biometric analysis of incipient speciation in the ringed snake, *Natrix natrix* (L.) *Experientia* **31**, 180–182. [183]

Thorpe, R. S. (1976). Biometric analysis of geographic variation and racial affinities. *Biol. Rev.* **51**, 407–452. [124]

Thorpe, R. S. (1979). Multivariate analysis of the population systematics of the ringed snake, *Natrix natrix* (L.). *Proc. R. Soc. Edin. (B).* **78**, 1–62. [118, 183]

Thorpe, R. S. and Leamy, L. (1983). Morphometric studies in inbred and hybrid house mice: multivariate analysis of size and shape. *J. Zool. Lond.* **199**, 421–432.

Thorpe, R. S. and McCarthy, C. J. (1978). A preliminary study, using multivariate analysis, of a species complex of African house snakes *(Boaedon fuliginosus). J. Zool.* **184**, 489–506. [118, 183]

[103] Thurston, L. L. (1947). "Multiple Factor Analysis: a Development and Expansion of 'Factors of the Mind'", 535 pp. Chicago University Press, Chicago, Ill.

[95] Torgerson, W. S. (1958). "Theory and Methods of Scaling", 460 pp. Wiley, New York.

[188] Townshend, J. L. and Blackith, R. E. (1975). Fungal diet and the morphometric relationships in *Aphelenchus avenae*. *Nematologica* **21**, 19–25.

Tukey, J. W. (1980). We need both exploratory and confirmatory! *Am. Statist.* **34**, 23–25.

[151] Udvardy, M. D. F. (1969). "Dynamic Zoogeography", 445 pp. Van Nostrand Reinhold, New York.

[186] Usher, M. B. (1979). The *Neptis* butterflies (Nymphalidae) of Ghana. *Syst. Entomol.* **4**, 197–207.

[144] Usher, M. B., Booth, R. G. and Sparkes, K. E. (1982). A review of progress in understanding the organisation of communities of soil arthropods. *Pedobiologia* **23**, 126–144.

Usher, M. B., Davis, P. R., Harris, J. R. W. and Longstaff, B. C. (1978). A profusion of species? Approaches towards understanding the dynamics of the populations of the microarthropods in decomposer communities. *In* "Population Dynamics", Eds Anderson, R. M., Turner, B. D. and Taylor, L. R., pp. 359–384. Blackwell, Oxford.

[2] van der Waerden, H. (1963). "Science Awakening", 306 pp. Wiley, New York.

[146] van Groenewoud, H. (1965). Ordination and classification of Swiss and Canadian coniferous forests by various biometric and other methods. *Ber. Geobot. Inst. ETH Stift. Rubel Zürich* **36**, 28–102.

[180] van Vark, G. N. (1974). The investigation of human cremated skeletal material by multivariate statistical methods. I. Methodology. *Ossa* **1**, 63–95.

[180] van Vark, G. N. (1976). A critical evaluation of the application of multivariate statistical methods to the study of human populations from their skeletal remains. *Homo* **27**, 94–114.

[117] Vogt, W. G. and McPherson, D. G. (1972). The weighted separation index: a multivariate technique for separating members of closely-related species using qualitative differences. *Syst. Zool.* **21**, 187–198.

[72] Waloff, N. (1966). Scotch broom (*Sarothamnus scoparius* (L.) Wimmer) and its insect fauna introduced into the Pacific North-West of America. *J. Appl. Ecol.* **3**, 239–311.

Warburton, F. W. (1962). The practical value of factor analysis in education. *Statistician* **12**, 172–188.

[103] Warburton, R. W. (1964). A reply to Mr Ehrenberg's criticism. *Statistician* **14**, 49–51.

[83] Ward, L. K. (1968). The validity of the separation of *Thrips physapus*, L. and *T. Hukkineni* Priesner (Thysanoptera, Thripidae). *Trans. R. ent. Soc. Lond.* **120**, 395–416.

[115] Warner, J. W. (1976). Chromosomal variation in the plains woodrat: geographical distribution of three chromosomal morphs. *Evolution* **30**, 593–598.

[149] Webb, L. J., Tracey, J. G., Williams, W. T. and Lance, G. (1967). Studies in the numerical analysis of complex rain-forest communities. I. A comparison of methods applicable to site/species data. *J. Ecol.* **55**, 171–191.

[113] Weber, E. (1959). The genetical analysis of characters with continuous variability on a Mendelian basis. I. Monohybrid segregation. *Genetics* **44**, 1131–1139.

[113] Weber, E. (1960a). The genetical analysis of characters with continuous variability on a Mendelian basis. II. Monohybrid segregation and linkage analysis. *Genetics* **45**, 459–466.

Weber, E. (1960b). The genetical analysis of characters with continuous variability on a Mendelian basis. III. Dihybrid segregation. *Genetics* **45**, 567–572. [113]

Welch, P. D. and Wimpress, R. S. (1961). Two multivariate statistical computer programmes and their application to the vowel recognition problem. *J. Acoustic Soc. Am.* **33**, 426–434. [50]

Wharfe, J. R. (1977). An ecological survey of the benthic invertebrate macrofauna of the Lower Medway estuary. Kent. *J. Anim. Ecol.* **46**, 93–114. [179]

White, M. J. D. (1973). "Animal Cytology and Evolution", 1000 pp. 3rd Edn Cambridge University Press. [110, 115]

White, M. J. D. and Andrew, L. E. (1959). Cytogenetics of the grasshopper *Moraba scurra*. V. Biometric effects of chromosomal inversions. *Evolution* **14**, 284–292. [110]

Whitehead, F. H. (1959). Vegetational changes in response to alterations of surface roughness on M. Maiella, Italy. *J. Ecol.* **47**, 603–606. [128]

Wilkinson, C. (1970). Adding a point to a principal coordinates analysis. *Syst. Zool.* **19**, 258–263. [186]

Wilkinson, E. M. (1951). Goethe's concept of form. *Proc. Br. Acad.* **37**, 175–197. [5]

Wilkinson, P. (1978). Graphical methods for representing computer classifications. *J. Ent. (B)* **42**, 103–112. [173]

Wilkinson, P. (1974). Numerical classification: some questions answered. *Can. Entom.* **106**, 449–464. [186]

Williams, D. A., Beatty, R. A. and Burgoyne, P. S. (1970). Multivariate analysis in the genetics of spermatozoan dimensions in mice. *Proc. R. Soc. Lond. B.* **175**, 313–331. [114]

Williams, W. T. and Lambert, J. M. (1959). Multivariate methods in plant ecology. I. Association analysis in plant communities. *J. Ecol.* **47**, 83–101. [146]

Williamson, M. H. (1978). The ordination of incidence data. *J. Ecol.* **66**, 911–920. [127]

Wolpoff, M. H. (1976). Multivariate discrimination, tooth measurements and early hominid taxonomy. *J. Hum. Evol.* **5**, 339–344. [180]

Worsley, A. (1981). In the eye of the beholder: social and personal characteristics of teenagers and their impressions of themselves and fat and slim people. *Br. J. Med. Psychol.* **54**, 231–242. [180, 183]

Yang, S. H. and Selander (1968). Hybridization in the grackle *Quiscalus quiscalus* in Louisiana. *Syst. Zool.* **17**, 197–143. [113]

Younker, J. L. and Ehrlich, R. (1977). Fourier biometrics: harmonic amplitudes as multivariate shape descriptors. *Syst. Zool.* **26**, 336–342. [173]

Zimmerman, J. R. and Ludwig, J. A. (1975). Multiple-discriminant analysis of geographical variation in the aquatic beetle *Rhantus gutticollis* (Say) (Dytiscidae). *Syst. Zool.* **24**, 63–71. [185]

Subject Index

A

THE
PALACE

THE
PALACE

PAUL
ERDMAN

Doubleday
NEW YORK
1988

All the characters in this novel are fictitious and any resemblance to actual persons, whether living or dead, is entirely coincidental.

DESIGNED BY PETER R. KRUZAN

Library of Congress Cataloging-in-Publication Data
Erdman, Paul Emil, 1932–
The Palace.
I. Title.
PS3555.R4P35   1988   813′.54   87-24523
ISBN 0-385-24488-6

PART ONE

1964–1969

1

Danny Lehman's shop was, well, kind of shabby-looking, but it somehow belonged because it was just on the fringe of downtown Philadelphia, which is also shabby. Danny Lehman did not exactly exude class either. Class is something you're either born with or you're not, and Danny most certainly wasn't, which did not bother him in the least. His father had been in scrap until he went bust in the early 1930s, and Danny had maintained the tradition: he had started out with a pawnshop and had then moved up to a coin shop. Scrap, used coins: no class.

Danny was plump, usually sported one day's growth on his chin, and liked green sport shirts. He liked to bowl. He liked bagels. He liked his mother. He liked girls, in fact he liked them a lot, but for some reason had never married one.

In the fall of 1964 the guy who owned the shop next door, a travel agency, suggested that it was time for Danny to broaden his horizons. He could work out a trip to Europe for him, booking all the first-class hotels at cut rates since it was off-season, and would not even charge any of the usual commissions. After all, they were neighbors.

Never one to pass up a deal, Danny took him up on the offer. Almost immediately after arriving on the other side of the Atlantic, Danny was impressed by two things. The first was how he got taken every time he had to change dollars into pounds, francs, or lire. It was the same in every hotel and restaurant bar none. By the time he hit Switzerland he had smartened up and decided to change his money at a bank *before* getting robbed at the hotels and the restaurants. He was disappointed to discover that the banks were offering rates that were almost as lousy. After all, they were banks, for God's sake, run by Episcopalians, or disciples of whatever type of God they worshipped in Zurich. Danny found out that Catholics were no better: in Milan they did the same thing; at the airport in Madrid, ditto. Danny figured that the money changers were taking an average of 5 percent in the middle for nothing more than giving him 375 marks in exchange for 100 dollars.

The other thing that astonished Danny was that nobody in Europe ever talked about the stock market. In Philadelphia, even in the bowling alley he frequented, everyone knew what the Dow Jones was doing, more or less. He could hardly believe it when somebody in Paris told him they didn't *have* a Dow Jones in Europe. What they did have, according to his French informant, a bartender at the George V who would talk to anybody to milk a tip, was a very highly developed interest in gold and silver. Coins maybe, Danny suggested, like the napoleon? It was the only French word he could pronounce with confidence. No, coins were for peasants and Americans, his financial adviser told him. The name of the money game in France, Switzerland, and Germany was bullion. That's where the rich Europeans had always put their dough. That's why they were still rich, and still able to frequent the bar at the George V.

Their reasoning was this, the erudite bartender went on to explain: every time there was a war, be it one of Napoleon's, World War I, or World War II, the governments of France or Germany or Italy would finance it by simply printing a lot of money. The inevitable result was that the value of everybody's paper money, be it marks, or francs, or lire, went all to hell. The smart Europeans, knowing that history constantly repeats itself, would begin trading in their cash for gold as soon as they saw the war clouds gathering. And when the war ended, in every instance they were among the few in society who had been able to preserve their wealth, in the form of that tangible asset gold, whose value had remained constant.

Later, after leaving a 25 percent tip, proving the bartender right once again, Danny figured that the only country that was at war right then was the United States in Vietnam. So maybe he should be thinking about getting into the gold business back home. The problem was that it was still illegal for Americans to buy gold in bullion form. All right, he concluded, then why not silver?

Danny asked around about who was number one in the silver bullion business in the United States. The answer was always the same: Engelhard Industries. When he asked foreign exchange, it was Deak-Perera. So with the help of a head-hunter, Danny hired one of the best traders away from each. Soon every smart travel office in Philadelphia knew that the place to send people who needed foreign money before going abroad was the American Coin, Metals, and Currency Exchange on Broad Street: the new name that Danny had given his shop. And as the storm clouds started to gather over America's financial system as a result of the strain of the Vietnam War—giving birth to a new clan of doom-and-gloom merchants who had visions of the almighty dollar ending up in ruins, and who sought safety in, yes, bullion—well, the word spread that you could get good deals in Philadelphia.

Danny had a cousin in Trenton, New Jersey, who worked for a bank there. When he saw what Danny was doing in Philadelphia he suggested that Danny open up an office in Trenton, too, which he would be glad to run. Danny did. His cousin concentrated on the Jewish trade and it worked. The place was in the black by the third month.

Then Danny's cousin told him about a guy he knew in Boston who was a stockbroker and who was perfect for their type of operation: he was a big gun in the Catholic Church up there, had built up a huge clientele among the parishioners. Danny's cousin figured that on the Irish trade alone they stood to do very well with a manager like that. Once again, his cousin was proven right.

In Chicago, Danny went the Polish-Catholic route; in Miami, he staffed the new place he opened up there with Latinos; in San Francisco, he hired some Orientals. Just as Aer Lingus and El Al had managed to build up the lucrative North Atlantic trade by appealing to ethnic loyalties, and by relying upon word of mouth among the faithful rather than expensive advertising, Danny Lehman built up his currency and bullion business by doing the same—and by offering rates that beat every bank and every currency dealer in the United States by 25 percent.

By the end of 1966, two years after Danny's historic trip across the Atlantic, the American Coin, Metals, and Currency Exchange had spread to seven additional cities. Danny Lehman's net worth was well into seven digits. In December of that year, Danny went to Miami. He stayed at the Fontainebleau Hotel. He had decided that it was time to stop thinking about his business for a while, so every day he went to the races at Hialeah, and every night he went to the jai alai fronton. He loved to watch the horses by day and the Basques by night, and he was excited by the atmosphere, by the crowd. For it was this gaming crowd, the gamblers and the bettors who seemed to enjoy life more than most people, which fascinated him. They lived joyously for the brief moment of victory. They overcame stoically the more prevalent brief moments of defeat. And they always came back. Danny never bet; it was a sucker's game, he figured.

On December 23 Danny succumbed to the inevitable and decided to check up on the local Miami branch of the American Coin, Metals, and Currency Exchange, which he had opened about eighteen months earlier, and which was on the main drag of downtown Miami, West Flagler. José, the manager, was a Cuban. Danny, not being an early riser, got there at about ten-thirty and had coffee with José in the back office. José said the place was making pretty good money in spite of the hard times. Actually, it was making damn good money. All of a sudden the rich Latin Americans had discovered how cheap the United States was compared to Europe, and were starting to come to Miami in droves to buy their mink coats, outboard motors, and Rolex watches. And they flocked to the American Coin, Metals, and Currency Exchange because the people there spoke Spanish and its exchange rates were better than those offered by any other bank or exchange office in town. The word had even gotten around São Paulo, Bogotá, and Caracas.

At around a quarter to eleven, a Cuban who worked at the coin and bullion counter entered the back office and mumbled something in Spanish, and immediately José got up, excused himself, and left, leaving Danny alone and puzzled. Danny went out to the shop proper and stood there watching a guy, definitely *not* a Latino, bringing in bags of coins, twenty-seven of them. Then José and the guy who was bringing in the silver—he could have been either Italian or Jewish—conferred, and a lot of hundred-dollar bills began to change hands. It took maybe ten minutes in all.

At eleven, José and Danny were again in the back room, each with new cups of coffee. Good coffee. Cubans know how to make coffee.

"Who was that?" asked Danny.

"Dunno" was the Cuban's response.

"Dunno?" asked Danny. "Then how come you didn't check out the coins, for Chrissake?"

"Don't have to," answered the Cuban. "They're always good."

Actually, there was not much to check. Either the coins were quarters minted before 1965, and thus almost pure silver, or they weren't. These were. With the silver bullion price on the rise, pure silver coins were starting to demand a premium, since they could be melted down and sold as bullion at a price higher than they were worth as coins in the monetary system. After 1965 the U.S. Government had started to smarten up and to use an increasingly smaller portion of silver per coin in the minting process; these new coins weren't worth melting down, at least not yet.

Danny mulled over the Cuban's answer for maybe a minute. "Always good, huh," he finally said.

"Yes."

"How often is always?" Danny then asked.

"Oh, about every ten days," answered the Cuban.

"How much's he bringing every ten days?" inquired Danny further.

"Oh, twenty-five, sometimes fifty bags."

"Always quarters?"

"Yes."

"Where's he from?" Danny asked.

José just shrugged.

"From around here?" Danny asked again.

"No," answered the Cuban.

They stopped here for a minute. Danny picked up his coffee cup and took a sip. So did José. They sipped some more. Then Danny continued with his interrogation. "So tell me this. Why'd he come here in the first place?"

"He came in with a customer. He had just one bag of silver the first time."

"A customer. Who's this customer?" asked Danny.

"A man who works at the jai alai place," answered the Cuban.

"What's his nationality?"

8

"Italian," answered José.

"When was that?" asked Danny.

"About six months ago," replied José.

"How much business in the meantime?"

"Let's see." José thought it over. "Something like four hundred thousand dollars."

"He let you know when he's coming?" Danny inquired.

"Well, he usually calls an hour beforehand," José answered.

"Who?" Danny asked.

"The same man" was the answer.

Danny paused and thought it all over. "Listen," he finally said, "next time he calls you I want you to call me immediately. Got it?"

"Yes, but I never really know in advance. Sometimes it's two weeks, sometimes maybe only five days," replied the Cuban.

"Doesn't matter. Just call me. I'm at the Fontainebleau. What time's he usually call?"

"Mostly between nine and ten o'clock in the morning."

"Okay. He calls you, you call me at the Fontainebleau, Room 756. Got it? Write it down."

José wrote it down. Danny took another sip of coffee, stood up, and left. Even on vacation, it was hustle that built a business. He went back to the hotel a happy man, an idea bubbling in his head.

◆   ◆

On January 4, around nine thirty-five in the morning, the Cuban called Danny at the Fontainebleau to tell him he had received word that a new silver shipment would probably be hitting the store before eleven o'clock that morning. It took Danny just twenty minutes to shower and shave, and another twenty minutes to take a cab from Miami Beach to West Flagler. He went to the back room, where both José and a cup of coffee awaited him, and sat there with his employee chatting about the cold weather, jai alai, and the races.

Once more the counter clerk came into the back room and, in a hoarse whisper, told his boss that the guy with the silver bags was out there again. This time Danny Lehman and both Cubans walked out together. Danny stood immediately to the right and a little behind his Cuban manager as the customer brought in one bag after the other from the van standing at the curb. Every time the tall man, who was somewhere around six feet four inches, maybe even six feet six inches, shoved bag after bag across the counter, he looked first at the bag and then at Danny, silently asking himself, it seemed, who this new guy was and what exactly he was doing there. When the delivery was completed, José went through what was obviously a practiced ritual of weighing each bag, then writing down the weight on a pad. The posted selling price on the board behind the counter indicated that he was buying at $1.57 an ounce, selling at $1.67. Then he moved to the calculator, where ounces were converted to dollars. Having done all this, he wrote down the final number, circled it, turned the sheet around, and showed it to the tall man. The customer looked at the number, picked up the piece of paper, folded it, put it in his shirt pocket, and nodded his head. The Cuban went to the small safe located immediately behind the counter, opened it and pulled out a steel box, brought it to the counter, and put it down. It was full of one-hundred-dollar bills. He took out six packages with wrappers around them and, one by one, ripped off the wrappers and counted out singles. He used up almost all six packages. Then he reached into the cash drawer and extracted a fifty, a twenty, a ten, a five, and a couple of ones and laid them down beside the hundreds. He shoved the entire amount across the counter to the tall man, who put the bills into a very large, old-fashioned folding-type briefcase. Then he nodded again and turned to leave.

The number on the sheet of paper was $37,587.37.

"Sir," said Danny Lehman, just before the tall man had reached the door.

He paused, looked back, stared at Danny, and asked, "Who are you?"

José intervened immediately. "This is Mr. Daniel Lehman, the owner of this company. He's from Philadelphia, where our head office is located. We have branches in eight cities along the East Coast."

The tall man nodded again.

"I'd like to talk to you," said Danny.

"About what?" asked the tall man.

"I'd be interested in perhaps suggesting a special arrangement for you; a volume discount, if you would like to call it that," answered Danny.

The tall man said, "I'll let you know," and walked out.

◆   ◆

Eight days later Danny Lehman was alone in his Philadelphia home watching the 76ers playing the Celtics when the phone rang at exactly ten o'clock. About two hours earlier he had hurried back from the bowling alley through a miserable, snowy Philadelphia winter night in order to catch the basketball game from the beginning. All he wanted at this moment was to watch the 76ers win, which they were doing. Then he wanted to take a good hot shower and go to bed. Now the damn phone was ringing. Danny Lehman, like everybody else, knew that ten o'clock phone calls rarely brought good news.

He turned down the volume on the TV, went to the phone, and barked, "Yeah," ready to slam the receiver down just as quickly as he had picked it up.

"Is this Danny Lehman?" came a voice.

"Yeah," repeated Danny. "What do you want?"

"I want to talk to you," the voice said.

"So talk."

"I've heard about you from a mutual friend to whom you introduced yourself in Miami a week ago. Do you remember?"

11

"Yes."

"I'd like to discuss the development of a business relationship, provided, of course, you're interested."

"I'm listening."

"Good. I'll be over in forty-five minutes." The phone went dead.

It was exactly a quarter to eleven when Danny's doorbell rang. When he opened the door, there stood a man about five feet six inches tall, the same height as himself. He was wearing a blue cashmere coat with a fur collar, and a fedora-type hat, and as he walked through the door into the hallway, one could not help but notice his very shiny, pointed black shoes. The man said, "My name is Joseph Amaretto. I'd prefer you to call me Joe."

"All right," said Danny, "Joe it is." When Amaretto started to take off his top coat, Danny reached out to take it from him. But his visitor said, "No thanks, I'll hang on to it. This won't take long. No more than fifteen minutes."

Left with little choice, Danny led him into the living room. "Do you think you'll have time for a brandy?" he asked.

"I've never refused one yet," Amaretto replied, flashing a big smile.

Danny indicated that he should take a seat and a few minutes later returned with the brandy bottle and two snifter glasses. He poured a short one and handed it to his guest. Amaretto nodded his thanks, drank it, and quickly handed his glass back to Danny for a refill. Having received it, he settled back on the sofa, indicating that maybe they should get down to business. The social part of the evening was over.

"How many silver coins can you handle a month?" was his question.

Danny looked at him, raising his hands in a gesture meaning the sky was the limit, and answered, "Any amount you can come up with."

"How about five hundred bags a month?"

"No problem."

"Do you have the facilities for picking up quantities this large on a regular basis?"

"Where from?" asked Danny.

"We'll get to that," said Amaretto. "First the answer."

"The answer is yes. Anywhere on the eastern seaboard," replied Danny.

"I'm not talking about the eastern seaboard."

"Well, I can't say for sure unless you tell me exactly where this stuff is coming from."

"First let's clear something up," Amaretto said. "We know a lot about you. We like what we've found out. We would like to enter into a business relationship which will go on for quite a long time. Everything would be absolutely confidential, if you understand me. I'd also hope that we could expand the framework of cooperation to areas beyond silver coins."

The sentence might have been somewhat garbled, but Danny got Amaretto's message. "Fine with me," he said.

"You understand," Amaretto repeated, "that confidentiality is a primary requirement from our point of view."

"Perfectly," said Danny, this time with more accuracy.

Amaretto returned to the brandy, which he now sipped rather than gulped. "Fine," he said after the third and last sip. "The pickup place can be either Las Vegas or Reno, take your choice. Five hundred bags a month. What would you pay for them?"

"The price would be calculated from the closing quote on the London Metal Exchange. That's a bullion quote. We'd adjust it for bagged coinage. I'd give you the formula for your approval. There would be a charge of ten cents per ounce for transportation, handling, and insurance."

"Hold on," said Amaretto. "These coins wouldn't be presorted. And there will be dimes and silver dollars, not just quarters."

"We'll sort them out," Danny confirmed. "No additional charge."

The Italian said, "You have a deal. We'll call you in a day or two about the same time of the evening—here at home, if you don't mind. I'll tell you the exact time and place where you should make the pickup. Okay?"

"Okay as far as I'm concerned," said Danny.

"Fine." Amaretto stood up and put on his coat, which he had kept folded across his lap. He reached down for his hat, walked to the front door, shook hands with Danny, and walked out into the night. Danny stood at the open door and watched him go down the path, turn right at the sidewalk, and then move off at a brisk pace. There was no sign of a car, taxi, or limousine; in fact, there was no light or activity whatsoever outside. Danny watched him move fifty, one hundred, then two hundred feet down the street, and disappear. Shrugging, Danny turned around and went back into the house.

◆   ◆

On January 21, 1967, a Railway Express van backed up to a warehouse in an industrial park in Reno to pick up three extremely heavy wooden crates. Three days later, they were dropped off at the warehouse of the American Coin, Metals, and Currency Exchange in south Philadelphia. After they were sorted and bagged, they were sold at retail throughout Danny's chain of outlets. The shipment brought in a quarter of a million dollars. Danny's profit amounted to approximately 17 percent of the sum. It was the largest single transaction that Danny had made in his business career. After all, he was literally a nickel, dime, and quarter guy.

2

Almost six months later, Joseph Amaretto returned to Philadelphia and again telephoned Danny, but this time at his office and in the morning. They arranged to meet for lunch at Bookbinder's, at the old Bookbinder's down near the water, not the other, newer one near downtown. Apparently Amaretto knew his way around Philadelphia.

Amaretto was with another man when Danny arrived. It was summer and Amaretto was wearing yellow slacks and a green sport shirt. But his colleague, who was introduced as Sam Sarnoff, had made no concession to the season: he wore a dark suit and an almost black tie with a gleaming white shirt. Sarnoff was also wearing cuff links studded with diamonds, which Danny figured were worth somewhere between twenty and thirty thousand dollars, per sleeve. Danny Lehman, like his father and grandfather before him, not only knew all about scrap metal and used coins, but also had an eye for precious stones. During lunch he questioned Sarnoff about the value of the diamonds. Sarnoff said, "Around fifty thousand dollars, both arms." He obviously appreciated the question, for it confirmed what Joseph Amaretto had said about Danny Lehman, namely, "The guy's got class."

The fish was excellent and the beer good. Apart from precious stones, the talk centered mainly on baseball: on the Phillies and the Dodgers. Danny Lehman rose even further in the estimation of his lunch partners when he confessed that he had been a secret admirer of the Dodgers since his boyhood days, when they played their home games not that far east of Philadelphia, and that he had remained a fan even though they were now based a few hundred miles west of Vegas.

It was a congenial sort of lunch, and it was not until a quarter to two that Amaretto burped, pushed back his chair, and said, "Danny, we've come here to break bread with you. We've been extremely happy with the way you've handled our silver coin business, and we want to express our appreciation for the confidential manner in which you've conducted these transactions. Our aim today is to tell you in person that not only do we want to continue our past business relationship with you, we'd now like to expand it. I think Sam would like to make a proposal to you. Sam?"

Amaretto's speech had sounded like a prepared introduction by a chairman of the board. That's exactly what it was.

"There is no reason to beat around the bush," began Sam Sarnoff, leaning across the table toward Danny. "What we have in mind, Mr. Lehman, is to seek your advice, your assistance, in redirecting the flow of funds arising from our silver transactions. To put it bluntly, our clients are experiencing difficulties in concealing cash receipts of this size. I'll explain it in a minute. But first"—and he paused—"first, we understand that not only do you deal in precious metals, but during the past couple of years you've also built up a business in foreign exchange. We assume that in such a business you must have extensive foreign bank contacts."

Danny looked at Sarnoff, nodded, and said, "Not exactly. They're more or less restricted to the Caribbean, although I'm not sure that's really any of your business." Then he stopped smiling. "I'm afraid, gentlemen," he said, while looking at Sarnoff, "that we're all going to have to be a lot more explicit with each other. Otherwise . . ." and his voice trailed off.

Sam Sarnoff looked at Joe Amaretto and Amaretto looked at Sarnoff, and their eyes must have confirmed an agreement to proceed. It was Sam who spoke. "Mr. Lehman, first I want to make it quite clear that we're acting as middlemen, or dealers, or perhaps a more precise word would be wholesalers. We're in essence a two-man operation. Now, what precisely do we do?"

"I've been wondering about that myself," Danny interjected.

"Why?"

"Just natural curiosity."

"Nothing more that that, I trust," Sarnoff said, glaring at him.

"What's that supposed to mean?" Danny replied, returning the glare.

"We're very, very allergic to people nosing around us."

"Who isn't?" Danny replied. "The Feds don't like anybody who deals strictly in cash. And I suspect that I've been in the cash business a lot longer than you guys, so cut the crap right now."

Again Sarnoff and Amaretto exchanged glances. Both nodded ever so slightly. Danny Lehman had apparently passed the final test. "You must know that skimming's a widespread practice in casinos in Nevada," Sarnoff said.

"When you talk about skimming, like everybody else I know that it goes on, but I'm not that familiar with Nevada casinos to know that it's widespread," Danny responded.

"Take our word for it. It is. And now we're coming to the heart of the matter, Danny, and why we have to be so careful about whom we deal with."

"We've already covered that," Danny said.

"All right. Our clients're involved in skimming in a big way. Their source: slots. Their problems: twofold. One, having skimmed, how to market the skim. And two, having marketed, how to hide the proceeds from the IRS and other interested parties. You follow me?"

Danny nodded.

"Okay. Thus far we've been devoting our attention exclusively to solving the first problem. Now what we propose to do today is mutually address ourselves to the second issue."

"Just how many suppliers do you gentlemen work with?" Danny asked.

"Eight," came the answer. "But the two principal ones—they're partners—contribute, I'd say, approximately 75 percent of the total."

Danny, frowning, mulled that one over, and then said, "They own these casinos?"

The visitors from out of town laughed. "Of course not. They *manage* these casinos or co-manage them," said Amaretto.

"So what you're saying," Danny said, "is that these clients have been hired to run these casinos and that they're running a skim on the side. And maybe the owner knows what's going on, and maybe he doesn't."

"Precisely" was the answer.

Danny Lehman was puzzled. "How do they do it? Skim, I mean, without the owners knowing?" he asked. "Don't they have auditors like in any other business?"

"I don't think you have to worry about that. That's a problem for the casino owners and, maybe, the IRS, right?"

"Sure, but you said that you have eight suppliers. Isn't it dangerous, if each one knows what the others are doing?"

Sarnoff gave the answer. "Your thinking's good, Danny, very good indeed. The point is that none of our suppliers knows that there are others. We deal with each one separately. As wholesalers, we collect the skim from Casino A, Casino B, and Casino C; we combine it, we box it, and we ship it to you for redistribution."

"Therefore," Sarnoff continued, "not only are they not aware of each other's existence, but also none of them knows of the existence of you."

Now a hint of a smile crossed Danny Lehman's face. He liked it. In fact, he liked it very much. "All right," he said, "you proposed that we address ourselves to problem number two. What is it?"

Sarnoff stretched his body into another position and leaned against the back of his chair. His hands lay flat on the table as if he were admiring his cuff links. He took a sip of his coffee and continued, "It's very simple. The men we deal with are all family men. They like to send their children to good schools. They like their wives to live in fine houses. They like to have motorboats. Some even like to have little airplanes so they can get away now and then and maintain their health. These things cost money. The problem, as I've already mentioned, is that the IRS is all over Nevada keeping an eye on these gentlemen. Thus they get the money from us, but *they can't spend it!*" Sam emphasized the last four words.

"So how can I help?" asked Danny.

"Well, like we said at the outset, we've heard about your foreign exchange operations, Danny," said Sarnoff. "You mentioned the Caribbean, which confirms what we've heard. In fact, we know that you go to the Bahamas quite often. We don't want to give you the impression that we are taking too deep an interest in your personal activities, but you know how it is: these things sometimes get brought to our attention, or at least to the attention of our friends. We figured from this that you might be able to help us, from down there."

They were very close to the mark. For what Danny had done was set up an arrangement with a bank down in Nassau, whereby it would conduct his offshore currency transactions for him, and channel them through what was essentially a mailbox bank of the same name that they controlled on the Cayman Islands, where the profits were left. That way there were no taxes due to anybody, anywhere, either in the Bahamas

19

or in the United States. The Cayman bank apparently had no offices of its own but had been merely domiciled in the office building of a solicitor who worked for the Nassau bank. The only employee, if one could call him that, was a part-time accountant, a British national—British because they were much less susceptible to bribes than the locals—who flew over to the Cayman Islands from Nassau twice a year in order to fulfill the minimum accounting and reporting requirements set by the Cayman Islands' banking commission. Who *owned* these banks was never made exactly clear to Danny, but from his point of view it didn't matter. What did matter was that the Bahamas had stringent bank secrecy laws. On top of that, there was really no way that Danny Lehman could ever be identified with that mailbox bank, since he had never been to the Cayman Islands in his life, and intended never to go there either. So he decided to take the next step.

"You're right. I do have financial connections in the Bahamas. In fact, I work with a bank down there which is certainly able and, depending on certain provisos, may very well be willing to help you."

"All right," said Sarnoff. He was now excited. "First let me tell you that we sure appreciate your openness. You're just the way Joe told me you'd be, isn't that so, Joe?"

Amaretto smiled.

"What we propose, then, is this: from now on—and we'll provide you with the exact figures each time—we want approximately 75 percent of the proceeds from future silver sales to be directed to that bank of yours in the Caribbean. Now and then we'll request that your bank make a series of loans to two particular individuals. Their names will be given to you at the proper time. We'll suggest that these loans should be long-term, say twenty-five or thirty years; that they should not be collateralized; and that the interest charged to the borrower should be very high, since it is deductible. We anticipate that probably after a few years our clients, having duly made their payments on interest and principal might, no, *will*, want to pick up perhaps half the interest they've paid, but all of their payments on principal in, say, Nassau. We suggest the payout might be in the form of 'consultancy fees,' or whatever type of bookkeeping entry the bank might choose to get it off its books. This is the way our clients would like to finance the foreign travels in which they increasingly expect to indulge themselves in their golden years. Now our questions are these: do you think that the bank in the Caribbean with which you have that special relationship would have the facilities, one, to provide these loans now, and, two, to handle such confidential travel assistance services in the future?"

Danny answered their questions with a single word, "Yes."

◆   ◆

The new arrangements were applied to the next shipment of silver from Reno. Twenty-five percent of the proceeds on the silver sales were returned to Amaretto and Sarnoff for cash payment to some of their Nevada clients; the other 75 percent was forwarded via Nassau to the bank in the Cayman Islands and credited to the accounts of the two clients.

Their names were Lenny De Niro and Roberto Salgo. Danny had to know because he set up their accounts. The signature cards went from the Cayman Islands to Nassau to Philadelphia to Nevada and came back, delivered by Amaretto in person. Before Danny forwarded them to Nassau he must have looked at the signature cards ten times. Danny reasoned that though these two men might be the biggest thieves in Nevada, there was no cause for him to worry. If they were ever going to present problems, they would be problems for Amaretto and Sarnoff.

Right?

◆   ◆

Over the next eight months the coin volume coming out of Nevada suddenly stepped up appreciably, as did the balances credited to De Niro and Salgo. Then in March 1968 the first loans were made to the two new Nevada clients of that bank on the Cayman Islands. Again, it was through Danny Lehman that the loan papers were forwarded, executed, and returned. The bank took a 10 percent fee up front and charged 16 percent interest on the thirty-year loans. The bank kept half of the interest and recredited the other half to its clients' accounts. The front-end fee the bank split with Danny Lehman. It paid it into his account in Nassau as a commission. Danny Lehman then did one more thing. He asked a friend of his to check around, for no real reason, perhaps, other than that the names Salgo and De Niro had set something in motion in his mind, to find out who was running the biggest casino operations in Nevada: the Sands, the Dunes, Caesars Palace, Raffles, the Circus; in other words the top operations. Mr. Lenny De Niro and Mr. Roberto Salgo, it turned out, were managing director and assistant managing director, respectively, of Raffles. Danny, who had never been in a casino in his life, also learned that Raffles, while not the biggest and not the best, was, perhaps, *the* casino in Nevada with the most unsavory reputation. A cautious man, Danny advised his friend to forget about the whole matter. The only two parties who could link Danny Lehman and the crooks who ran Raffles were Amaretto and Sarnoff, and an executive of a bank located on an obscure island in the Caribbean. Surely none of them were likely to talk. He was safe.

About three months later, Amaretto paid another midnight call to Danny Lehman at his home in the suburbs of Philadelphia. He delivered two large packages, each containing one million dollars in cash; paper money, mostly fives, tens, and twenties, with some hundreds. They were, he said, intended for deposit in the De Niro and Salgo accounts respectively, and he had been told that further such cash deposits would be forthcoming on a periodic basis in the future. In all cases they would be hand-delivered by either Mr. Sarnoff or Mr. Amaretto.

In early 1969 Danny Lehman added up the cash flow, less

10 percent, that had passed through him to Mr. De Niro and Mr. Salgo: it amounted to just over twelve million dollars, or six million per annum, on average. Obviously they had moved well up from merely skimming slots.

3

On March 12, 1969, the board of directors of Raffles Inc. met in the company's corporate headquarters in the new part of Los Angeles known as Century City. The boardroom itself was wood-paneled, the board table solid oak, and the artwork on the walls subdued. The directors of the Wells Fargo Bank would have felt perfectly at home there. But it was the board of a Vegas casino, Raffles, that gathered there this day, and the seven members of that august body of that not quite so august corporation faced only two matters on the agenda. The first was a report by management on the company's activities and financial results for the fiscal year ending December 31, 1968; the second was an address by the corporation's general counsel on matters that had not been more closely defined on paper.

Where most companies are concerned, it is usual that the members of their board of directors gather for regular meetings at a reasonably early hour in the morning, say nine o'clock on the East Coast and maybe closer to ten o'clock on the West Coast. Not the board of Raffles Inc. It met at five o'clock in the afternoon. For most of its members tended to suffer from severe hangovers in the morning hours; then needed a late lunch to get them going; then a picker-upper around four-thirty—all this in order to brace themselves sufficiently for the quarterly fulfillment of the obligations incumbent on them as holders of high corporate office. Thus on March 12 the meeting had been scheduled for five o'clock, but owing to late arrivals it was not until around a quarter to six that the chairman of the board picked up his gavel and smacked the oak table. This is something that board chairmen rarely, if ever, do to call a meeting to order. It was, however, a practice that Mario Riviera had decided to introduce in 1967, since it was the closest he would ever get to achieving, or at least publicly emulating, that station in life to which he most avidly and secretly aspired: being a judge. Once the members were finally seated, Chairman Riviera motioned to the two men who had been left standing at the foot of the long oak boardroom table to join him.

The two exchanged glances and just stood there for a while before slowly, deliberately, and, it seemed, provocatively moving forward. When they finally made it to the head of the table, instead of taking chairs they just stood behind the chairman with looks of surly arrogance on their faces. It was they, not the chairman, who now dominated the boardroom, their presence reeking of menace and latent brutality.

Turning to them, Riviera said, "Gentlemen, if you don't mind? I'd appreciate your sitting here, Mr. De Niro, and you there, Mr. Salgo." "Here" and "there" were to the right and left respectively of where Chairman Riviera was sitting. When the maneuvering was finally completed, a hush settled over the room. Chairman Riviera lost no further time. "I call this meeting to order. Our managing director, Mr. De Niro, has the floor."

De Niro rose. "I've been asked to come here to report on the results for the preceding fiscal year. The results were not good. In fact, they were very bad. We lost almost two million dollars. The final figure will be available from our accountants in due course. The loss this year will probably be higher." He sat down.

The first reaction around that oak table was profound silence. Then a movement of eyes began. First they looked across the table, then moved up in the direction of the chairman, then down to the colleagues sitting near the foot of the table. Nobody looked at De Niro. The basis for all this silent eye-moving was a desire among the members of the board of directors of Raffles Inc. to find one among them who could come up with an appropriate response to the words that had just been uttered by the thug now in their midst: Lenny De Niro.

The spokesman who emerged was from St. Louis, the city where the Western Pension Fund of the Teamsters' union was headquartered, and what that union's representative now said was to the point: "How the fuck is that possible?"

Lenny De Niro immediately got to his feet again and, without the slightest hint of apology in his voice, said, "I don't particularly like to be addressed in that tone. Understood?" He glared at the man from St. Louis for a full five seconds. Then he continued, "All right. The explanation is very simple. Business is lousy for everybody. It's just worse for us. If I knew why, I'd fix it, wouldn't I?"

What De Niro was obviously trying to indicate was that if market conditions in the gaming industry in the state of Nevada were so adverse during the fiscal year of 1968, how could a member of this board of directors have reasonably expected him to pull off a goddamn miracle? The lack of any signs of agreement around the board table seemed to indicate that such reasoning, such a line of logic, such an excuse, might have been appropriate had Raffles been part of the shoe industry, even the trucking industry. But the gambling industry?

It was that same man from St. Louis, a Teamster, and thus closely affiliated with the trucking industry, who again spoke up. "I know the guys at the Sands," he said. "I know the guys at the Dunes. They're not losing money, Lenny. Something's wrong here. How the fuck can you lose money when every game in the house's rigged against the jerks that come into your place?" He was a huge, beefy man, and his words came out in a rasp, the product, no doubt, of smoking cheap cigars, two of which were sticking out of the breast pocket of his off-the-rack brown suit.

De Niro answered in a steady voice, "I don't know exactly what you're trying to imply, George. I hope nothing. The whole country is in a recession, not just Nevada, not just Vegas. If you think you can do better, be our guest. Right, Roberto?"

Roberto Salgo just glared at the bigmouth from Missouri. At that point the door opened and in came a man in a pin-striped suit wielding a large attaché case, and obviously in a state of extreme agitation. This was a highly welcome interruption for Riviera, who, after all, had been responsible years ago for bringing in De Niro and Salgo as the operating heads of the casino. So it was with relief in his voice that Riviera stood up, both to offer due recognition of the entry into the boardroom of the corporation's general counsel and, at the same time, to say, "Gentlemen, I think the first part of the agenda has been covered. Let's move on to the second."

When the last syllable had left the mouth of the chairman, De Niro and Salgo looked at each other and then defiantly around the board table. Nobody, not even the director from St. Louis, was going to object to the chairman's proposal. So De Niro and Salgo got up and, with measured steps, disappeared out the boardroom door.

Their next stop was Harry's Bar, where both tossed down two back-to-back bourbons in silence, and simultaneously lit up two fine Montecristo Cuban cigars. Then, as if it was part of a regular ritual, both reached out and shook hands. After all, it had been no small accomplishment: they had managed, once again, to get away with a year of theft on a truly grand scale, an achievement which, by itself, gave no cause for special recognition in the United States of America in the 1960s. Tens, no, hundreds, of thousands had done the same. But what gave them cause for self-congratulation in that bar in Century City on March 12, 1969, was that they had stolen such vast sums, not from mansions in Beverly Hills or from customers of brokerage firms in New York, but rather from some of the biggest crooks not only in the United States, but in the entire Western world.

Upstairs on the seventeenth floor of the same Century City tower building that housed Harry's Bar, the general counsel had moved to the head of the table and taken the chair so recently vacated by Lenny De Niro. In contrast to his predecessor, he had arrived armed not only with mental notes, but with pound upon pound of documentation, which he proceeded to remove from his oversized attaché case. The very last thing he extracted from it was one single sheet of yellow paper.

Since Chairman Riviera did not seem to be in the mood to make any more introductions that day, the lawyer simply took command by holding up that piece of paper and talking. "Gentlemen, these are notes reflecting the precise key words of an oral and still totally confidential statement which was made to me two days ago by the Gaming Enforcement Division of the Casino Control Commission of the Sovereign State of Nevada. It's not exactly good news. The chairman of the Gaming Enforcement Division said, and I quote, 'It's known to us that three of the men currently on the board of directors of Raffles Inc. are acting in a "fiduciary capacity" for organized crime syndicates from three separate geographical areas in the United States.' End quote."

The room was once again very quiet. The dapper attorney then stretched his thin, rather bluish lips in the semblance of a smile and said, "That was not good news, was it? No. But now I have slightly better news." He paused for effect. "The chairman of the Gaming Enforcement Division also said that he did not intend, in his words, 'to reveal the identity of these individuals at this stage of the investigation.' He in fact further stated, and I quote, 'Maybe they must never be revealed by us.' End quote.''

The mood in the room seemed to improve; it were as though a light had been switched on. But then the lawyer continued, quickly dimming that light.

"As always, I'm afraid, one must pay a price for better news, because, and now listen carefully, you guys," he said, slowing the pace of his delivery, and articulating each word with a threatening emphasis, "the only way that the names of these individuals will not be brought before the public—that is, the only way that Raffles Inc. and its one and only operating unit, the Raffles Hotel and Casino of Las Vegas, Nevada, will not end up being worth zilch—and again I ask you to listen very carefully, is if these three unknowns will voluntarily agree to divest themselves of their holdings in Raffles Inc. This has to be done within a reasonably short time, to be specific, within sixty days of this board meeting."

The lawyer then looked around the room and said, "All right, we need three volunters. Who are they?" No one in the room moved.

"Let's try another approach, a fresh approach, one which I can only hope and pray will work for the sake of all of you. I stress: *all* of you. Because, believe me, *my* interests in this entire situation are limited and becoming more limited every minute. Let me suggest that perhaps the three unknowns in this room could, without prejudicing their positions because what transpires in this room this afternoon would go no further than the confines of this room—"

At this point the man from the Teamsters' union in St. Louis said in a voice that was equally quiet and threatening, "Sidney, get to the point."

"I will," the lawyer replied, "in my own way.

"Now," he continued, in an even more deliberate manner as if to rub it in, "it seems to me that the gentlemanly thing for

these three particular individuals who are acting in a fiduciary capacity for outside interests to do would be to *buy out* the other four partners sitting around this table, the clean ones if I may so describe them; to buy them out at a fair appraised value. I'll arrange for the appraisal to be done. And to be done not within sixty days, but better yet within thirty days. Then these three unknown individuals who are acting in, and I again quote, 'a fiduciary capacity for organized crime' could fight their own fight with the fucking Casino Control Commission. Any thoughts on this so far, gentlemen, whoever you are?"

Silence.

"All right," said the lawyer, "I quit." He put the single piece of yellow paper back into his massive attaché case. He rose, nodded first to the chairman and then to the other members at large, strolled to the door, and disappeared.

The dilemma facing the men remaining in the boardroom was the fact that all were, in some way, shape, or form, operating as fiduciary agents for people who, by almost any definition, would be regarded as participating in organized crime. While all obviously knew that about one another, none of them knew what the Nevada Casino Control Commission's Gaming Enforcement Division knew. So why volunteer to divest if their name was not on the hit list?

Raffles Inc. was still alive, if not exactly well. Who could know what it might be worth if this current little problem was solved? To divest now, which was really just a nice word for selling out for a song, would be crazy. So there had been no sense in anybody standing up the first time, had there? The same logic applied to the second so-called solution that their ex–general counsel had proposed, namely, his suggestion for an internal takeover. Why would anybody throw good money after bad if he was among the three on the hit list in Nevada? They would close down Raffles Inc. anyway, wouldn't they?

Now, Chairman Riviera might not have had too great a knowledge of how to run a corporation, or too much style in running a board meeting, but he was able to recognize a situation that desperately called for a solution. He explained himself very clearly and succinctly. "We've got to find somebody who will buy this fucking company," he said, "and take *all* of us out." He then added, "Quick."

The man from the Teamsters' union decided to seek further clarification. He did so because the Teamsters had seven million bucks in Raffles Inc., in the form of unsecured loans. If Raffles Casino got closed down and became an untouchable, and if, as a result, this seven million never got repaid, it would mean personal trouble for him. For not only had he personally sponsored this loan in his capacity as a high official in the union's pension fund, which acted as the lender, but he had also personally taken a finder's fee on the loan from the borrower, from the hand of Chairman Riviera to be precise, of a quarter of a million dollars. If Raffles went into Chapter 11, and people started to sniff around . . . well, he had already concluded that there was no sense sticking his head in the sand; if he was going to be thrown into the Detroit River he might as well have it done in warm weather. So he decided to speak up then and there.

"Riviera. Who wants to buy a casino that's losing millions of dollars a year? Who wants to touch a situation which even the company lawyer has taken a walk from? Like nobody in his right mind! And so who's going to pay me back my seven million bucks? If the company can't, I'm afraid you guys, let's face it, have a moral responsibility here, especially you, Riviera." He hated to push a guy like Riviera this way, but there was not much left to lose, he reasoned.

"I've got a man who can sell this place," Riviera now stated, totally ignoring what had just been said. After all, who could blame a guy for being afraid of Jimmy Hoffa?

Led by the smartass Teamsters man, they all laughed.

"I'm telling you, I've got this guy who's a Hollywood agent. He's always selling crap to suckers."

The other board members looked at him. Hollywood agents were the sleaziest of all merchants of deadbeat properties to be found anywhere on the face of earth. Everybody knew that. Middlemen from Lebanon, purveyors of tax shelters from Texas, even Swiss lawyers could not compete. So if there was a last chance for the sale, it had to be through a Hollywood agent.

They were right, although at first it didn't seem so.

4

The agent that Chairman Riviera had in mind was Mort Granville. Mort Granville ran so true to form as a Hollywood agent that Riviera had to make five futile attempts to get hold of him before Granville condescended to return his call. One problem with Hollywood agents is that they have no choice but to act in a totally stereotypical fashion; otherwise they lose their mystique, and if they ever lose that, there'll be nothing there. Like Oakland.

Or to put it another way, *everybody* is overqualified for the job of Hollywood agent. So when Riviera did finally get hold of the wonderboy of Wilshire Boulevard, not only was he asked to join him for lunch the next day, a great honor in and of itself, but he was commanded to join Mort in his booth at the rear of the patio of the Polo Lounge of the Beverly Hills Hotel. Riviera ordered a Cobb salad and iced tea, so Granville, feeling that such understatement could hardly be left unchallenged, did the same. They had barely ordered when Riviera blurted out the truth: he needed a buyer, quick, for Raffles Inc., and the sellers were not too fussy. The operative word was quick!

After Riviera repeated the word a second time, Granville motioned to the Philip Morris–type bellhop and had a phone brought to the booth. It was barely one o'clock in California and so he got his man in New York without any trouble or delay. The conversation, however, lasted two minutes. Mort ended it by slamming the receiver down without saying good-bye. That was the way he ended a lot of phone conversations.

"No dice," he said to Riviera. Then Granville continued, "Since you sounded so desperate I decided to start immediately at the bottom of the barrel. That was a guy from New York," and he went on to name one of the most prestigious investment houses on Wall Street. Mort had once met the guy at a party and then fixed him up with an aspiring singer from Hong Kong who had proceeded to feed him Chinese food and sleep with him, intermittently, for three days. He had dizzy spells for weeks afterward. They were due to his being allergic to MSG, he found out later with great relief. The next time he was in L.A., Mort had fixed him up with an Italian actress. She had given him the clap. But the guy still stayed in touch. Weird, even for an investment banker. The point was that Mort, despite these past "favors," had never asked for anything back. Thus the phone call.

"What'd he say?" inquired Riviera.

" 'Are you kidding?' that's what he said," the agent replied. "He also said, 'They're shutting Raffles down.' " Then Granville added, "Riviera, you might have told me about that shutting-down aspect of the situation. It's pertinent, you know."

"What now?" Riviera asked, ignoring as usual what he didn't want to hear.

"Don't ask me," replied Granville.

At this point the agent looked at his watch and, as if acting upon a printout that had appeared there, jumped to his feet, telling Riviera that he was already late for his next appointment. Riviera was left with the check and with the reality that the word must have already gotten out that Raffles Inc. was indeed dead, because if Mort Granville walked away from a deal that theoretically would have brought him 10 percent of anything, after just one phone call, there was probably no hope left in this world.

For Riviera this was not good because, in contrast to the other members of the board, he did not act in, as the Casino Commission guy had put it, "a 'fiduciary capacity' for organized crime." In this case there was no fiduciary necessary. He *was* organized crime, where the southern half of California was concerned. Therefore the looming debacle would mean more than a mere loss of money; it would involve an irrevocable loss of prestige among his peers, and that could mean a lot more than the lousy five million dollars he had invested in the fucking casino in the first place. The fact that a shit-faced agent had just walked out on him, and just after a shit-faced lawyer had done the same, could hardly be viewed as anything less than a portent of what lay ahead.

Riviera should have had more faith in his fellowman. After all, the human race is not made up exclusively of agents and lawyers. Yet.

◆   ◆

The next day Mort Granville, being an agent, and therefore being in touch with all that was trendy, decided to get into silver. Sure, he had heard that a lot of guys were doing it: like the anchormen around town; production types at the studios; entertainment lawyers; i.e., men who were always state-of-the-art–oriented. But it was not until that moment that he decided to go for it.

The final decision did not come as a result of any revelation as he drove his Porsche back to the office from the Beverly Hills Hotel. Rather, it had been spurred by such a mundane thing as a full-page ad he had seen that morning in the financial section of the Los Angeles *Times* announcing in a somewhat gaudy manner the grand opening of the first West Coast office of the American Coin, Metals, and Currency Exchange. By coincidence, about an hour later, when his secretary had brought in the mail, it had included an invitation to attend a cocktail party at the new coin place, which was going to be hosted by the man who owned the entire nationwide operation, a certain Daniel Lehman from Philadelphia. The explanation for the invitation, Granville figured, was that somebody must have sold the coin and metals guy a list of the names of L.A.'s emerging elite. In fact, his name came from a listing of everybody in L.A. who owned a Porsche.

As far as Mort Granville was concerned, with these new offices just three blocks up Wilshire Boulevard from his own office building, if it turned out to be a bummer, he would waste no more than a five-minute walk up Wilshire Boulevard and a five-minute walk back. Even an abortive trip by foot might do his body good because of the deep breathing he always practiced when walking. It was a good day to breathe, one of those days without even a trace of smog in the air, a day when you could walk up Wilshire Boulevard, see the mountains in the distance, and *have* to say to yourself with true conviction, "Thank God I live in Los Angeles!"

◆   ◆

When Mort Granville got to the new branch of the American Coin, Metals, and Currency Exchange, the first thing he did was look at the store front. Not knowing a damn thing about coins, metals, or foreign currencies, what he looked at was the color scheme, and he liked it. There were those old collectors' gold coins, American eagles, displayed elegantly beside some bluish currency which, upon closer examination, turned out to be German marks. Mexican silver coins were laid out around some red and very large bills. When he squinted and took a closer look, it seemed to him that they were Swiss francs. What caught his eye most were the bars of silver bullion, the status investment Mort Granville had concluded he should seriously consider making, because though he was still "just" an agent, there was very little doubt in Mort's mind that within a matter of years or perhaps months he would become a producer, then a "major" producer. After that, his posting as head of a studio could not be far off. Looking ahead to the time when he would be running Columbia, or at the very least Twentieth Century-Fox, it would be rather nice if when moving into his new offices he could have various bars of silver, preferably very large bars of silver, to put on the bookshelf behind his desk.

With this vision of the future in mind, Mort Granville entered the premises to be greeted by a fairly subdued crowd of perhaps a hundred people. He gave them a quick scan and came up with nobody of importance. Perhaps it was still too early for the entertainment lawyers and the financial planners.

It was then that a short, chubby, balding man grabbed his arm and introduced himself. "You look lost," he said. "I'm Danny Lehman. What do you think of this place? Interesting, isn't it? I own it and I'm damn proud of it. Let me show you around." It all came out in one breath as if memorized.

Mort Granville did not appreciate the laying on of hands, but somehow he immediately liked the little guy. Be kind, he thought. "Splendid place, Mr. Lehman," said Granville, as they moved from counter to counter, display case to display case. "You are not by any chance connected with Lehman Brothers in New York, are you?" This guy sure didn't *look* like it, but you could hardly go by that in L.A.

"No connection. And call me Danny," Lehman replied. "You're a coin collector?"

"No, I'm an agent."

"A real estate agent?"

Now, that was a very low blow, but when Granville looked at Danny's face he saw a type of innocence which indicated that this remark had hardly been meant in a vicious way. "No, actually I represent people here in the industry: like entertainers, actors, actresses, screenplay writers, you know, that sort of business. My name's Mort Granville."

This truly impressed Danny Lehman. For despite the fact that he was now a relatively rich man he had never in his life met anybody from this type of industry. This time his face expressed an obvious pleasure at having the opportunity to mingle with a man who mingled with the stars. Who knows, he thought to himself, maybe someday this simple boy from Philadelphia might be mingling with the same sort of crowd.

What he was thinking showed. Mort Granville, who appreciated any admiration from any source, decided to humor the little man further. "It seems that you're off to a good start here, Danny. How many shops have you got like this?"

"This is the seventeenth."

The agent's condescension visibly lessened. "And how long have you been in the coin business?" he now asked.

"About ten years" was Danny's answer.

"What's your turnover? You don't have to answer, but I'm curious about your type of business."

"About sixty million dollars this year, I estimate," answered Danny, nonchalantly.

"Sixty million!" Then as a further comment he whistled quietly through his teeth.

"Let me tell you, Mort, it's been a good business, but it's going to be a hell of a lot better in the future."

"Why's that?" asked Granville.

"There're two reasons. The war in Vietnam means the dollar's got to start going to hell sooner or later. Then everybody'll want to get out, won't they? Some'll want real estate, some'll go for other tangible things: paintings, that sort of crap. Some'll go to Switzerland to buy gold illegally. But the smart ones'll come to my place and legally buy Swiss francs, or German marks, or Japanese yen. The best of all, though, is going to be silver, especially if you're a bit of a gambler."

"Gambler? What do you mean by 'gambler'? Can you explain that gambler bit on silver?"

Danny did so at length, explaining that a lot of crazies were starting to get into this silver business. Not pension funds or Wall Street types, or banks, or insurance companies. The people who were going into the silver business were almost all individuals. And they were all doing it on an emotional basis. They were convinced that they, and they alone, knew what was coming. And the only way to survive it, they figured, was to trade in money for the real thing, for real coins, real bags of silver, real bars of bullion. The *real* crazies among the crowd were going into futures, buying silver on margin. Not the kind of margin you have to put up for stocks, but only 10 percent or even 5 percent margin. Leveraging to the hilt. That made for wild fluctuations in the market because one day they would double their money, even though the price had only moved up 10 percent. Then they would dump it and sell out, and the same thing would happen in the other direction: the price would move down 15 percent and in the process wipe out ten thousand speculators, all this within one week.

"So it's really just a roller coaster," Danny pointed out. "The stakes and the action are a lot bigger than you can find anywhere over there in Nevada. But that's what makes for horse races, which I love. That's also what makes the profits for people like me who're in the silver business, because it's like having a new type of bait attracting the same old suckers." Danny paused for breath. Then he finished his spiel. "Everybody's basically a gambler," he said, "and what we've got here is a new type of casino. You're a smart guy, you know in the end it's always the house that wins. Well, in the silver business I run the house, so I make a lot of money."

God must have sent him here, Mort's internal voice was saying. And if God also sent *me* here, He would want me to be open and honest. So Granville asked, "Want to buy a casino?" These were words that would lead to a total revolution of the gaming industry in the United States of America, but it did not seem that way at first. For Danny Lehman's reply to Mort Granville's question was "Nah, not really. I visited Vegas a few weeks ago for the first time, and you know something? That place has no class. It seemed to me, just looking around, for about forty-eight hours I guess it was, that those places are run by a bunch of amateurs. No public relations. What do you think of when you think of Las Vegas? I'll tell you: call girls, bad drinks, windowless rooms, sleaze. Right?"

Granville agreed.

"That's no way to push a product or to run a place of business, is it? People like clean places. Look at my place, how clean it is. Got me?"

"You may be right," said Granville. "But just because that's the way things are over there right now, that doesn't mean it can't change." Mort Granville could feel there was something here, maybe something big, something that had to be followed through as only he could do; something that would then end up as a really viable deal of which he would get 10 percent. Intuitively, to keep it going, his finger suddenly lashed out and pointed at a large silver bullion bar. "How much does that cost?"

"Four hundred dollars, give or take," came the answer.

"I'll take two," said Granville and whipped out his checkbook ready to close the deal on the spot. After an exact quote had been given by the counterman, Granville filled out the check and handed it over to Danny with his calling card. "Hold the silver for me, will you," he said. "I'll have it picked up later." Not too many people got mugged on this stretch of Wilshire in broad daylight, but you never knew. Better his secretary than him.

As Mort Granville turned to go, Danny Lehman's hand once again came down on his sleeve, but this time very, very lightly. "Just one question, Mort," said Danny.

"Yes?" and now that tingling that Mort had felt just a few minutes ago became more intense. "Yes?" he repeated.

"What's the name of that casino you have for sale?"

"Raffles," answered Mort.

The owner of the American Coin, Metals, and Currency Exchange nodded and said, "I think you might hear from me."

◆   ◆

It was three hours later when the party was finally over and Danny Lehman was able to head back to his hotel in a cab. It did not take him to the Beverly Wilshire, or the Beverly Hills Hotel, or even the newly opened Century Plaza. No, he roomed at the Westwood Holiday Inn. That was his style. Danny knew it and it did not bother him in the slightest.

Raffles. He was more than tempted and he admitted it. But it was risky. The temptation arose, naturally, from the fact that he knew something about Raffles that nobody else did: De Niro and Salgo. But the risk came from the same knowledge. He, Danny Lehman, was acting as a fence for the not insignificant amount of silver coinage that Lenny De Niro and Roberto Salgo, the co-managers of Raffles, had filched from their own casino over the years. If that was not bad enough, that type of grand theft was peanuts compared to what had started coming in recently in the form of bills. If the middlemen, Joe Amaretto and Sam Sarnoff, had lied to him and had indeed informed their clients about the identity of "their man in Philadelphia"—their man who was fencing the skim; their man who had also set them up with that bank in the Cayman Islands which was enabling them to make fools of the IRS—then he, Danny Lehman, might disappear for a minimum of thirty years, or he might just disappear, period, depending upon who got to him first. At some point a go or no-go decision would have to be made. It would involve the biggest such decision he had ever taken in his life. If he went ahead, he might really become somebody. Or, let's face it, he might blow everything.

But what was there to blow, actually? They've never let you have a real crack at anything. Coin stores! Big deal. When everything went bad in Philadelphia after his father's death and he'd had to spend his Easter school vacations with his grandma in Atlantic City, she did not talk to him about *coin* stores.

Together they used to walk down the boardwalk, trailing behind the really rich people as well as the prostitutes and con men who were trailing them. "One day *you'll* be leading the Easter parade in Atlantic City," she would tell him.

Danny, though a very successful coin, metals, and currency man, had not restricted his education in money matters to these three areas alone. He, like almost everybody else in the sixties, had been lured into the stock market: first into mutual funds and then into opening a trading account at Merrill Lynch. He was familiar with such concepts as mergers and acquisitions and, in fact, had made good money on some merger arbitrage deals—Fred Carr's takeover of National General Insurance, Loews buying Lorillard, ITT bidding up the price of Hartford Insurance before absorbing it—so such determinants of the market value of a company as annual sales, net asset value, or price-earnings ratios were nothing new to him. Since he was an independent businessman, it was the last measure which always impressed him most. Who cared about assets if you could not make money out of them? Who cared about sales if they did not yield profit? No, when you bought something, you bought profit. Not just potential profit, either, in Danny Lehman's scheme of things; none of that blue-sky stuff. If you bought something, you wanted it to make money now and you wanted it to make a hell of a lot more money later. That's what capitalism according to Daniel Lehman was all about. He intuitively wanted an immediately positive cash flow, but *that* one even he'd never heard of yet.

So he mulled all this over in his head as he sat there on the sixth floor of a motel near the San Diego Freeway in the Westwood section of Los Angeles. The crazies these days were paying prices for companies amounting to twenty-five times earnings. Even dogs were selling at ratios of fifteen to one. Raffles, from the transparent eagerness of Mort Granville to get him interested, was obviously not just a dog, but a very sick dog. Therefore, ten times recent earnings would be the maximum to which a reasonable buyer would go.

He knew that six million dollars per annum, on average, had been stolen from Raffles during the past couple of years. It would seem logical that if Raffles was on the block to such a degree that an agent like Mort Granville would offer it to a man he'd met only five minutes earlier, you had to conclude that the casino was not only in deep trouble because it was losing money, but probably also in jeopardy where its license was concerned, because the whole place must be run by crooks from top to bottom. But, he reasoned, who cares about the licensing status? The guys who owned the place obviously wanted to get out, and to get out before the ax fell. Once they were gone, there would be no reason for the state of Nevada to quarrel with the new owner, since there was nothing wrong with Danny Lehman's reputation. On second thought, there *would* be nothing wrong with Danny Lehman's reputation so long as the De Niro–Salgo–Amaretto–Sarnoff relationship with himself remained as it was at this particular moment in time—unknown.

But back to the money angle. How big a loss was the casino showing officially? A million? Two? It was probably two million, maximum, Danny concluded, because Salgo and De Niro were not dumb enough to steal so much that the financial viability, the very existence of the casino, would be endangered as a result. That would amount to the killing of their golden goose.

So what were the *real* profits of Raffles? And based upon those profits, what was the value of Raffles? The calculation that Danny came up with was this: six million dollars' skim less the average losses per annum of the casino for the past two years, say two million, equaled four million, less the tax that would have had to be paid after depreciation, loss carryovers, and all, say a half million, equaled three and a half million, times ten, equaled thirty-five million dollars, which, realistically, would end up at, say, forty million. That's what Raffles Inc. was worth to Danny Lehman, subject to a look at their books, of course. Which raised a new problem: if he went back to that guy Granville and asked him for specifics on the financial status of Raffles, that is, if he asked for copies of the financial audits of Raffles during the last two or three years, invariably when this request was passed on to the people who owned Raffles his name would come up. In spite of the shape they were in, they were hardly going to hand that stuff out to just anybody. But then would come the moment of truth: if his name came up and rang no bells anywhere, fine. Then he would make an offer.

However, and this was a very big however, if the bells *did* go off, he would have to pull out of Los Angeles and stay out of Los Angeles for a long, long time. But think it all the way through, he said to himself. Who could possibly know of him and thus be able to ring those bells? Answer: four guys, none of whom had anything to do with the ownership of Raffles and none of whom wanted their boat to be rocked either. Since neither Amaretto and Sarnoff nor De Niro and Salgo would want to risk having him entangle them in anything, and so long as their paths, which had remained separate in the past, would also remain separate in the future, once this current brief interlude had ended, well, nothing would happen to him or them. Right? So . . .

◆   ◆

At nine o'clock the next morning Danny Lehman rang Mort Granville at his office. Naturally Granville did not take the call, but he returned it six minutes later, a fact which if it had ever gotten out would have ruined the man in that town forever. Danny's request was simple. "Get me the financial statements."

The answer was equally simple. "I'll have a messenger over at your hotel with them within the hour. Where are you staying?"

The voice on the other end of the phone went a little quiet when Danny mentioned the Holiday Inn as the place to which the delivery should be made. But then the cloud seemed to lift, since Mort Granville's twisted mind immediately came to the conclusion that he was dealing with a very cagey operator who understood the impact of understatement. First Riviera's Cobb salad and iced tea; now the Holiday Inn. Was this a new trend that he had failed to spot?

5

Exactly three days later the board of directors of Raffles Inc. reconvened. That something truly extraordinary was afoot was apparent from the time of day that this event took place: ten o'clock in the morning. The strain showed on the paunchy, sallow faces of the men there assembled. But in contrast to that late afternoon meeting a few days previously, there was a mood of guarded optimism in the air.

The chairman had personally phoned each of them to say that he thought he had found a "solution." Since nothing more was said, the normal reaction would have been unbridled skepticism. But they all knew that this time the chairman's reputation was on the line as it perhaps had never been before, and that if he was going to interrupt the daily pattern of their lives to the extent of calling a board meeting in the morning, then there must be something of true substance in the offing.

This guarded optimism, however, disintegrated immediately and totally when, after they had all duly gathered, the chairman walked in with nobody else but that hotshot Hollywood agent, Mort Granville, at his side. To them Mort personified precisely the opposite of substance. And when they examined the agent more closely and noticed that he had what could perhaps best be described as a perpetual manic grin on his face, the mood of the crowd in that boardroom was on the verge of turning ugly.

For they all knew Mort. How? Because Granville, along with the rest of them, was a regular customer at that part of one of the bars just off the casino floor of Raffles known as "hookers' corner." Mort would invariably turn up once every other weekend, usually on Friday night when a new act was opening in the big showroom featuring one of Mort's clients, or a friend of one of Mort's clients, or somebody Mort would have liked to have as a client. The last class was, of course, the largest, for in spite of what Mort Granville thought about himself, the truth of the matter was that in the galaxy of Hollywood star agents, Mort Granville was not a supernova; rather, in the eyes of some, he more closely resembled a black hole.

"Gentlemen," said the chairman, "I think you all know Mort," and he paused here, hoping that no one would groan aloud, which nevertheless two did. The effect on Granville was nil; the manic grin persisted. "Mort," continued the chairman, "lay it out for them."

"I got a buyer, fellas," said Granville. "Let me tell you who he is: a guy from Philadelphia who owns a bunch of coin shops. I had him checked out. He's clean. He's got plenty of borrowing power. He's waiting outside and he's prepared to make an offer."

"What kind of an offer?" asked the Teamsters' union representative. "Nothing down and a hundred years to pay?"

Granville stared him down for a few seconds and said, "Why should I explain further, asshole? Let this man do it himself. His name is Danny Lehman. I'll get him."

Granville went to the door, opened it, stuck his head through, and motioned to Danny, who had been sitting there in the outside office eyeing the cleavage of the receptionist. It didn't bother her somehow, as this paunchy, balding little guy turned her on. So while she was fully aware that the man sitting across the room from her was copping, if not a physical feel, at least a series of highly targeted mental feels of that part of her anatomy she was most proud of, she refrained from shooting him a withering glance. Instead, she began sorting and resorting everything on top of her desk, a pattern of moves designed to display her anatomy from the most flattering angles.

Her thoughts were actually not too far off the mark, for Danny Lehman had been impressed by this leggy, bosomy blonde, just as he was impressed by Century City, by Los Angeles, by this whole new world of California. To the guy from Philadelphia it seemed almost unreal that these people from this town were willing not only to accept him as an equal, but to bow and scrape. To his credit, Danny Lehman was already perfectly aware of the reason why, the one and only reason why: money. He had money, they knew he had money, and they wanted his money. Therefore, they would do any fucking thing they had to do to get it: butter him up, deceive him, cajole, beg; you name it, provided that the end result was his money ending in their pockets. Which was absolutely fine with Danny Lehman, because already years ago he had figured out that anybody who had any illusions about life's being some kind of romantic, mystical journey was just asking for it, asking to be had. What life was really all about was money: making money, accumulating money. It all came down to the simple fact that all the best things in life, including romance, were for sale; and without money, no sale.

Danny had dressed up for the occasion. He had on a complete suit, a new shirt, a new tie, and new cuff links, which he had acquired the previous evening for a grand total of one hundred and forty-nine dollars. He walked to the head of the boardroom table, bowed his head slightly to the men present, and then shook hands with Chairman Riviera. The chairman beamed and seemed on the verge of putting his meaty arm around Danny's shoulder, as a further sign of his benevolence, before thinking better of it and saying, "It's a true pleasure to meet you, Mr. Lehman. We've heard a lot about you and we like what we've heard." Especially that he had a lot of money. "We're all eagerly looking forward to hearing what you have to say."

With that, Danny had the floor. "I guess you all know why I'm here," he began. "I want to make you an offer to buy this company. Mr. Granville here told me it's for sale. He's given me the recent financial statements and I think I know enough about the situation to come right to the point."

He then reached into the inside pocket of his suit jacket and pulled out a check. "Here, to show my good faith," he said, "is a cashier's check in the amount of five million dollars. I had it issued by my bank in Philadelphia in favor of Mr. Granville's company, which, he has informed me, will be acting as the agent in this matter. As I understand it, these funds will be put into escrow today at the Union Bank here in Los Angeles, provided we can agree on a deal." He paused, looked around the table, and sensed the latent greed present there. "This is my offer: forty million dollars." He paused again. He knew that he'd stunned them. It was only seconds later, however, when it was he, Danny Lehman, who would be stunned. In fact, beyond any doubt he became the most stunned man in the room. For during the pause he had created by throwing out the figure of forty million dollars for a casino in Vegas that was losing money, was losing its customers, and was about to lose its operating license, the door to the boardroom opened yet again and two men, both in their mid-fifties, entered the room.

For a moment they just stood there while the chairman, obviously angry, stared at them and seemed on the verge of losing control. Then he backed off. "Would you mind taking a seat back there. We'll get to you in a minute. All right?"

The two intruders chose to remain standing. It seemed to

dawn on the chairman that their lack of manners might offend the man who had just made an offer of forty million bucks for what up to now had been an unsalable property. So rather than risk giving the impression that they were anything but just one big healthy, happy, and prosperous family, he quickly addressed himself to the visitor from Philadelphia.

"I do hope, Mr. Lehman," he said, "that you will pardon this interruption. However, these two gentlemen are the managers of our casino operation in Las Vegas. That's Mr. De Niro on the left and Mr. Salgo on the right. They are, if I may say so, the best—the very best—in the business. And very punctual normally. Their plane must have encountered fog."

Visibility that day at both the Vegas airport and LAX was a hundred miles. But this guy, Riviera figured, was from Philadelphia, so how would he know that? The chairman now beamed at his two revered employees and said, "Gentlemen, this is Mr. Daniel Lehman, who has come here to discuss the possibility of assuming the ownership of our corporation. This would mean, of course, that should we achieve a meeting of minds today, he would replace me as the chief executive officer of this company. In other words, he would be your new boss."

Danny watched the two men with the utmost care. His hope was to detect absolutely no sign of recognition. For had there been just the *slightest* hint, the merest *passing* flicker of mental recollection of the name Daniel Lehman, then it was time to stop fishing, to cut bait, and to get out. Quickly!

But there was nothing. Absolutely nothing. In spite of this apparent reprieve, Danny could hardly help but focus his mind on the fact that the situation that had now arisen was nothing less than outrageous. For there, no more than thirty feet away from him, stood two men, De Niro and Salgo, who he knew for a fact had been involved in a massive looting of the revenues of the corporation he was now trying to buy. What was even more grotesque was that he, Danny Lehman, was the man who not only had been fencing that loot, but who had also, through his Bahamas–Cayman Islands connection, arranged for the return of the stolen money to the thieves in the form of "legitimate" loans. And yet there they stood: calm, self-assured, cocky; proof positive that the middlemen, Amaretto and Sarnoff, had done precisely what they had promised him, namely, keep his name, his identity, completely concealed. The question now was: if he moved toward consummation of this deal; if he gained control of Raffles Inc.; if he therefore became the boss of Lenny De Niro and Roberto Salgo—what would be the reaction of Joe Amaretto and Sam Sarnoff? He been through this thought process before, Danny recalled. Then he had concluded that it was an imponderable but hardly insurmountable obstacle. Now, he reasoned, it would once again very quickly and very simply reduce itself to money. If money could buy Raffles Inc., money could just as surely buy Amaretto and Sarnoff.

But to ensure that the waters that seemed still would remain still, Danny decided to address a few choice words to those two men who, when one came right down to it, represented not only a threat to this deal but quite obviously a threat to the future of Danny Lehman as a surviving member of the human species. So he looked toward the far end of the room and made a statement. "I'd like to go on record right here and now that, should this transaction go through, I fully intend to keep you both on as the senior management of our casino operations. In fact, I'd greatly welcome it if both you, Mr. De Niro, and you, Mr. Salgo, would not only stay on in your current capacities, but would also agree to enter into the negotiation of a two-year employment contract which would guarantee to me that a smooth and mutually profitable transition could take place between one ownership group and the other ownership group."

Danny was very nervous, and the way his words had just flooded out reflected that. But nobody seemed to notice. Where De Niro and Salgo were concerned, his words evoked no response whatsoever. No smiles of gratitude; no signs of eagerness to please the new boss. Nothing. They just continued to stand there stony-faced and silent, and vaguely threatening.

Well, fuck them, Danny thought, and turned to whisper a few quick words in the chairman's ear. The words took immediate effect, for Riviera then said, "Lenny. Roberto. I think that Mr. Lehman has made things very clear. The matters that remain concern the board alone. Do have a good trip back to Vegas." Given little choice, the pair left the room having spent a total time there of just over three minutes. Everyone seemed relieved.

The rest of the meeting was perfunctory and tedious. All that really happened was that Danny Lehman said he intended to give the sellers a one-year note for the balance of the purchase price. He insisted that the note be nonnegotiable, since, as he put it, he did not want to find himself owing money to anybody but people of the highest repute. He then suggested that it might be appropriate if he were to leave the room to allow the board to consider his offer and to formulate an appropriate response.

Ten minutes later they called him back in. Their response was simple: too little up-front cash, especially if that note was nonnegotiable; and the maturity of the note itself was way too far out: six months was the maximum that would be acceptable. In summary, what they wanted was fifteen million dollars cash now and the remaining twenty-five million in 180 days. Instead of starting to haggle, Danny referred the matter to their mutual agent. Mort came up with the not exactly original suggestion that they might split the difference down the middle, that is, ten million cash and a nine months' or 270 days' note. Both the seller, as represented by Riviera, and the buyer, in the person of Danny Lehman, pounced upon this marvelous idea with an alacrity that would have appeared suspect to even the most naïve of observers. For both the buyer and the seller were convinced, no, more than that, *they knew for a fact*, that the other party was being taken to the cleaner's.

To seal the deal before anything new could possibly come

up, Danny Lehman immediately handed over his five-million-dollar check to the agent and stated for all to hear that his lawyer would show up in two days from Philadelphia, with another check for an equal amount. He then suggested to Riviera that, following this, his attorney would contact the Raffles attorneys and start working out the details. Riviera wrote down the name and number of Raffles Inc.'s new law firm for Danny. That was it. Danny shook hands all around and left.

Once the door had safely closed behind him, everybody in the boardroom started to talk at once. The impossible had happened. The Teamsters man even went up to the chairman and pounded him on the back. All the while Mort Granville just stood to one side, his manic grin becoming ever broader as the realization sank in that within a matter of days he would become a millionaire. Then, just for a fleeting moment, one that lasted barely five or six seconds, the grin subsided and a cloud passed across Mort's face. The cause of this very temporary consternation was a question which, in a truly inexplicable fashion, had come into Granville's mind; namely, was it fair to take a nice, dumb little guy like Lehman for such a ride, one that would inevitably wipe him out?

But the response to this disturbing thought was immediate and obvious: nobody had forced Lehman to do this and maybe, who could tell, the little guy might get a little fun out of it during the next couple of months before either the losses or the authorities closed the place down. Although, Granville concluded, upon further reflection it was highly doubtful whether during such a brief period he was going to be able to get forty million bucks' worth of fun!

◆   ◆

Danny had gone directly to the bank of elevators, passing the leggy, bosomy receptionist without looking at her, without even giving her a passing thought. Likewise, seventeen stories down, he had hurried past that Italian restaurant with its cute little bar without even noticing that it was there. For Danny's mind was engaged, fully engaged, with one subject and one subject only: money; his money, in fact all of his money and a lot more. Not that there was any doubt in his mind that this was anything but the best deal that he had ever made in his life, or perhaps would ever make in his life. No, he was not worried about the forty million dollars he had to put on the line. What he was already subconsciously trying to work out was what this forty-million-dollar investment would bring to him during the next two, five, or ten years. Danny Lehman sensed that in this particular instance, past was by no means prologue. If, no, *when,* this was all worked out, his past would be discarded forever. Forgotten. Buried. But it would be an orderly burial, starting, he thought, right now.

He found a cab across the street in front of the Century Plaza Hotel. Once at the Westwood Holiday Inn, he got on the phone to his lawyer in Philadelphia. The lawyer listened for a while; first about the need, the immediate need, for another five million dollars in cash. The news produced no words, just a grunt. Then he was told about the thirty-million-dollar note. That produced an even more emphatic grunt. When, finally, the attorney was told that the note was due in 270 days, he spoke for the first time. "Danny—I'm saying this as a friend and your lawyer for the last ten years—are you sure you know what you're doing?"

Danny ignored the question. He reminded his attorney that he was fully equipped with his power of attorney and that he should go back to the bank and do what he had to do, i.e., get the second five-million-dollar cashier's check, and then get on a plane for Los Angeles. There was really no sense in any more talk on the phone. "Just get on the plane when you're done," he said, "and take a cab to the Holiday Inn on Wilshire in Westwood."

The lawyer managed to get a 5:30 P.M. plane the next day out of Philadelphia and arrived in Westwood around nine o'clock that evening, California time. When he walked into Danny's hotel room he first dropped his overnight bag and briefcase to the floor and reached into the inside pocket of his jacket for his wallet, from which he then proceeded to extract the cashier's check and hand it over to Danny.

"Danny," he said, "let me tell you straight and emphatically: I think you're making a terrible mistake here. You're throwing away, on some whim, everything you've built up. Frankly, I don't get it." There was true worry in the man's voice, and the reason for it came out in the words that followed.

"I know you're a clever businessman, we all know that. But you must realize I had to give the bank everything you've got as collateral, including your house, to get this second check. The bank likes you, likes your business, but after this, let me tell you, you now have zero credit available to you in Philadelphia, and my guess is that the same holds for any bank in the United States. You're borrowed to the hilt. You understand that, don't you?"

"It figures," Danny responded.

"Okay. So now let's go on. You've got to come up with thirty million dollars for that note, and not in ten years, or in five years, but in 270 days. Now tell me, how the hell do you expect to be able to do that?"

"I'll find a way," answered Danny. "That's my job. Your job is to close this deal and to close it quickly."

"All right," said the attorney. "You're the boss, Danny. But I'm telling you, you'll get ruined as a result of this. I look at you as a friend, and I have for years. I hate to see this happen to you. Why? Can you at least explain to me why you are doing this?"

Danny refused to answer. Although he knew, in a way, in a way that was becoming clearer by the hour. If Riviera, Mort Granville, or hoods like De Niro and Salgo could do it, then Danny Lehman could do it a hell of a lot better and a hell of a lot quicker. He had probably already wasted five years of his very valuable time, out of sheer ignorance of how easy it was! No lawyer could possibly understand that. Nothing was ever easy in their minds. So Danny just stood silently as his lawyer stooped to pick up his briefcase and his overnight bag.

"What's the next step and when?" the lawyer asked. He then added, immediately echoing Danny's thoughts, "These things tend to get complicated, you know."

"I'll make the arrangements," said Danny, "and then call your room."

◆   ◆

As it turned out, there were no complications. The new attorneys of Raffles Inc. drew up the contract, which had but one new element in it, namely, that the buyer would be regarded as being in total default should the note not be paid in full within thirty days of maturity. Danny's attorney, Benjamin Shea, did everything possible to get this taken out in the hope of leaving the whole default question as vague as possible. But it was that or no deal.

At this juncture in the negotiations he spoke on the phone with Danny and pleaded with him to take the second option—no deal—so that they could get the hell out of there and go back to Philadelphia where they both belonged. They could still get out more or less intact, he explained, because there was no question that he could retrieve the ten million bucks from that escrow account. Maybe the Los Angeles attorneys and that agent would try to play games, but he would get it back. Then they could walk away from this thing before promising something that they simply could not fulfil.

But Danny refused to listen. "I'll take care of that note," he said. "Please do what I'm asking, Ben. I know what I'm doing. And if it goes wrong, you're on record as saying it would."

Benjamin Shea was undoubtedly a friend, and considered him one also, Danny reasoned, but fear of malpractice had already spread from the medical to the legal profession, a fear that transcended all other emotional attachments, even blood relationships. Hence, lawyer off the hook, the contract and note were signed, and the resignations of the board of directors of Raffles Inc., the en masse resignations, were submitted and notification thereof forwarded to the Nevada Casino Control Commission. At the same time, the credentials of the new owner of Raffles Inc. were forwarded to the commission, along with the names of the three men who would form the nucleus of the board of directors of Raffles: Daniel Lehman as chairman; as vice-chairman, Benjamin Shea, lawyer from Philadelphia; and as company secretary, of all people the agent from Los Angeles, Mort Granville. Finally, an application was made for an operating license for the casino in the names of the new parties.

It was decided that it would be better if the latter submission to the Casino Control Commission be made in person, so it was on March 27, 1969, that Benjamin Shea went to Las Vegas to perform that final act which gave Daniel Lehman control of the Raffles Casino and Hotel of Las Vegas, Nevada. After the Nevada officials had gone through Shea's submission, they were so happy with the solution that was immediately getting Raffles out of mob control that they gave him a temporary operating license on the spot and, in essence, guaranteed that he would receive a permanent license in no more than sixty days. The lawyer had wanted Danny to come along with him in order to, as he pointed out, "at least look over the place that you have bet your life on." But Danny refused. So at the same time the lawyer was catching a PSA flight from Los Angeles to Vegas, Danny Lehman was boarding a United flight to Philadelphia.

◆   ◆

By nine o'clock the next morning, Danny had made a series of telephone calls: to the Provident National, to the Philadelphia National, and to the Mellon National Bank over in Pittsburgh. In all three cases he arranged an interview with one of the bank's senior loan officers, and since he indicated that some urgency was involved, he managed to get both meetings with the Philadelphia banks set up for that same afternoon, while the meeting with the man from Mellon was agreed upon for lunch the next day in Pittsburgh.

By one o'clock that next day Danny knew exactly where he stood. He had just paid ten million dollars down and given a thirty-million-dollar note for a property that no bank in Pennsylvania would accept as collateral for a loan of any type or any size. In none of the three cases did they even get to any discussion of property values, of profitability, of cash flow. Once the niceties were taken care of, the conversation that followed over lunch with the senior loan officer from the Mellon Bank in Pittsburgh was typical.

"Mr. Lehman, if you don't mind let's discuss business. I'd like to say that we have run a check on your operation over there in Philadelphia and we like what we have seen. Now, what can we do for you?"

"I'm trying to finance an acquisition," Danny had replied.

"We do that kind of business, of course. What exactly are you trying to buy?"

"Well, it's a little outside the field I've been in during the past ten years."

"I see."

Pause.

"It's a hotel complex."

The man from Mellon responded, "W'have done a lot of hotel financing, in fact more than a few Hiltons in the state of Pennsylvania. We've even financed a few abroad. Where is the hotel?"

"Well, it's in this country. In the West."

"I see."

"To be more precise, it's in Las Vegas. Nevada," he added, as if to pinpoint that city more exactly in case the banker was not quite sure where it was.

"What's the name of the hotel?"

"Well, it's actually not *just* a hotel. It's also a casino. Raffles. Raffles Inc. I've arranged to purchase Raffles Inc."

"Sorry, Mr. Lehman, I'm afraid that's out."

"But let me tell you—"

"Sorry, Mr. Lehman, as I said, that's out. The Mellon Bank does not finance casinos or the acquisition of casinos. For obvious reasons."

◆　　　◆

The day after Danny's meeting in Pittsburgh, Benjamin Shea was back in Philadelphia to report that everything had been accomplished with remarkable smoothness in Las Vegas.

"The meeting with the Casino Control Commission people could not have been better, Danny," Shea said, as they sat in the dark living room of Danny's house. "They were not just helpful, but downright solicitous."

"How come?" Danny asked.

"You want me to tell you straight?"

"You always do."

"All right. It had nothing to do with their welcoming you to Vegas because of your being Mr. Clean or something. To be sure, it was made quite clear that they approved of you, me, and the entire new slate of directors. But let's face it, Danny, they reached that conclusion pretty damn fast. And that's what bothers me. I could not help but detect a sense of *urgency* in getting the transfer of ownership of Raffles over and done with as quickly as possible. And now my opinion as to why. Ready?"

"Go ahead."

"Because they know there's something basically wrong with the casino itself. Financially wrong. And they don't want Raffles to get into a jam and cast a cloud over the viability of the entire gaming industry in the state, scaring off investors. The whole financial condition of Nevada is directly linked to the continuing success and further growth of the gaming industry. And they're planning on you using your money—money which you have not yet got, I might add—to bail out Raffles and thus them."

"So you think I've bought a lemon."

"A very big forty-million-dollar lemon."

"Maybe. Maybe not."

"Look, Danny. I'm not sure how, but I'm willing to try to get you out of this deal before it ruins you."

"Thanks, Benjamin. But I'm staying in. End of discussion. Okay?"

The two men just sat there in silence during the next few minutes. Then Shea spoke again. "Okay, Danny, you're the boss. But the least we should do is get to work and at least try to finance that note as soon as possible. Do you want me to try to start scheduling a few things?"

"No," said Danny. "I'll take care of that myself."

For the next thirty-two days Danny Lehman never showed up in his office on Broad Street. Instead, he kept to the neighborhood in which he lived, a working-class neighborhood, just like the one he'd grown up in. His house was even furnished the same way as his parents' had been, with heavy furniture, thick curtains, and an old-fashioned kitchen. His cleaning lady came in every afternoon except Sundays to take care of the dishes, if there were any, to make Danny's bed, do the laundry, dust. Despite the fact that he was "staying home," Danny seldom saw her. He was out taking in a matinee at the movies, or having a beer or two with the boys at the bar next door to the bowling alley. At night he watched sports on television. Not once did he return his attorney's phone calls, at least a dozen of which had been registered on his answering service as the days and weeks passed. Danny had gone into a defensive crouch: thinking, weighing up, calculating, hoping, but coming up with nothing. For the first time in his life Danny was going through the process of thinking before acting, and it was getting him nowhere.

6

At the end of April 1969, the doorbell rang at Danny Lehman's house at ten o'clock precisely. Danny, as he had been doing night after night, was sitting in his living room, alone, naturally, and thinking. That note was due in 230 days. The doorbell rang again and again. Danny, who had not even checked with his answering service for five days, was determined to let the thing ring until whoever it was went away: probably his lawyer coming to tell him yet again what a fool he had been. After the sixth ring, Danny decided that this problem also was not going to go away by itself. When he opened the door, he was not overly surprised to see it was Joe Amaretto.

"Please excuse me, Danny. I really didn't want to intrude on your privacy. But I've tried to phone at least seven or eight times. . . ."

"No excuses necessary," Danny said. "You know you're

always welcome. Come on in. Let's start with a cognac." This time there was no overcoat or hat to be taken care of. It was an early spring in Philadelphia, and a touch of warmth still lingered in the evening air. The room remained silent, neither man saying a word, as Danny went through the ritual of getting the cognac glasses and then filling them a lot fuller than he normally would have done. It was Courvoisier, a gift. He sat down across the coffee table from his guest and asked, "Do you want to start, Joe, or should I?"

"Maybe I'd better," said Amaretto.

He reached for the huge briefcase he always carried and laid it on top of the coffee table. At first he was going to open it, but then decided to just let it lie there. "There is another half million here, Danny, all big bills this time," he said, "to be processed as usual, with the exception that the entire amount is to be loaned to a new guy. I've got his name written down here." He reached into his jacket pocket and then handed over a small, folded piece of paper, obviously some kind of hotel stationery. Danny unfolded it. The name of Mr. Rupert Downey had been typed on it. That was all. Danny then refolded the paper and handed it back to Amaretto, who, for the moment, appeared not to know quite what to make of this. But he immediately found out.

"I'm sorry," Danny said, "but you'll have to find somebody else to help you out on this one, or, for that matter, any future ones. The last shipment was the last shipment, if you know what I mean. I'd have contacted you, Joe, but I've been busy and I've had things, you know, on my mind during the last month. So I'm really sorry that you had to make this trip to Philadelphia and waste your time with those phone calls, and then have to come over here in the middle of the night, but—"

"We understand, Danny," said Amaretto. "You see, we've heard."

This was hardly a surprise to Danny. Even *The Wall Street Journal* had run a half column on the sale of Raffles.

Amaretto paused for a moment before he continued.

"We've heard, Danny, but frankly we don't quite understand." Then he added quickly, "When I say 'we,' don't misunderstand what I'm talking about: I mean just the two of us, me and Sam Sarnoff, because believe me, Danny, this thing has remained strictly, I mean strictly and exclusively, confidential."

Danny broke in, "I know, I know, Joe. I appreciate it, believe me, I appreciate it. But what's this about not understanding?"

"Well, if you've gotten control or are getting control of Raffles the way the word has it, how come you're sitting here in Philadelphia? How come, you know, you're not in Vegas, taking over that place from those goddamn crooks?"

"It's too complicated to explain at the moment, Joe, but as long as we are on the subject, what are you going to tell Salgo and De Niro? And this new guy. What was his name again? Rupert Downey? Who the hell is he anyway?"

"We don't know and frankly don't care. It's just a name that Salgo and De Niro gave us."

"All right, but I'll ask you again: what are you going to tell Salgo and De Niro?"

"The truth. That our distributor in the East has folded up, gone out of business, and that we've tried to find a new one, unsuccessfully. That we've concluded that it's probably going to take a while to find a new one."

"Any thoughts on how long?"

"Oh, probably months," replied Amaretto.

Danny nodded. "I like that answer. I like that answer a lot."

Then Amaretto became very serious, leaned forward, and asked, "Now, what about those guys in the Caribbean that have that bank down there that's made those loans to De Niro and Salgo? If push comes to shove, are you going to be able to keep their mouths shut as to where this money came from?" There was a touch of fear, not just anxiety, in Amaretto's voice now.

"I think so," said Danny. "In fact, I'm pretty damn sure so, but you can never be certain."

Amaretto winced. "Maybe you should talk to them," he suggested, "soon."

"You're right," Danny replied. "I'll do it tomorrow, early tomorrow."

"Our names never came up with those guys down there, did they?" Amaretto asked.

"No, never; never mentioned," replied Danny.

"And it'll stay that way, I hope?"

"It will definitely stay that way."

"So I think we understand each other. I think we agree, don't we?" Amaretto asked.

"I think so. I think we've never met, isn't that right?" asked Danny.

"That's exactly right."

Amaretto got to his feet and then picked up his oversized briefcase, still unopened, still apparently very heavy. "I'd better take this with me," he said, grinning for the first time.

"What are you going to do with it?" asked Danny.

"Give it back to De Niro and Salgo," replied Amaretto. "It's their problem now because I think I know what I'm going to tell them. In fact, I *know* what I'm going to tell them: that not only have they lost their distributor in the East, but they're going to lose their agents in the West. Danny, Sam and I are going to get out of this business right now."

"Take it easy," Danny then said, sensing that the man's fear was developing into a latent panic; he'd seen De Niro and Salgo and he understood why. "To make it easier for you to get out I'll do that last deal for them. Leave the money here."

"Thanks," Amaretto replied, the relief flushing his features. He eased the briefcase back onto the table. As they walked out of the living room toward the front door Amaretto suddenly stopped and said, "Danny, I've got to ask you something. It might sound like prying, but I'm going to ask it anyway."

"Go ahead."

"Well, you know De Niro and Salgo are going to keep stealing, don't you?"

"Of course."

"So how are you going to stop them?"

"I don't know yet."

"Now listen, Danny, I'm going to give you an opinion for what it's worth, even if you didn't ask for it."

"Go ahead," said Danny.

"The only way you're going to be able to stop it's to kick out not only De Niro and Salgo, but also the next six guys under them. That's one big pack of thieves."

"You may be right."

"The problem is, Danny, you try that and they'll get you, you must know that they'll get you. They are so entrenched there and they're so fucking greedy and so fucking arrogant that they think they own the place. They're so fucking stupid that they don't even know that if they keep stealing at the rate they've been stealing, then the place is going to go belly up. They can't figure that out. But believe me, if you try to reason with them, or kick them out, or, worse, if you try, you know, to get them into trouble with the authorities, they'll *kill* you, Danny. I swear to God, they'll *kill* you. And if they find out about us, they'll think we put you up to this and they'll kill *us.*" It was the first time Danny had ever heard Amaretto speak with such desperation in his voice.

"Now let me tell you something else," Amaretto continued. "As I just said, I've decided that Sarnoff and I, about a week or so from now, are leaving. We're not only leaving Nevada, we're leaving the United States for a while. And when we come back to Reno, we're going to get into an entirely different kind of business, and we're never going to set foot in a casino again. You'd be well advised to do the same, Danny. Because you can't beat these guys. Nobody can."

With that he turned, opened the door, and left. He also left a chill behind that had nothing to do with the weather. It was caused by the long shadow of two men in Las Vegas.

◆   ◆

Just before noon the next day, Danny called the First Charter Bank in Nassau and contacted his man there, the one in charge of, as they so nicely put it, "special projects." He started off by explaining that something new had come up, and asked whether they could chat about it on the phone. He was informed that he would get a call back in about thirty minutes.

When the call did come precisely thirty-two minutes later, his Bahamian partner apologized profusely but said he had reason to believe that there was a tap on the bank's lines, a completely illegal tap, one that had been put in under the auspices of both the Securities and Exchange Commission and the Internal Revenue Service. The problem, he said, stemmed from some of the trust arrangements they had set up with a string of American clients, especially a bunch from Northern California. It appeared that a hotshot district attorney in San Francisco had gotten some sort of a court order that allowed them to go after these clients any way they wanted. American judges, it seemed, could unilaterally extend their courts' jurisdiction to wherever they wanted, including the Bahamas. Cheeky, but not something the Bahamas was likely to go to war over. Thus the precautions.

It was somewhat startling to hear all this explained in an impeccable Oxford accent, one that Danny knew for a fact was by no means phony, since his Bahamian partner, by the name of Montague Davies, was in fact a product of both Eton and Oxford—a law graduate of Oxford, to be more precise. The man had decided to pursue his postgraduate career in a warmer climate and from Nassau's Bay Street had proceeded to specialize in the establishment of legal façades for the most sophisticated types of white-collar criminals.

"Now, what's the problem, Mr. Lehman?" he asked, thus turning from apologies to business as usual, at least business as was usual for him.

First, Danny explained, he had one more loan to be arranged for those parties in Las Vegas, the ones from whom the funds had been coming and to whom the same funds had been going back via that subsidiary bank in the Cayman Islands, the bank through which he, Montague Davies, always made the "arrangements." It would be the last one. Whether or not Montague Davies, the First Charter Bank of the Bahamas, and/ or the First Charter Bank of the Cayman Islands wanted to continue a relationship with those clients was, of course, entirely their business. All that Danny was concerned about, all that he really wanted, was the ironclad assurance that in the future, as in the past, no mention whatsoever would ever be made of his name.

"It's understood, old chap. Consider it done," answered the Englishman.

"But if they ask at some time?"

"The Bank Secrecy Act forbids us from making any unauthorized disclosures to third parties regarding the affairs of our clients. We regard you, Mr. Lehman, as a very valuable client and have for years. We are most appreciative of the high volume of foreign exchange business that you direct to us, and hope that it will continue in the future. Does that put your mind at ease?"

"Completely. Thank you," said Danny. "Now for something else, as long as we're both on the line. I'm in the process of buying a hotel-casino complex in Las Vegas. To be precise, I've really already bought it with ten million down and a thirty-million-dollar note, which will come due in just under eight months. It's a 6 percent note by the way. I want to raise some money to cover that note. I'd agree to what I think you British call a 'floating charge' on the entire property. What I'm looking for is a loan for around five years. I'd be willing to pay a very nice finder's fee to a party such as yourself if you could introduce me to a lender. I'd also be willing to pay a nice setup fee, front end, to such a lender, and I'd be prepared to go as high as 10 percent interest, even 12 percent."

"I think I fully understand what you're after," said Montague Davies, a bit too dryly for Danny's liking.

Undeterred, Danny plunged on. "Now, I guess you know why I'm asking you. I've found out that no bank, in fact no lender I know of in this country, will touch anything to do with the gaming industry. So my question is really this, Mr. Davies: is your bank interested?"

"Theoretically, yes. But it's too big for us, Mr. Lehman. We are not that large an institution. We're a merchant bank, not a commercial bank. We're not Barclays or National Westminster, you see. Our principal activity involves arranging financing, not providing it. In any case, we certainly would not be in a position to lend thirty million dollars on any one transaction, all the more so in the United States and especially to a firm in the type of business that is subject to rather extraordinary surveillance. We have, as I have already explained, enough trouble in the States. We have to think of the clients we've already got. And we wouldn't want to do anything that would, you know, provoke the Americans even further, so that they might tend to come down on us like a pack of wolves. You must understand that."

Danny said, "Yes, of course, I certainly do. Well, I'm sorry I brought it up but I thought that maybe—"

Then the Englishman cut in. "However, I do think that perhaps we still may be able to be of some service to you in this matter. We do have associates, friendly parties, with whom we work very closely in our little bank in the Cayman Islands, the one that has been able to service your clients, or should I now say ex-clients, from Las Vegas. They are people who, I believe, understand the industry that you've become involved with. I think they would at least hear you out on a proposition. Are you interested in pursuing this, Mr. Lehman?"

"Who and where are they?"

"Well now, I don't think I can exactly tell you *that,* but I will tell you *this:* they are represented in some matters by a man, an attorney, in Miami. If you'll agree I will let him know about this

68

matter and will give him your name. You understand: your name will be mentioned. And if he is interested he will call you. Is that agreeable?"

"Yes," answered Danny.

"Just one thing, one more thing," the Englishman said. "I believe at the outset you did mention that there would be a substantial finder's fee involved should we be able to develop something for you. I trust that you will remember us should something happen?"

"Don't worry, I will," replied Danny.

"Thank you so much, Mr. Lehman. As usual it's been a pleasure doing business with you. Goodbye, sir, and good luck!"

The line went dead.

7

Ten days later, during the first half of May, Danny got the phone call. It was brief and to the point.

"I'd like to speak to Mr. Daniel Lehman," said the voice at the other end of the phone.

"Speaking."

"A mutual friend in the Caribbean told me about your special project. We're interested in it. Are you free tomorrow?"

"Yes."

"Good. Take Eastern Flight 741 to Fort Lauderdale tomorrow morning."

Pause. "All right. Where do we meet?"

"Don't worry, I'll find you." Click.

The big break? The much sought-after partner . . . ? Danny chose not to spell out the word.

◆   ◆

The airport in Fort Lauderdale was not much to look at in 1969. For the most part its buildings looked like leftover World War II barracks, and the baggage claim area was nothing more than an open shed. About the only place today where you encounter similar outdoor baggage claim areas in the United States is Hawaii: Maui or Kauai.

Flight 741 of Eastern was not even half full, since May was definitely off-season. Within twenty minutes of landing, the sixty-odd people who had been fellow passengers with Danny Lehman all seemed to have found their baggage, and after another five minutes they had all disappeared into cars, cabs, vans, and buses. This left Danny a completely solitary figure standing uncertainly on the curb adjacent to the Eastern baggage carrousel. Another five minutes passed, and then another. Stoically, Danny just stood there. Then a brown Chevrolet Impala suddenly appeared at the curb in front of him. A man leaned over from the driving seat and asked rather quietly through the open window on the right, "Are you Lehman?"

Danny answered, "Yes."

"Might as well get in, then."

Danny first opened the back door and threw in his overnight suitcase, closed the door, and then climbed in beside the driver. The man held out his hand and said, "My name is Saul Meyers. A pleasure."

Danny took his hand, and said, "It's mutual." Then he added, "I appreciate your picking me up personally."

"Well, I don't do it often, I can tell you that. But for you, Mr. Lehman, nothing but the best." He gave Danny a wolfish smile. *That* got Danny worried.

Then they drove off. They headed toward the Port Everglades area and then turned south on Highway 1, drove for maybe four or five miles, and then turned left, heading toward the ocean. In the final stages of the trip they passed through an uninhabited wilderness area. It seemed rather strange for these parts. Danny assumed it was probably a swamp that had been partially filled in and then abandoned as too remote for commercial property development. The road itself was almost brand-new. It was paved, and it was leading toward what appeared to be either a medium-sized high-rise hotel or a condominium complex—one that stood in rather splendid isolation on the very edge of the swampy meadow and the beach.

As they approached the structure, it became apparent that it was completely surrounded by a high fence topped with barbed wire. The newly paved road narrowed down between cement walls, ending at an imposing barrier flanked on both sides by guardhouse structures, both manned. As they drew up, uniformed guards with revolvers at their hips came out, approached the car, and carefully scanned the interior of the vehicle from both sides, in spite of their obviously knowing whose car it was and who was driving it. Then one of them asked, "Are you all right, Mr. Meyers?"

The answer: "Fine, everything's fine."

Both guards nodded. One went back into his bunker and the barrier lifted. The other waved them through. They were met at the entrance to the building by a doorman who immediately summoned a young Latino to take care of the car. They walked through the revolving doors and again were met by two uniformed and armed security people. Again the same question was asked: "Everything all right, Mr. Meyers?"

And it drew the same answer: "Fine, everything's fine."

They entered an extremely elegant hallway, the floor Italian marble, the ceiling high, and antique furniture lining each wall. After about twenty paces they went up ten steps and passed into what appeared to be a lounge area. The centerpiece was a huge circular seat designed to take at least thirty guests. It was now two-thirty in the afternoon and there was absolutely nobody in the huge room except the bartender clad in a white bow tie and a red jacket—maybe his idea of what a bartender would wear in an elegant English club, something that was as far removed in atmosphere and decor from this building between Fort Lauderdale and North Miami Beach as any place possibly could be. But he didn't know that.

"You must be thirsty," stated Meyers, who up to now had said very little to Danny Lehman except for making the usual remarks about the flight, about Eastern Airlines, about how early it had gotten hot that year and yet, in spite of the heat, for some reason the humidity had stayed fairly low. Florida talk.

"Vodka and tonic would go down great," said Danny.

Mr. Meyers held up two fingers to indicate that they would both have the same. Once the drinks had been served, Meyers said to the bartender, "Jimmy, I'd appreciate it if you'd hunt down a couple of cigars for me. You know the brand I smoke, and take your time, take your time."

Jimmy nodded, took a couple of swipes at the bar with a towel, and then disappeared.

"Well, Mr. Lehman, I suggest that you call me Saul. May I call you Danny?"

"Certainly."

"Okay, Danny, I know what you're here to talk about. You've bought yourself a casino, if I've been informed correctly?"

"Yes."

"In Vegas, they tell me."

"Right."

"Why would anybody be selling a casino in Vegas? Almost all of them're money machines, aren't they?"

"Almost all of them. That's right."

"But not this one?"

"Not this one, or at least on the *surface* not this one."

"I see. And you know what's going on under the surface, is that right?"

"Correct."

"What's the name of the casino?"

"Raffles."

"It figures. That's what we thought. That dumb bastard Riviera wouldn't know how to make money if somebody handed him the goddamn United States mint. Okay, tell me more."

Danny Lehman told him more. In fact, he told him everything, all except for the details about *how* he had come to know that the profitability picture of Raffles was radically better than it appeared on the surface. He waited for Saul Meyers to insist upon his going into it further, but he found out that there was nothing to worry about.

"How you know all this, Danny, is your business. All I can assume is that you are telling us the truth, because it won't do anybody any good, especially yourself, if you're not, will it?" It was put as a statement, not a question. Danny had noticed that a lot of big-money guys had a habit of putting statements in the form of questions. "Now you want us to refinance you so that you can meet your obligations on that note, which must be coming due now in just under eight months. Am I right?"

"Yes."

"Okay. Now I'm going to offer a suggestion. Take it as just that: a suggestion, and not an offer. Okay?"

"Fine," said Danny, sensing that this was either going to be "it," or . . .

"Now, assuming that your figures are right about the basic profitability of the situation out there at Raffles, and for the moment leaving aside any discussion of *why* such a large proportion of the profits has been, shall we say, diverted, but further assuming that you must somehow know how you're going to solve that problem, we'd be willing to consider helping you. However, I think you must know that we regard this as a real high-risk situation."

"I recognize that," said Danny.

"The key part of this is that we might be prepared to lend you, let's say, thirty or thirty-one million dollars for five years at 10 percent per annum interest, involving a straight-line amortization," continued Meyers. "But listen carefully: I think that rather than looking at this as a loan, you'd be better advised to consider it a convertible debenture or, let's say, a loan with warrants attached. Let's be even more precise. If you can't manage the situation, and if you can't make any of the payments—and we'd expect payments every six months, again on a straight-line amortization basis plus accumulated interest—we'd want to have the option to take over your complete equity interest in Raffles for the remaining face value of the note. Do you understand?"

"Not exactly."

"All right, let me run through it again. Let's assume we make the loan tomorrow. Let's say that six months from now you were able to make the three-million-dollar payment. Then six months after that you also made the second three-million-dollar payment. But six months after *that* you couldn't. Follow me? Fine. Well, that would then mean that we'd take over your interest in the casino and you'd be out the original ten million bucks cash down payment, plus the two three-million-dollar payments that you made to us. Do you understand?"

"Yeah, I understand. I don't think that I'd be willing to go for a deal like that."

"I'm not asking whether you would or wouldn't. I'm not done. Do you want to hear me out further?"

"Go ahead."

"We'd expect that even in the event that such a default occurred, you'd remain on as the titular owner and chairman of Raffles, along with your entire board of directors. And that you all would stay on for at least one full year until arrangements could be made by my principals to prepare for an orderly transition of personnel which would meet the needs of the Nevada Casino Commission. And that from beginning to end you'd keep all this to yourself." He stopped there.

"Look," he continued, "we don't want anything to go wrong. What lender does? But if you had to stand in for us for a while, then don't worry, Danny, we'd pay you well, take care of you for a long time. But the fact of the matter is that you'd be out at least ten million bucks of your own money and probably a lot more. As I said, it's a high-risk situation."

"Anything else?" asked Danny.

"Well yes, it's not something that would be part of any contractual agreement, you understand. It'd be more a gentleman's agreement to the effect that should everything work out for you, as we all hope and expect it to, with the result that in five years you are home scot-free, then if at some time, who knows when, I or some of my principals would come to you and ask for a favor, and at the moment I have no idea what that might be, but if we came and said, 'Hey, Danny, you know we scratched your back a little bit back in '69, so maybe you could help us out a little bit in '76,' well, we wouldn't want you to forget us."

"Makes sense," said Danny. "In fact, it'd be understood, but let me ask you a few things."

"Anything."

"If I decide more or less to accept your proposition, how soon could these funds be made available?"

"Within thirty days. Maximum."

"And who'd be the lender?"

"I don't know, but most probably a corporation domiciled in the Netherlands Antilles. That would be the same party that would get the option, or warrants, or whatever, depending upon how we formulate this."

"Why the Netherlands Antilles?"

"Because of the double taxation agreement with the United States: no withholding tax applies on interest or dividend payments made from the United States. That keeps the IRS out of it."

This was all news to Danny Lehman, but Saul Meyers was rising steadily in his estimation. "And who owns this company?" Danny then asked.

"A Dutch bank."

"I see. Then you mentioned, if I heard correctly, thirty-one million dollars."

"Yes."

"But I only need thirty million."

"No, you'll need thirty-one million, and I'll explain why," continued Saul Meyers. "There'll be the finder's fee for our mutual friend in the Bahamas. I think five hundred thousand dollars would probably be fair. Okay?"

Danny nodded.

"And I think that the same amount would probably be fair where my particular services are concerned."

Danny again nodded.

"Don't worry," Meyers added, "my principals will be fully aware of both of these commissions. I think they'll be appreciative of the fact that you saw fit to cover these expenses, rather than calling upon them either to assume or to partially assume the usual cost of doing this kind of business."

Meyers paused and then asked, "So what do you think?"

"I think I'm going to have to think about it."

"Fair enough," replied Meyers. "How about another drink?"

Right then the red-coated bartender called Jimmy came back with two Partagas cigars and handed them to Meyers. "Do you smoke these things?" asked Meyers.

"Today, yes."

The two men lit up, set to work on fresh vodka and tonics, and got to talking about the upcoming Joe Frazier fight, when two bronzed young women entered the room and came up to Meyers. "I'd like you to meet two friends of mine," he immediately said to Danny. "This one's Laurie and that's Rita.

"Say hello, girls. This is my friend Danny."

They both gave him a big smile. Laurie took the barstool on Danny's left while Rita took the one on Saul Meyers's right. Meyers then ordered four more vodka and tonics. Danny found himself talking about Philadelphia, about Joe Frazier, about jai alai. And the girls seemed to think that their old friend Saul Meyers had finally come up with a really nice guy for a change. Then, around four o'clock, Danny glanced at his watch.

"What's that all about, Danny?" asked Saul. "No sense looking at the time. What I thought was that the four of us could go on to dinner and make an evening of it. There's lots of room here at the inn for you, Danny."

Laurie, who by that time had firmly hooked her arm around Danny's, gave him an especially hard squeeze when the last words were spoken, as if to emphasize that nothing in the world could give her greater pleasure than to spend the next twenty-four hours with him. But Danny Lehman shook his head. "Thanks, Saul, thank you, girls, but I'm afraid I'll have to take a rain check on that. And believe me, I want to take a rain check on everything!" And now he squeezed Laurie whatever-her-name-was with a strength born of sincerity. "I've got some business to attend to that's already waited too long," he added. "And like I told you, Saul, I'll be thinking about it."

Meyers was disappointed. "Don't take too long, Danny. We don't like to be kept waiting, you know."

Fifteen minutes later Danny was back in the brown Chevrolet Impala, but this time it was the young Latino parking attendant who was at the wheel. He drove him in complete silence to the Fort Lauderdale airport, dropping him off, upon Danny's instructions, in front of United. Danny got out, and then opened the back door of the Impala in order to retrieve his bag. He thought he saw two men in a car that had just slowly passed the Impala looking at him through the back window with more than just incidental interest. As he walked into the terminal there was an alarm bell ringing in the back of his mind. But he ignored it, and it silenced itself as he found himself hurrying for a flight that would take him to the Miami airport, where he would have to change to get United's Flight 201, headed for Los Angeles, but with a stop in Las Vegas, where Danny Lehman intended to get off. If anybody *had* had him under surveillance up to this point, they would sure as hell have a tough time following him any farther.

At the check-in counter they told him that they expected boarding to start in approximately five minutes. So Danny decided to make a quick phone call. When he got hold of the reservations desk at Raffles, he asked for a top-of-the-line room, told them he would probably be there for three nights, and asked them to leave a message for either Mr. De Niro or Mr. Salgo, or preferably one for both. They should be informed that a Mr. Daniel Lehman expected to be arriving at McCarran Airport on United Flight 201 at around eight o'clock, and that he requested the pleasure of their company for either a late dinner or a drink.

Flight 201 from Miami was a dinner flight, but Danny waved off the stewardess's offers of both food and drink. He had already had enough to drink at that bar during the past couple of hours, and his nerves were such that he definitely was not hungry. So while the rest of the people in the first-class section were eating, drinking, and reading, Danny just sat there thinking. He liked that guy in Fort Lauderdale: Saul Meyers had a nice way about him. He liked those girls, too, especially that one that he had drawn—Laurie. Laurie. He had to keep her in mind.

But going back to that Meyers fellow, the one thing that had seemed puzzling was why Meyers had never really pressed him on how, or why, he was so sure he could make a go of Raffles. Meyers must have found out that the casino was in trouble, not necessarily in trouble with the Casino Control Commission, but in financial trouble. Yet Meyers seemed to have had no qualms about the thirty—or rather, thirty-one-million-dollar loan.

And then it dawned on Danny. Out of outrage over his own stupidity he banged himself on the head with the palm of his left hand, causing the stewardess who was walking up the aisle to give him a strange glance. It was obvious that they *knew;* they *knew* about De Niro and Salgo. After all, Meyers's principals, whoever they were, were friendly parties, close to that bank in the Cayman Islands that had made those phony loans to De Niro and Salgo. So that meant the club of insiders who knew the truth about what was going on at Raffles had expanded from Amaretto and Sarnoff to himself, then to the English banker in Nassau, and now, quite obviously, to the friendly parties, whoever they were, associated with the Cayman bank, and finally to their attorney in Fort Lauderdale, Saul Meyers. The club that had started out as a cosy little three-man conspiracy now seemed to be developing a membership that would soon challenge the Trilateral Commission in numbers. The more Danny reflected on this, the less comfortable he became.

Maybe this whole trip had been a mistake from beginning to end. Maybe he shouldn't have raised this issue with that Englishman in Nassau in the first place, causing him to bring up the name of Raffles with those principals of his, who, in turn, had brought in Saul Meyers, who might or might not be having him tailed. Which raised yet another question: if *they* knew all about De Niro and Salgo, was it not inevitable that De Niro and Salgo would somehow find out all about *him*? Could it be that word had already somehow started to leak back to the crooks running Raffles, word that this guy Lehman already knew too much? And if it had, was this the most clever thing for him to be doing right now?

The stewardess was coming back up the aisle again and this time saw what seemed to be an expression of deep concern on Danny Lehman's face. They had encountered a little turbulence over New Orleans, and maybe, she reasoned, the man's a little queazy about flying. Normally she avoided conversations with single males beginning to approach middle age like the plague, but somehow this one, although a little pudgy and obviously not too tall, seemed like a nice guy. And there was something sexually attractive about him. It was impossible to say what, but there was no doubt in her mind that he would be one hell of a lot of fun in the sack.

There seemed to be something about Danny Lehman that even attracted women back when he had brushed them off. They didn't seem to mind! So repeatedly during the rest of the trip she dropped by to offer him a drink, to suggest a snack, to give him some peanuts, to ask if he wanted something to read. But Danny just sat there. He never read. Well, almost never.

Now and then he read the sports page of the *Philadelphia Bulletin.* He also read *TV Guide.* But that was about it. After all, you could hardly make any money while reading, could you? So why waste the time?

The next time the stewardess stopped at his row, she bent down to ask, "Mr. Lehman, are you sure you won't have something to drink? We still have quite a bit of time until we arrive."

"Okay," said Danny. "Bring me a beer, any kind of beer."

After she brought him the beer she sort of draped herself over the back of the empty seat beside him and asked where he was going to stay in Vegas. When he told her Raffles she raised her eyebrows, since he hardly looked like the type who would go to such a place.

"How long are you going to be there?" she asked.

"Oh, I guess three nights," Danny answered.

"Actually, you know," she said, "I go on to Los Angeles. But then I've got the next couple of days off." She continued, "Come to think of it, I haven't been to Vegas in over half a year."

All of a sudden Danny felt expansive. "I'll tell you what I'll do," he said. "You give me your name and address, and not this week, maybe not even this month, but pretty soon you'll hear from me and you can go to Raffles for a day, or two days, or even a week anytime you want to. Bring a friend; girlfriend, boyfriend, who cares. And it'll be on the house. How's that?"

The stewardess now undraped herself. "Gee," she said, "how could you arrange that?"

"You'll find that out when you hear from me."

She never did. Danny *meant* what he said. But soon, very soon, there were other things on his mind.

◆   ◆

When Danny emerged at the arrival gate at McCarran Airport he looked around, half expecting somebody to be there. Nobody was. Down at the baggage claim area again, he took a couple of scans. Still nobody. The queaziness that had attacked his stomach in the airplane over New Orleans gradually started to return. And fifteen minutes later it got a lot worse. For when he walked into Raffles—skirting the casino floor and going directly to the reception counter—and mentioned his name, the girl looked at him in a rather odd way. Then she went to the back of the office, returned, and said, "Mr. Lehman, I am afraid we don't have a reservation in your name. How do you spell that again?"

"L-E-H-M-A-N, first name Daniel."

"Well," she said skeptically, "I'll check again."

When she came back she had a little white slip of paper in her hand. "Sorry about the delay, Mr. Lehman. Now we've found it."

He filled out the registration form, informed her that he would be there for probably three nights, gave her his American Express card, and waited while she made an imprint from the card. He walked over to the bellhop who was ready to take him to his suite. "I'm expecting a message," Danny told him, "from one of the executives here at the casino. Either Mr. De Niro or Mr. Salgo. Where would such a message have been left?"

"Let me check, sir," said the bellhop. He went back to the reception desk and returned immediately. "I'm afraid there are no messages, sir."

Danny followed the bellhop over to the elevators at the far side of the casino and went up all the way to the second floor. The room he had been given was immediately adjacent to the elevator bank. As they walked in, from the smell of it, it was also right above the kitchen. When he looked out of the tiniest of windows, all he saw was an immense ventilator funnel. He also heard the blast. He turned immediately to the bellhop and said, "This won't do."

The bellhop just shrugged, indicating he had nothing whatsoever to do with such matters.

"Stay here," Danny said. "I'll call down to reception and get this changed." He called down and was informed that the hotel was fully booked. His was the last room they could make available, and it was implied that he should appreciate the fact that Raffles had decided to make any space available for him at all.

He gave the bellhop a five-dollar bill and sat down on the bed. All of a sudden he felt hungry. He'd had neither lunch nor anything to eat on the plane. Maybe, he thought, this was a sort of compulsive hunger because of the situation that seemed to be developing. But still, it was better to eat something. Danny was no health freak, but he believed in taking care of himself. So he picked up the phone and called downstairs again, this time to room service. A salad and a steak was all he wanted. Plus some coffee. That was it.

Ten minutes went by. Twenty minutes. Half an hour. No food. He called down to room service to complain. They assured him that somebody would be up within five minutes. Half an hour later, he called down again, got the same assurance, upon which he told them to take the goddamn food and shove it. This time he slammed down the phone. If it had not all been so damned ominous it would have been funny, he thought. There he was, the guy who owned this fucking hotel lock, stock, and barrel, and he couldn't get a decent room, not to speak of a decent meal. In fact, he couldn't get a meal, period, decent or not!

Now what? he asked himself. De Niro and Salgo, that was what. Get to them, fire them, kick them out *right now*, that was what. But first he had to find them, which would hardly be possible if he sat around that room any longer, listening to the ventilator while inhaling the fumes from the kitchen.

He found out immediately that it might not be that simple. For as he closed the door on Room 202, embarking on what he had already mentally labeled his search and destroy mission, he noticed two men standing in the area in front of the elevators immediately adjacent to his room. They watched him. They were both big, very big, and both had on blue suits. They were not talking; just standing there. In fact they were just standing there looking at him. He rang for the elevator and as he went in he noticed that both were continuing to track him. Then one reached for the walkie-talkie unit that was clipped to his belt and started talking in a low voice. When Danny reached the main casino floor, a full story below, he decided to have something to drink, this time something a little stiffer than beer. So he left the elevator and headed for the bar next to the baccarat tables.

The casino floor, he noticed as he walked, wasn't exactly crowded. At least half of the blackjack tables were not operating. There were people at maybe one out of five slot machines. As he approached the baccarat area, he saw that it was totally deserted. To get to the bar itself you had to take a step up, passing through a gap in the brass railing that separated this particular bar area from the main casino floor. Some sort of headwaiter was standing there. He asked Danny whether he preferred the bar or a table. Danny thought a table would be better, and commented to the effect that this bar seemed to be by far the most popular area in the whole casino.

"Well, that's not too hard to understand, maybe," the man answered. "This place is what they generally refer to as hookers' corner. Look around and you'll see what I mean."

After Danny had taken his place at a very small table against the wall he did look around, and it didn't take a great deal of imagination to figure out where that name had come from. There must have been at least fifteen young ladies there; some of them, Danny noted, were very young indeed, but this was somewhat compensated for by the fact that there were also a few who appeared to be awfully old indeed.

The crowd in the casino seemed to be picking up a bit. In fact, immediately to the right of the bar area there was even some action at the baccarat tables. Danny ordered a bourbon and water, and as the waitress brought it back he noticed that the two gorillas he had seen outside his room on the second floor were now talking to the headwaiter. The three of them were standing about ten yards away. Twice while they were talking all three looked directly at him. Of that he was sure.

While this was going on, it seemed that all fifteen ladies, young and old, in the bar area were simultaneously sizing him up. And every time there was eye contact between one of them and Danny, however inadvertent, immediately the lady in question flashed as big a come-on smile as she could muster. Danny decided that it was probably not the worst idea in the world, but also decided to spend a little time picking and choosing before making up his mind.

So for a while he just sat there surveying the vast casino floor. The crowd he estimated at probably six or seven hundred people. He checked some of the physical characteristics of the place, and it did not take much more than a cursory glance to note that the carpeting was shabby, that the pit bosses were bored to death, that most of the waitresses one could perhaps best describe as slovenly. Out of the corner of his eye Danny noticed that the bar's maître d'hôtel, or whatever one called such a man in such a bar, was slowly making the rounds of the place with a word here and a word there to each of the hookers. Once he had moved on, as if by magic they had transferred what was left of their interest in Danny Lehman to other targets. Eye contact was reduced to zero. Then, as if to add further insult to injury, the maître d' slapped down a RESERVED sign on all three tables surrounding Danny's, leaving him there in splendid isolation, or perhaps "quarantine" would have been a better word.

The maître d' then disappeared, probably to report back to the gorillas. To say that Danny was starting to develop a severe case of paranoia would have been a definite understatement. For right then he remembered hearing that above the ceilings in these casinos there were catwalks from which everybody in the place could be observed through peepholes, or through the phony mirrors up there. He envisioned De Niro and Salgo just above him, sitting there and laughing their asses off. Let them laugh, Danny thought, as long as they're laughing about what's happened so far.

Then he was momentarily diverted, because even in that crowd, which was now up to probably a thousand customers, you couldn't help but spot the statuesque figure of the woman approaching: in her early twenties, at least six feet tall, and dressed in a perfectly cut black suit and white blouse, with stunning hair, she approached the bar area with a stride that indicated that she knew who she was: the most attractive woman, by far, in the casino.

When she took the single step up into the bar area, the slit in her skirt opened to reveal legs which, although Danny just caught the briefest glimpse of them, sent a stinging sexual shock through his system. The fact that they were black heightened her overall aura. She was about to take a seat at the bar when her eyes moved, met Danny's, and stopped right there. He maintained the contact and so did she. With no hesitation whatsoever she came right over and sat down. "Honey," she said, "you seem to like to be alone. If I'm bothering you, just say so."

"You know something?" Danny exclaimed. "You are the most gorgeous creature I have seen in one hell of a long time." She smiled a broad smile, reached over, and patted his hand, then left her hand on top of his and said, "Honey, you don't look so bad yourself. My name's Sandra Lee. What's yours?"

"Danny. Danny Lehman. That's my name, and I sure as hell would like to buy you a drink." He waved a waitress over. She seemed fidgety as hell as she took Sandra Lee's order for a champagne cocktail, but nevertheless she brought it back without saying a word.

"I haven't seen you here before," the black girl said. "Where are you from?"

"Philadelphia."

"Did you come in here on one of those junkets?"

"No, by myself," Danny answered.

"Uh, I like that." The hand that had been holding his now

moved below the table to make an exploratory feel. Danny, who was hardly inexperienced in these matters, felt a shiver that he hadn't felt in weeks, no, months; in fact, since this whole damn thing with Raffles had started in that new shop he had opened on Wilshire Boulevard. He had been so mesmerized with the forty million bucks he had put on the line that he had reverted to a state of total inactivity in other areas, which he now realized had even extended to sex.

"Look," he blurbed out, "I hope you don't mind my saying so right away, but I would sure like to fuck you!"

"Honey," she answered, "I think you found yourself the right girl. You're staying here, aren't you?"

"Yeah."

"Show me your room key."

He fished it out and laid it on the table.

"You stay right here, Danny boy," she then said. "I'll be ready in a minute, okay?" She then got up and headed, Danny assumed, toward the ladies', to make sure that everything was in place that she normally had in place when she got down to business.

She was gone at least five or six minutes before she returned, walking this time with a brisker and somewhat firmer stride. She sat down, leaned as close to Danny's ear as possible, and said, "Man, I don't know who you are, and I don't know what you're up to, but you're in trouble, baby!"

"You got the word, too," Danny replied, also in a very low voice.

"I got it," she said.

"So why're you still sitting here?" Danny asked.

"Because nobody tells Sandra Lee what to do, that's why."

This time it was Danny Lehman who moved his hand under the table to grasp her thigh through the slit in her skirt, which, upon further exploration, seemed to extend almost up to her hip.

"Would you like a word of advice?" she now asked.

"Yes," Danny answered, and he really meant it.

"Get the hell out of here!" she warned.

"I'm not sure that would be the end of it," Danny replied.

"You're probably right," she said. "When Sid and Gino have an interest in you . . ."

"The two gorillas in the blue serge suits?"

"That's them," she answered. And they both just sat there for a minute.

"I've got an idea," she said.

"Whatever it is, I'm for it," Danny replied. For he had already concluded that he was in a no-win situation.

"Come on, honey," she said, and suddenly got up, grabbed her purse, grabbed Danny's hand, and began to lead him out of the bar and straight across the middle of the casino floor. They were quite a pair: the tall, gorgeous six-foot-two-inch black whore moving through the crowd as if she owned the place, and the five-foot-six-inch plump, balding guy trotting along beside her. Out the main casino doors they went. The black girl handed the parking attendant her valet ticket, the car was whipped up in front of them within a matter of minutes, and she climbed in on the driver's side while Danny joined her from the other side. His door was not even closed when she put the car in gear and they took off with a roar of the engine and a screech of tires. She was proud of her white Thunderbird convertible.

"Hold on," she said, as they turned out onto the Strip and as she gunned it up to a speed of at least fifty miles an hour within thirty seconds. Ten minutes later they approached the downtown area of Las Vegas; she turned into a cross street and then very soon they were in a rather run-down residential neighbourhood.

At this point he finally asked, "Where are we heading?"

"My place, honey. It's like I told you: I don't like people fucking around with me, so if it can't be your place it's going to be my place. Okay?"

Damn, Danny thought, two of the most enticing pieces of

ass being offered to him within a span of less than half a day, and he was going to have to pass on both of them. Life was becoming increasingly unfair. Then he said, "Sandra, look, sometime later I will explain all this, but would you mind just taking me to the airport instead? I don't like to cut and run. I've never done it in my life. But I don't like *these* odds one bit." She slowed the car down, looking over at him. Danny continued, "I'll take care of these monkeys in my way and in my time. Until I'm able to, I sure as hell don't want to get you in trouble."

She had now stopped the car at the curb. "I already am," she said, "but don't worry; Sandra Lee can take care of herself. The one I'm worried about is you. Why, for God's sake, are you fooling with this crowd?"

"Because they've got something I own and I want it."

"What can be worth that much?"

"It's not what's *it's* worth. It's what *I'm* worth. If I can't beat these stupid crooks at their stupid games, then I can't beat anybody. But suspect I can, and I think that after I do, it's going to be *them,* and a lot of lot of other people, who are going to think twice before they start fooling with *me.* Do you understand?"

"I think I do," Sandra Lee replied. "Though I don't know why at this point."

"Well, next time, and by God there definitely will be a next time, I'll explain further. At your place, my place, or both."

She'd heard that one before. Yet . . . She gunned the car into a U-turn. Fifteen minutes later they turned right into the access road to the airport terminal. Danny extracted five one-hundred-dollar bills from his wallet and reached over to put them on Sandra Lee's lap.

"What's that all about, honey?" she asked.

"Just a little gesture of thanks."

She turned to him and flashed an enormous smile. "Danny Lehman," she said, "you've got class. I like you. I like your guts. I like what you just said a few minutes ago."

He smiled back at her and laid a hand on hers. "You'll hear from me. You can count on it."

She would, and she would not be disappointed.

◆   ◆

When he got to the American Airlines counter, walking slowly and deliberately and looking straight ahead the entire way, he asked for the next flight east. When the girl at the counter suggested he would have to be a little more specific, he just told her to tell him when the next American flight was headed east. She didn't even give him a funny glance, since in Las Vegas you got all kinds of queer requests. "There's one leaving for Chicago in exactly twelve minutes. Gate 7. You want a ticket?"

He gave her his American Express card and, after a wait which seemed interminable, was finally handed a ticket.

"You'd better run," she said, "fast."

When he got to the gate they were closing down the counter. He waved his ticket at them, and they waved him toward the tunnel leading to the waiting DC8. *Now* he took one last glance back into the terminal. Two very large men in blue serge suits were trotting in the direction of Gate 7, though still a good fifty yards away. Immediately after he stepped into the plane the stewardess slammed the door. He sat down in the very first row and fastened his seat belt, and almost immediately the plane jerked as it was towed away from the gate. About four minutes later they were airborne.

Five minutes after that, Danny Lehman's breathing gradually returned to normal, but not his mental state. "I'm going to beat those bastards if it *kills* me," he said in a low mumble to himself. Then for some reason the old Jack Benny joke occurred to him, the one about "Your money or your life," coming from a stick-up artist holding a gun to Benny's head. "Lemme think about it," had been Benny's reply.

Well, before he, Danny Lehman, could give the same answer he first had to have the money. But that meant going the

Fort Lauderdale route, the route that led from Saul Meyers to Montague Davies in the Bahamas to the Cayman Island bank and those close associates of Davies and Meyers who knew all about the casino business.

When you went that route, what was the final destination?

◆   ◆

That was precisely the question put to him the next day by his attorney and the still-reluctant vice-chairman–elect of Raffles Inc., Benjamin Shea. "You know the answer," Shea said immediately, and added, "so don't do it, Danny." And before Danny could get stubborn and begin to plant his feet in concrete—Shea knew his client—he continued, "I've got excellent contacts at Mercier Frères."

"What's that?" Danny asked, despite himself.

"An investment bank."

"Forget it, Shea. Banks don't lend money against casinos. We've been through that, remember?"

"It's an *investment* bank, Danny."

"So what?"

"As you know, they invest *in* companies, raise money *for* them, and not just lend money *to* them."

"So?" Danny wasn't in the greatest of moods.

"Well, it can't hurt to talk to them."

"About what?"

"The thirty million you need."

"I need a loan, not a partner."

"You think what they're offering you in Fort Lauderdale is just a loan?"

◆   ◆

Mercier Frères was not just "an" investment bank. It had been started in Paris in the nineteenth century and by now had been established in both London and New York for nearly a hundred years. Though Mercier had been founded later than Lazard Frères, it was that merchant bank which Mercier had tried to emulate from the very beginning. Now, almost a century later, the partners of Mercier were still keenly aware that their bank was Avis and Lazard was Hertz, but none would ever admit it out loud.

Henry Price was one of their three managing directors, the one responsible for all of their operations in North America. He was Groton, Harvard, a Fulbright scholar, on the board of governors of the Council on Foreign Relations, the board of directors of IBM, the board of elders of the Episcopal Church. *Summa summarum,* he numbered among the one hundred most powerful men in the United States.

Price was forty-nine and was, by any standards, a very fine physical specimen. When he was in France, which was often, invariably someone there would remark that there was an uncanny resemblance between him and Giscard d'Estaing. Both were very tall, and rather thin, and looked down upon the world with a great deal of skepticism that was often taken, and correctly so, for disdain. But where Giscard would look down upon his world of politics from the Élysée Palace on the rue du Faubourg St.-Honoré, Henry Price ruled his world of finance from the offices of Mercier Frères, at times from their original site on the Boulevard Haussmann, more often from the merchant bank that they maintained under the same name in Moorgate in the City of London; but usually from the offices on Wall Street, from which Mercier Frères ran their American investment banking activities.

Mercier Frères was not as fast-growing as Salomon Brothers, not as large as Goldman Sachs, not as snooty as First Boston, and certainly not as aggressive as Lehman Brothers. Lazard Frères was the house that it resembled most closely. And, inevitably, the next comparison that was made was between the two "stars" of these two institutions: Felix Rohatyn of Lazard and Henry Price of Mercier. Rohatyn was the more flamboyant of the two, the one who enjoyed publicity. He liked to see his name on articles expressing rather radical opinions concerning the economic future—or lack thereof—of the United States on the editorial pages of *The Wall Street Journal,* which had the circulation, or in *The New York Review of Books*, which lacked circulation but made up for it by having the type of readers that Rohatyn felt were sufficiently bright to appreciate him. It was said that Rohatyn had political ambitions; that he wanted to become the architect of America's economic and financial policies, just as Henry Kissinger now was of its foreign policy. The similarities did not end there. Both Kissinger and Rohatyn came from Central European origins; both were first-generation immigrants consumed, probably to a very healthy degree most of the time, by ambition.

Not Henry Price; ambition, politics, and Central Europe were all foreign to this man, but especially ambition. At forty-nine, if he was consumed at all it was by a desire to gradually abdicate his position in that American corridor of power which, like the Eastern shuttle, runs from Boston to New York to Washington. Spurred on by a disastrous divorce, he was determined to rebuild his private life into one that would offer everything that was new, different, and satisfying. He had recently acquired a farm in Virginia and at the annual meeting of the principals that year had stated that while he would remain on as managing director at Mercier indefinitely, he would henceforth devote his attention only to special projects; to new ideas.

So on May 27, 1969, he received Daniel Lehman and his attorney, Benjamin Shea, for lunch in one of the dining rooms on the tenth floor of the Mercier Frères building on Wall Street. He did so because Shea was a nephew of one of the firm's other partners, but also because of the nature of the proposal.

He regretted his decision the very first moment he saw Lehman. Which was a pity, because the project, a casino, was both new and intriguing. Nobody at Mercier Frères had ever considered getting involved with the gaming business. In fact, as far as Price knew, nor had anybody at Morgan Stanley, or Lehman Brothers, or Goldman Sachs. Even the brethren in London who would try anything—the more daring merchant banks such as Hambros; Hill, Samuel; Guinness Mahon; or even S. G. Warburg—had never thought of casinos. In itself, this made the idea intriguing.

The menu that day consisted of melon, cold salmon, and a chocolate mousse. When port was offered with the melon, it was obviously a first for Danny Lehman. It was, therefore, also obvious to Henry Price that it would be a waste of time to try to engage the man in small talk. So he got right down to business.

"I assume that Mr. Shea has told you about our firm, Mr. Lehman?" he asked.

"Yes. You raise money, you don't lend it."

"Precisely. We help corporations place their new issues of stocks, bonds, notes. And so forth. Usually we place these instruments with the general public. Sometimes we do private placements. What exactly did you have in mind?"

"I need thirty million dollars."

"For how long?"

"Five years."

"If I understand your situation correctly, you need the funds to complete the purchase of a casino in Las Vegas."

"That's right. Raffles."

The cold salmon arrived. So did a bottle of Sancerre. Price grimaced when Danny downed most of the wine in one gulp.

"I'll tell you bluntly, Mr. Lehman, that we have never been involved with casinos before. So it would be very difficult for me to present a case to my colleagues unless I could demonstrate that the borrower is extremely sound. After all, the question that is always asked first is: how will the loan be repaid? I assume you've brought along the financial statements." The last sentence was addressed to Benjamin Shea, but it was Danny who answered.

"They won't tell you much."

"Really. And why not?"

Lehman could hardly explain that the true earnings of Raffles were much higher than those reflected in Raffles' P&L statements, since, for years, management had been stealing millions. So he said, "Because financial statements cannot tell you about the future. And that's what counts."

"I agree. Tell me about that future."

Danny did, in a monologue that lasted twenty minutes. The chocolate mousse arrived, and Danny was still going on.

"When the Atlantic City casino gets operational, the profits are going to be phenomenal. Do you know how many people live within a hundred miles, a hundred and fifty miles, two hundred miles of Atlantic City? Tens upon tens of millions! I know, since I've done my research, Mr. Price. Benjamin, show him the numbers."

Lehman's attorney withdrew a fifty-page report from his briefcase and handed it to Price, saying, "Mr. Lehman asked me to commission this demographics study."

"You see?" Danny said, as Price started to flip the pages. "And a large percentage of those people can hardly wait to start gambling *legally*. That's the point. Today they have to travel thousands of miles to Nevada in order to do that."

While Danny continued to speak, he watched Henry Price as carefully as the investment banker had been watching him. After Price had handed the study back to Shea, Danny could see the banker's eyes begin to glaze over. He should have known better than to come here. A Henry Price was as much out of his depth in Danny Lehman's world as Danny was in Price's. He should have been blunt with Benjamin Shea before this unfortunate meeting had been scheduled in the first place. White shoes and casinos did not mix, and they never would.

◆   ◆

Lunch ended abruptly when Henry Price looked at his watch. It seemed he had a plane waiting for him. He mumbled something about a farm in Virginia. By two-thirty Price was airborne in his firm's Learjet. An hour later he disembarked at National Airport, where a limousine was waiting to take him to his farm, a three-hundred-acre spread near Middleburg that he had bought six months ago for $1.7 million. Henry tried to visit it every other weekend. For three reasons: the solitude, the horses he stabled there, and the woman who usually spent her weekends with him.

Although *he* referred to Natalie Simmons as his fiancée, she never called him anything but a "good friend." He was New York, she Georgetown; Middleburg was his escape from Wall Street, hers from the art gallery where she worked in Washington, D.C., a city where they also spent weekends together.

It was during one of his stays in Paris last year that he had met her on the curb outside the entrance to the Ritz. It was April and it was pouring. He had just missed his limousine, which had to circle the block one more time. She was frantically trying to get a cab and failing miserably, beaten every time by the Parisian mob.

So, out of nothing more than sympathy for a compatriot, he offered her a lift. The last time she could recall even considering a "lift" was in high school, but this handsome man in his slate-gray topcoat, who carried a rolled umbrella and spoke with a mid-Atlantic accent, could hardly be suspected of any motives other than kindness. And the alternative was simply giving up meeting her girlfriend for hot chocolate at Rumpelmayers, and she was damned if she was going to let a bunch of rude Frenchmen do *that* to her.

A couple of hours later, as both were returning to the hotel, they ran into each other again in the lobby. They had a drink in the bar, dinner at Lasserre, and champagne at the Crazy Horse Saloon, and, to the utter surprise of both of them, ended up in her bed at three in the morning. She kept her white nightgown on the entire time; he apologized when it was over. They tried it again the next night and this time her hand actually guided him into her. She did that very rarely. He had hardly expected it. The result was that both of them climaxed very quickly, and in contrast to the previous night, he stayed for breakfast and they shared both the *Herald Tribune* and *Le Monde*, and, eventually, their own war stories.

The biggest war in his life had been with his wife and it had just ended in divorce. That was one reason why he was spending more time than usual in Paris: to try and forget her once and for all. Natalie's disastrous relationship had also just ended, fortunately, as she explained, this side of marriage. But it had been a very close call.

After Vassar, she had come over to Paris to study the history of art at the Sorbonne. The semester had barely begun in 1967 when she had met, and fallen completely in love with, a young man from Basle who was working in Paris for CIBA, the Swiss chemical concern. After one month they shared his apartment, and three months after that he had proposed marriage. Upon his insistence they had gone to the American embassy in Paris to determine what his status would be should she decide to return to the United States "someday," as he had put it. No problem, they'd been told. He would automatically be issued a green card upon producing documentary evidence that they were indeed man and wife. It would be a matter of weeks. That evening they celebrated, leaving her so exhausted that she slept in the next day while he went to work. Sex had never been her strong point anyway, although he seemed to think of little else. Out of boredom, she took a look through his desk. And there was an entire folder of his correspondence with Du Pont. They were highly interested in him and willing to pay him a salary that was, she knew, triple what he was getting from CIBA. The catch was that they required that he have permanent residence status: a green card. The last letter in the file was one in which he explained that the green card was no longer a problem. He would have one within thirty days.

She had gathered together her belongings within thirty minutes and walked out. She had not seen the jerk again.

After Natalie had finished her story, they had gotten dressed. Price was about to leave for the office when she asked him to take the day off and join her for an afternoon at the Louvre. He immediately agreed. Her passion for art, he soon found out, had become very specific: women painters, and the appalling lack of appreciation they had been shown, even in France.

They spent most of the next two weeks together, and by mutual agreement decided to see if it would work beyond Paris. She'd had enough of the Sorbonne, Paris, and, in fact, Europe, and would be returning to the United States in the summer. To do what? Maybe work in a gallery in New York, or Boston, or Washington, D.C., with the idea of one day owning her own. She wasn't quite sure. She eventually ended up in Georgetown, in a small town house that she bought with funds from her trust. Her great-grandfather had been one of the founders of General Mills, so money was no problem. Her college friends who met Price at the occasional dinner party she gave in Georgetown that summer wondered what this twenty-four-year-old woman saw in a man of forty-eight. She knew. He filled a major vacuum in her life, as she did in his, and neither had any ulterior motives whatsoever, neither money, nor status, nor, most important, green cards. The arrangement even allowed her to play housewife now and then, a role she loved to play as long as it lasted only forty-eight hours maximum. When that mood hit her, she usually arrived at the Virginia farm for their prearranged weekend well before Price did, and worked with the kitchen staff in planning the meals.

So she was waiting on the veranda when the limousine pulled up. The first thing Price did before kissing her was to hand her a bag full of art books from Scribners and a box full of chocolates from Belgium.

"For the farmer's wife," Price said.

"Not quite," she said, and then added, "but you are a dear, Henry. And you are also early."

"It's because I walked out of my own luncheon at the bank."

"What drove you to that?"

"It was a who, not a what: a little grubby fellow from Philadelphia. Ghastly man."

"What made you invite him in the first place?"

"The usual: his attorney is a nephew of a partner."

"What did he want?"

"Money. To buy a casino. In Las Vegas."

Natalie Simmons responded, "From *you?*" She giggled. "How ridiculous."

"Well, not entirely," Price replied. "He's got an idea. About Atlantic City. He became almost lyrical when he got onto the subject. He thinks it's a natural place for casino gambling. He's figured out how many people live within a hundred miles, two hundred and fifty miles, of the Boardwalk, that sort of thing."

"How many do?"

"About fifty million. Even more, I think. In any case, a hell of a large chunk of the American population."

"All dying to gamble?"

"Exactly."

"And he would make Atlantic City into a Las Vegas East?"

"Precisely."

"Even the people who live in New Jersey don't deserve that!" she said.

"That's debatable. In any case, they don't deserve *him.*"

"Hardly the Right Stuff, eh?"

"Exactly. In fact, the perfect description of the man."

"Never mind," she then said, reaching over to pat his hand, "you'll never see him again."

"That's for damn sure," Price responded.

◆   ◆

This was precisely the same thought that was going through Danny Lehman's head at exactly the same time. He had just arrived in Fort Lauderdale courtesy of a National Airlines flight from La Guardia. There was also a limousine waiting for him. In it was Saul Meyers. He was sitting in the back of the black Lincoln that was parked at the curb next to the baggage carrousel. "So you've made up your mind," Meyers said as Danny climbed in and joined him. Danny nodded.

"We thought you would," Meyers continued, and added, "You won't regret it."

He made it sound believable. Almost.

PART TWO

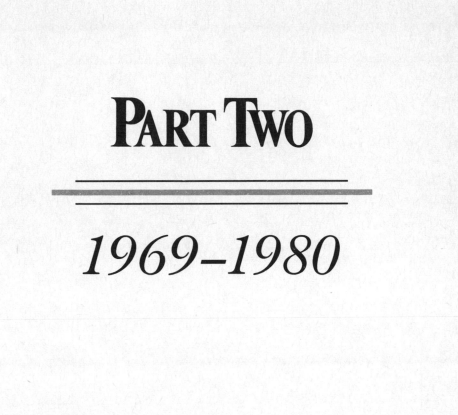

1969–1980

8

Two weeks later, on a day that was hot as hell in Philadelphia, Danny Lehman got to the office just before eleven o'clock. He was sweating profusely as he stepped through the doors of the American Coin, Metals, and Currency Exchange World Headquarters, as the not too discreet sign said on the building that housed it. The only problem was that the building, and the company, in fact the whole damn thing, now essentially belonged to the bank who, against that collateral, had advanced him the ten million dollars he had used as a down payment on the casino. Though, to be sure, now that Fort Lauderdale had come through he had been able to pay off the note and now owned that casino lock, stock, and barrel.

In retrospect, buying it had been easy; taking possession was proving less so. Much less so. And if he did not get possession soon, very soon, the Fort Lauderdale mob would take possession of him. But he had made a decision in principle: to wait it out. To wait for an opening. He had no intention of walking into that den of thieves again unless he could walk in with artillery heavy enough to blow them away. The situation was taking its toll. Danny's appearance had changed markedly during the past month. He was fifteen pounds lighter and 10 percent balder. He told everybody that the weight loss, at least, was fully explainable and was due to his having taken up tennis. He had not only taken up tennis; he had become a tennis fanatic in thirty days flat, spending up to three hours a day taking lessons, practicing alone, and even venturing into a few matches with other neophytes. He knew it was escapism, but what the hell.

"Mr. Lehman, somebody is trying to get hold of you. Frantically! We tried everywhere. The service said you left an hour ago." It was his general manager, normally not an excitable type.

"I played a little tennis before coming in," Danny replied. He was still sweating to prove it. "Who is it?"

"A Mr. De Niro. That's all he said. He'll call again in five minutes, you can be sure."

"All right. I'm going up to my office. I'll take care of it." He said it calmly, but his stomach was churning at a furious pace. De Niro calling *him!* This could be it. He worked out of a small office suite on the third floor, reachable by either a creaky old elevator or a very steep staircase. Danny took the stairs; it built up tone in his leg muscles. His secretary was standing inside the door of the tiny reception area waiting for him. "That man is on the phone again," she said.

"I'll take it inside. No visitors until I say so." He picked up his phone while still standing. "Yes?"

"Is this Lehman?"

"This is Daniel Lehman, yes. Who's this?"

"Lenny De Niro. And Roberto Salgo is standing right beside me."

"What do you want?"

"There are some other people standing right beside me, too. From the FBI."

"I see. And?"

"They're trying to raid the place!"

"And?"

"For Chrissake, man, it's your place. I'll put them on the phone and you tell them to get the fuck out of here and to stay the fuck out of here until our lawyers get here to protect our rights."

"What lawyers are you talking about?"

"Look, Lehman, you own this place. You've got lawyers. So tell these guys to get the fuck out of here or you're going to—"

"Hold on there, De Niro. I'm not going to do anything of the sort. If the FBI's raiding the place, they must have a reason. When you find out what the reason is, call me back. Okay?"

And Danny hung up. Had his new "associates" intervened?

Half an hour later, his secretary stuck her head around the door, very carefully. "It's Las Vegas again," she said.

"De Niro?"

"No. A man who says he's from the FBI," she replied, and her expression showed she was scared. She had heard, everybody in the building had heard, that Mr. Lehman had gotten involved with some casino in Nevada and that something had gone very wrong. She liked her job and she liked her boss. It seemed as if both might be in danger.

"Put him through," Danny said.

"Mr. Lehman?" It was a deep voice that came over the phone.

"Yes."

"My name is William Smith. I'm the special agent in charge of the Las Vegas office of the Federal Bureau of Investigation." He spoke with a southwestern type of accent, probably from Oklahoma. It was the same sort of semi-drawl that airline pilots affect. "Mr. Lehman," Smith continued, "we'd appreciate it if you'd come to Las Vegas right away."

"If you think it's necessary," Danny answered without hesitation.

"We've established that a flight is leaving Philadelphia in just about an hour from now, and from what we've been told, there's space on it."

"Which airline?"

"TWA Flight 62."

"I'll be on it."

◆ ◆

When the plane was about forty-five minutes away from Las Vegas, a voice over the loudspeaker asked that Mr. Daniel Lehman please identify himself to one of the stewardesses. When he did, she told him that the cockpit had been informed by the "authorities," as she put it, that they wanted him to disembark first. Two men would be waiting for him. When the door slid open at the airport and Danny stepped out, there were three men, not two. The oldest among them stepped forward, reached into his pocket, and produced what looked like a small leather wallet. He opened it and showed Danny his badge.

"Mr. Lehman?"

"Yes," replied Danny. The badge bit had impressed him.

"I'm Special Agent Smith. William Smith. Let's go."

Smith did not waste any time. He had no sooner started walking down the long corridor leading to the terminal exit when he started working on Danny. "We assume you're here, Mr. Lehman, in order to cooperate with us?"

"Your assumption's correct," Danny answered.

"What is your relationship with Mr. De Niro and Mr. Salgo?" the special agent asked.

"None. In fact, less than none," Danny replied. "As you must know, I assumed ownership of Raffles a number of months ago. But up until this point I have not interfered in any way with the management of the casino here in Las Vegas. This was not by choice, Mr. Smith. To put it very bluntly, the management, specifically De Niro and Salgo, locked me out."

The agent nodded his head as they continued to walk at a fairly brisk pace, and said, "That confirms exactly what we've heard."

Danny was glad that he was in such good shape as a result of his tennis playing and stair climbing. Once outside, Smith and his colleagues went directly to a gray Ford parked rather majestically where no car was ever allowed to park. They were watched by an airport cop who was obviously resentful of this higher police authority which had temporarily occupied a piece of his territory. The two younger FBI men got into the front seat, while William Smith and Danny Lehman took over the back.

As they pulled away, Smith turned to Danny and said, "Now let me explain the exact nature of our problem. We've obtained a search warrant for your place from a federal judge, but it has been successfully challenged, at least temporarily, by the shyster lawyer, if you'll excuse the expression, that De Niro and Salgo came up with about one minute after they talked to you on the phone. In fact, thinking of it now, they must have found him before that, because De Niro had barely hung up before this guy walked into the executive offices of the casino. Well, we really had no choice but to walk out."

"Why?" Danny asked. "And what were you searching for?"

Smith chose to ignore the second question. "The warrant is not specific enough and, frankly, it's because we *think* we know what we are after and even where it is, but we're not sure."

Danny was getting increasingly confused the more Smith spoke, and it showed.

"Don't worry, Mr. Lehman, we've got the place watched from the front, from the back, and from the top. We know every senior employee in the place; every one of our men has studied their pictures, and, as far as we know, so far nobody has gotten out with anything. Now, that does not mean that if and when we get in we're going to come up winners. There are probably a lot of good places to stash something in a casino. Nonetheless, we are pretty sure that if we have access to the entire casino, and sufficient time, we'll come up with what we want, which will be all we'll need to rid you of De Niro and Salgo for a long time."

"What exactly is that?" Danny asked again.

"I think that at this point, lest somehow what we do and say here may later be interpreted in court as entrapment, or in case it may be implied that for completely selfish and personal motives the two of us got together—colluded—in order to nail De Niro, Salgo, and all their pals—for those reasons I think it best I do not answer your question at this point. If I do, maybe this whole thing could get thrown out of court later. So if you don't mind going on faith a little bit longer—"

Danny broke in. "What do you want me to do?"

"You own the place. If you invite us in, there is no power on earth that can stop us. Right?"

Danny thought it over. "I guess so." He did not appear convinced.

Special Agent Smith then tapped the shoulder of his colleague sitting beside the driver in the front seat, and said, "Harry, let's have it." The young man produced a piece of paper and handed it back over the seat.

"This looks a little formal," Smith said to Danny, "but as I have already indicated, we can't be too careful in these matters. Now, all that this essentially says is that you, Daniel Lehman, as the principal owner, chairman of the board, and chief executive officer of Raffles Inc., which owns the casino and hotel in its entirety, hereby authorizes the Federal Bureau of Investigation to enter your premises and to have complete access to the entire property for a period of forty-eight hours starting . . . and we will fill in the date and exact time right now if you agree to sign this, Mr. Lehman."

Danny reached inside his jacket pocket for his pen. "Where?"

"Between the red X's."

Danny signed.

◆   ◆

When they pulled up in front of Raffles it was not the doorman who came forward but two more FBI agents dressed in almost exactly the same way as the two young men who were now emerging from the front seat of the gray Ford. As they walked up the broad stairs to the doors at the main entrance, they were met by an elderly guy, in his sixties, who held up his hand like a traffic cop and said, "Smith, stop right there! If you've got a new warrant, show it to me. If you haven't, back off!"

Smith said, "I've got something better than that. This is Mr. Daniel Lehman. He owns this place. Here, read it." He handed over the piece of paper that Danny had signed in the car. When the man was done, Smith didn't wait for him to give it back; he grabbed it back.

"Okay, fellas," he said to the four young FBI types, who were now all standing in a huddle off to the right, waiting. "In we go."

As they went through the door, Smith asked Danny, "Where's the counting room?"

Danny replied, somewhat shamefacedly, "Frankly, I don't know." So everybody stopped on the terrace overlooking the casino floor. Smith then said to one of his younger men, "Go down there, get a pit boss, and bring him back." The agent returned with a man who stood there with a completely blank look on his face.

"Could you please tell me where the counting room is?" Smith asked.

The man's arm went up, he pointed to the rear of the casino, and he said, "All the way back"; then added, "Anything else?"

The FBI man did not bother to answer this question; he just said to his entourage and Danny Lehman, "Follow me."

The counting room was indeed right at the rear of the main casino floor. It was really nothing more than a big square box slapped against the rear wall, and apparently constructed not of Sheetrock and plaster but of materials of substantially greater width and strength. Upon closer inspection, what it really resembled was a kind of blockhouse, which the interior decorators had tried their best to disguise by decorating it in garish colors, and with even more garish murals. Mounted into the front walls were thick glass windows with what looked like bank tellers standing behind them. The entrance to this bunker was a single door off to the left, a door that also had a narrow, thick plate-glass window mounted at eye level. It was obviously the only means by which one could gain access to the counting room.

Special Agent Smith stepped forward and thumped on the door with his right fist while at the same time tapping as sharply as possible with his left knuckles on that plate glass.

A woman's startled face suddenly came into view.

"Open up!" yelled Smith. He held up the leather wallet containing the badge and emblem of the Federal Bureau of Investigation of the United States of America, flat against the windowpane.

"I can't," came back the woman's muffled voice, the voice of a person who was scared out of her wits.

The agent now took Danny Lehman's arm and drew him to his side. "Tell her," he ordered Danny.

"My name is Daniel Lehman," he said, "and I am the new owner of this place. In that capacity, ma'am, I request that you open up this door immediately."

By this time there were tears starting to run down the cheeks of the face that was staring out at them. "I couldn't," she whined, "even if I wanted to. I don't have a key to unlock this door."

"Who does?" yelled Smith.

"I'm not authorized to say," she answered.

This made more than a little sense, since the knowledge of who had the key that would open the door that led to one of the largest accumulations of cash in the United States would inevitably have been too tempting for some of the more daring crooks of this world to resist using it for selfish purposes.

The agent now turned back to his entourage and said, "Jack. Harry. You stay right here. Nobody goes in. Nobody comes out. The rest of us are going upstairs." Smith now knew where he was going. He led them around about a quarter of the circumference of the casino main floor to an elevator that had been placed very discreetly down a short corridor. It was a single small elevator with no sign indicating where it led to. It was so small, in fact, that there was barely room for the four men in it.

Smith punched the button marked "3." When they got out, there was a sign with an arrow indicating that the executive offices were right ahead. Smith said to Danny, "I think it would be appropriate if you would lead. Just go straight down this corridor to the office at the end. My bet is that De Niro and Salgo are both in there."

When Danny got to the end of the corridor, instead of knocking on the door he just opened it and walked in. The FBI man had been right. De Niro sat behind the huge desk at the far end of the room; Salgo was sitting on a sofa off to his right. Both had in front of them large glasses filled with what was obviously a lot of whiskey and very little soda. De Niro stood up in violent anger and let fly, ignoring Danny. "Smith! You've been told once to get the fuck out of here, you and all of your creeps. Apparently none of you understand English too well, so I'm going to have to tell you again: get the fuck out of here. Now! If you don't I am going to arrange to have you removed, physically removed, right now. Do you understand that?"

Smith didn't say a word.

De Niro turned to Salgo. "Where is your goddamn lawyer?"

"He said he wanted to wait downstairs at the front entrance because he thought this might happen," Salgo responded.

"Well, get the senile bastard up here," De Niro commanded.

"Oh no, Salgo. You sit right where you're sitting." Smith stepped forward and produced the same piece of paper they had shown the attorney downstairs. It was becoming increasingly apparent that the attorney had been wise enough to leave the casino as unobtrusively as possible and head back to his office, if he had an office.

"Read it," the special agent said as he gave De Niro the document that Danny had signed in the car. De Niro read it and handed it over to Roberto Salgo. In the meantime, nobody in the room said a word.

Then Smith broke the silence. "We'll need all the keys."

"Why?" De Niro asked.

Then it was Danny Lehman who spoke. "All the keys. All of the casino's keys, which are now my keys; and specifically right now, the keys to the counting room."

"We don't have any keys," De Niro answered.

One of the young FBI men stepped forward and whispered something in Smith's ear. Smith listened with a slightly impatient frown, looked sharply at De Niro, and asked, "Who's your chief accountant here?"

De Niro just looked at him.

"All right," said Smith, "if you want to play it that way it's fine with us. Everybody stays here." With that he left the room and a minute later came back with a young, extremely good-looking secretary in tow. Smith pointed at the telephone on De Niro's desk and said, "Young lady, I want you to call the chief accountant and tell him to come to this office. Right now."

The secretary was obviously a tough nut, since in contrast to the woman down in the counting room, she was not intimidated in the slightest. Nevertheless, she did as she was told. Then Smith told her to go back to her desk, which she did. A few minutes later a very agitated man in his late forties walked in. He looked at De Niro. He looked at Salgo; he looked at Danny; and then at the FBI crew, not knowing exactly who he should address his first words to. He didn't have to worry in that respect, since the chief FBI honcho moved right in front of him and, with his face about six inches from the newcomer's face, said, "That man behind me is Mr. Daniel Lehman. He owns this place, as you know. He wants the keys, all the keys to everything in this fucking casino; he wants them right now and he specifically wants the keys to the counting room. Where are they?"

"Rupert," said Salgo, "not one syllable to these guys." He punched each word out with increasing emphasis.

Special Agent Smith ignored Salgo, stepped back from the newcomer, and addressed Danny. "Mr. Lehman, I would like you to meet the head of the accounting and auditing department of this casino, Mr. Rupert Downey."

An alarm bell went off in Danny Lehman's head. Rupert Downey? Rupert Downey? He knew the name. How in the world was that possible? Then he remembered: this was the guy who had received the last loan that was processed from his home, through the Caribbean, and back to Las Vegas. The money had come from the briefcase full of cash that Amaretto had left on his coffee table that night. Easy, Danny told himself. As he watched Rupert Downey's face, there was nothing, absolutely nothing, to indicate that Downey, who was now starting to tremble, had even the faintest clue who he was. This was another one he owed Joe Amaretto and Sam Sarnoff.

Just to make damn sure, Danny stepped forward, shook hands with Downey, and said very quietly, "I'm Danny Lehman, Rupert. I think you understand what's happening here. I want you to know that I'll try to be as helpful as I possibly can when this whole thing is over. So I would greatly appreciate it if you'd agree to Mr. Smith's request."

Now Downey's face began to twitch. He didn't dare look in the direction of either Salgo or De Niro when, in a voice even lower than that used by Danny, he answered, "Come with me." Danny followed him out of the room. Everybody else stayed. When Danny returned with Downey he pointed to a bundle of keys the accountant was carrying. "I think I've got exactly what we need."

Smith looked at De Niro and then at Salgo and said, "You've got a choice: you can stay here or you can come with us." Both men decided to go with the FBI.

It took the elevator two trips to transport everybody down to the casino floor, but once they were reassembled they moved like a wedge in the direction of the counting room. When they got there, Downey stepped forward to unlock the door and then stepped back to allow Smith and everybody else, including De Niro and Salgo, to move in ahead of him. Smith had said he knew what he was after and where it was. Obviously, they were now all inside the "where."

The counting room itself was the size of a small bungalow. It had been partitioned off into three sections. The first was the entrance area into which they had now all stepped. Straight ahead of them was the second area, a room about twenty feet by twenty feet. Everything inside was totally visible, since the upper halves of the partitions were all glass. In the center was a large table. Around the table stood five women. In the middle of the table was a pile of cash. One woman had a stack of bills in front of her and was counting them at an amazing speed. When she was done counting she handed the bills to the woman beside her, who did the same. Apparently they counted out loud. Then the bills were handed to a third woman, who ran them through a counting machine, which also wrapped them. Then a fourth woman took over, but it was not quite clear what she was doing; it appeared that her function was to take the information that was printed on the wrappers around each of the stacks of bills that had just emerged from the counting machine, and to record it in a ledger. The fifth and final woman was obviously there to put the now counted, wrapped, and recorded stacks of bills into containers, which were then sealed.

On the wall behind the counting table was a row of metal racks, obviously mobile, since all were mounted on wheels with rubber tires. The racks had two tiers. On each tier there were three metal boxes, and these boxes were, in turn, chained to the rack itself. The first woman approached the racks and produced a key that opened the padlock on the chain attaching the metal boxes to the metal racks. She then lifted a box off the rack and brought it to the counting table. She turned it upside down, and out of it tumbled another large pile of money, which the woman on her left now started to count, and thus the whole procedure started all over again.

Everyone just stood there in silence taking it all in until Smith turned to Rupert Downey and asked, "Where are your safes? Are they in there?" And he pointed to the counting area they were all watching.

"No," replied Downey, "in the other room."

"All right, that's where we'll go."

It required a new key to get into the other room, and Downey had it with him. This room was slightly larger than the one where the counters were so busy. It represented the nerve center of the casino, housing both the money bank and the data and information bank supporting the gaming operations. There were rows upon rows of filing cabinets, containing detailed financial information on all clients of any significance. At the front was a type of banking counter with thick bulletproof windows facing the casino floor. There stood the tellers who were visible from the main floor. There were two slits in the glass: one at the bottom of each window and one cut into the middle of the glass, allowing for verbal communication between the men standing behind the counter inside the counting room and men who approached the counting room from the outside—usually one of the pit bosses, one of whom was standing there now.

"What's he doing?" Danny asked Rupert Downey.

"That's where the pit bosses bring the markers," the auditor said. "The people in here then check our files to see if the customer's credit is still good. In essence a marker is just an IOU. They, plus the cash and the chips, constitute our 'inventory,' all of which is stored in this room."

"But exactly *where* are they stored?" Smith asked.

Downey pointed to the right side of the room. "There." He indicated a wall of steel doors.

Smith now looked at Danny and said, "I think that's what we want to look at." He then turned back to Downey. "How do you get into those steel cabinets?"

Downey replied, "With a key and a combination."

"Well? Let's get started," Smith commanded.

Downey moved forward, produced yet another key, inserted it into the lock, turned it, quickly spun the dial next to the key five times in alternate directions, and then slid open the large steel doors. Behind these doors, on the left, were two huge safes mounted one on top of the other, and, to the right of them, four rows of safety-deposit boxes and drawers, just like those one sees in the vaults of banks.

"How do you open those safety-deposit boxes?" was the next question that Smith put to Downey.

"It takes two keys, as usual. Our key and the boxholder's," Downey replied. "I'll demonstrate it," he continued. "Box number seventeen," he said, "is my box."

Downey reached into his pocket for his key chain. Then he sorted through the keys he had brought down from the office until he found its match. He inserted the two keys into box seventeen. He then drew out a metal tray and started to hand it over to Smith, who backed away from it as if the auditor were offering him a box full of rattlesnakes.

"No, no," Smith said, "this has to be done correctly."

Downey turned to Danny and then back to the FBI agent, and said, "Sir, I don't mind at all if you look through this. These are just personal documents, that's all. I would feel relieved if you did so."

Smith continued to look reluctant but nodded his agreement. Downey put the metal tray on a table in the middle of the room. Smith stepped forward to the table, lifted the lid of the tray, and began very gingerly to finger through its contents. It was quite obvious that there was nothing in the tray except insurance policies, some bank statements, and what looked like a will. When Smith, in a practiced drone, described what he was finding in the tray, it was not only disappointment that registered on the faces of the men from the Federal Bureau of Investigation, but suddenly a growing concern. Danny sensed that perhaps this whole thing was in the process of blowing up in their faces.

This might be very embarrassing for the FBI if it proved to be the case; for *him,* however, it could have truly disastrous consequences. For Danny could not help but remember the last words Amaretto had spoken to him in Philadelphia. "If you try to get them into trouble with the authorities," he'd said, "they're so fucking greedy and so fucking arrogant that they'll kill you, Danny. And then they'll kill us."

If the other safety-deposit boxes contained nothing more than this, if De Niro and Salgo were not caught red-handed, they could simply walk away from it all, and who on earth could stop them?

The agent in charge of the Las Vegas office of the FBI was watching Danny, and he must have sensed the train of thought in the man's mind. So once again he took charge and now addressed himself to De Niro, who was just standing there, smirking. "All right, Lenny, your turn."

"For what?"

"Look, do you want to do this the easy way or the hard way? We look now or we look later. You know damn well that I can get a search warrant, if not within an hour, at least within days, and in the meantime, if necessary, Mr. Lehman is going to stay right here and is even going to sleep here should it take that long; isn't that right, Mr. Lehman?"

Danny nodded his agreement. Why not? His neck was already in the noose.

"So," Smith continued, "it's up to you; make up your mind."

De Niro continued to smirk. Suddenly he said, "All right," to everybody's surprise and steadily mounting apprehension.

"Which is your box?" asked Smith.

"Not box," answered De Niro. "Drawer. And not drawer. Drawers, plural."

"Okay. Which ones, plural. Answer!"

"One, two, and three." And they were big drawers. That produced a noticeable increase in tension in the room. Nobody had that many insurance policies.

De Niro continued, "Roberto there, he's got three, four, and five, don't you, Roberto?"

Salgo nodded. That took care of all the drawers right there.

"You've got your keys with you?" Smith asked.

"Yes."

"Well, use them," Smith said.

De Niro looked at Smith and then spoke in a sharp voice. "Let's stop right here for a minute. I'll do this my own way, okay? In my own time. Right?"

Danny's apprehension mounted. Either De Niro felt it necessary to prove what a dangerous bastard he was through bravado to the end, or he, Danny Lehman, had pulled the pin on a grenade that was going to be a dud. Worse yet, it might prove to be a grenade with a timer. Nobody, especially De Niro and Salgo, was going to get blown up here today. But when the timer ran out in maybe a month or a year, it was going to have a very lethal impact on two other men in that room: Rupert Downey and Danny Lehman.

Nobody challenged De Niro, so he continued. "I want to explain something first. My colleague, Mr. Salgo, and I have indulged ourselves in a little bit of the action here during the past couple of months. I know, we all know, that's against the rules, but as you all also know, nobody pays any attention to those rules. What I'm telling you now is that Roberto and I have spent more than a few hours at the baccarat table and we've been lucky. Very, very lucky. And if you don't believe me, you can check it out with the guys running those tables. Their names are . . ."

It's a dud, was Danny's first panicky thought. These guys simply can't be beaten. The FBI's conclusion was obviously exactly the opposite. In a thickly sarcastic Oklahoma drawl, Special Agent Smith interrupted De Niro. "Okay, okay, Lenny, we can get to the names later."

De Niro said, "Roberto, give me your keys." Salgo did as he was told. De Niro said to Downey, "Give me your fucking key, too, asshole." Downey also did as he was told, his face absolutely ashen.

"I want everybody to remember that I'm doing this voluntarily," De Niro then stated. Having had the final word, he stepped forward, manipulated the keys, pulled out drawer number one, and put it on top of the table. Then drawers two and three. Then four, five, and six. He gave Downey back his key with the words "Here, asshole," and smiled at the rest of the group. "Gentlemen," he said, "be my guests."

Danny Lehman had, to put it mildly, a lump that felt as big as a basketball in the pit of his stomach. This whole thing, he thought, is going to be a bust. De Niro is making fools out of the entire Federal Bureau of Investigation. I never should have listened to those guys in the first place when they asked me to come here from Philadelphia. Those boxes are no doubt either empty or full of old newspapers.

Smith looked grim. He walked over to the table and raised the lid on the first drawer that De Niro had pulled out of the steel cabinet. He looked into it, and what he did next startled everybody in the room: he took the drawer, raised it, turned it upside down, and with a quick move slapped the metal bottom. What tumbled out were bills, stacks of bills, all hundreds and fifties. He took the next drawer and did the same thing with the same result. As he grabbed the third one, one of his younger associates said, "Stop it for Chrissake, Bill, that's fucking evidence!"

Smith looked even grimmer now. He addressed De Niro: "Okay, asshole, we've got *you*. Nobody, but nobody, will buy that horseshit about the baccarat table, and I don't care if every employee in this place swears to it. You're even dumber than we thought." Up until that point all of the employees in the counting room had tried to ignore as best as they could what had been going on, but now their activities ceased entirely. Smith reached inside his jacket pocket and produced what appeared to be a printed card. "Abiding by the Miranda decision . . ." He started to read De Niro and Salgo their rights.

When he was done, he motioned with his head, with the result that two of the younger agents stepped forward and handcuffed De Niro and Salgo. "Take them downtown," Smith said. "All right, one more thing," he continued, after the co-managers of Raffles Casino had been led off. "I want to go into that place where those women are counting that money and look at something. Who's got *that* key?"

One of the men who had been standing at the counter trying to go about his own business now stepped forward. "I've got the key, sir, but I'd like to ask that you alone step in. You can understand that we have to be very, very careful about these things."

"I understand that," Smith said. "We'll do it your way."

Everybody shuffled back out into the small hallway in front of the glass-enclosed room that contained the five women, the metal boxes, and the piles of cash. Smith went in alone. Although it was impossible to hear a word of what he said, it was quite obvious what his instructions were. All action stopped in there; the five women and Smith got together and started to pull the metal racks holding those chained metal boxes away from the wall. The FBI man crawled behind them on his hands and knees; he seemed to find what he was after almost immediately.

When he came out he just said to his colleagues, "It's exactly what we thought. I think we can go back to the office now, fellas." To the employees he said, "Okay, everybody, back to work. Tell them"—he pointed to the women in the counting room—"they can do the same."

Smith turned to Danny Lehman. "I suggest you come with us so we can give you a full explanation. I think you'll find what we are going to tell you, how can I put it, of great interest, great interest."

Danny said, "Do you mind if I do that a little later, say, tomorrow?"

Smith smiled at him. "No problem, Mr. Lehman. In fact, I thought you might want to have a little time to sit down and start to enjoy things around here. It's all yours now."

Danny returned the smile. "How about if we begin right now with my inviting everybody, all of you guys, for a drink. All the liquor around here belongs to me, too."

Smith laughed and actually patted Danny on the back. "Mr. Lehman," he said, "we'd love to more than I can tell you, but I think you understand we can't do it during business hours."

Danny got the message. "Mr. Smith . . . or may I call you Bill?"

Smith answered, "Bill it is."

"Why wait until tomorrow? How would it be if we reconvene here after business hours, say around eight o'clock tonight, so that you can start at the beginning and lead me all the way through this thing."

Smith liked the idea. "It's a deal. I'll call you when I get here." But then he thought of something. "Where will I be able to reach you?"

"Hookers' corner," Danny answered.

Smith grinned. He liked that. He then issued one final command. "Mr. Downey"—the accountant now looked as if he needed a doctor—"I think you'd better come with us, too."

Downey looked at Danny as if asking for help. Danny was embarrassed. First he looked away; then he walked away. And as he was walking, he was hit by a very disturbing thought: there but for the grace of Amaretto and Sarnoff go I!

9

Danny headed for the elevator to the executive offices on the third floor. The smartass secretary was there to greet him when he reached the door that led to the general manager's office.

He looked at her for a few seconds and then said, "I'm going to ask you to do something, and if you want to do it, fine; if not, that's also fine. All right?"

The woman nodded uncertainly.

"I want you to round up some boxes: wooden boxes, cardboard cartons, whatever's around. Okay? Then I want you to round up a couple of girls to help you. Is there a secretarial pool around here somewhere?"

"Yes, sir, it's in the back."

"Good. Get two or more girls from the typing pool, and then I want all of you to collect everything, and I mean everything, you find in De Niro's office, and everything you find in Salgo's office, and put it in those boxes. Do you understand?"

"Yes, sir," she answered.

"Now, before you do that, I want you to call the head of security around here, whoever he is, and tell him to come up here. Could you do that right away?"

The secretary made the telephone call and then disappeared to attend to her various roundups. Danny decided to take the chair behind the secretary's desk and, while he was at it, to take a quick run through her papers. None proved very interesting. About five minutes later, a rather heavyset swarthy man came hurrying up the corridor. He saw Danny behind the desk and didn't approve of what he saw.

"What are you doing sitting there, and who the hell are you anyway?" he demanded.

"My name's Lehman, Danny Lehman. I own this place. What can I do for you?"

The impact of his words was immediate and devastating. "I'm terribly sorry, Mr. Lehman. I'm head of security. You asked for me, apparently."

"I did. I also asked some of the young ladies up here to pack up all the belongings of De Niro and Salgo. I want them stored somewhere secure under lock and key, and when it's done, I want you to tell me where the place is and then give me the key."

"When?" the man asked.

"Right now. Wait in De Niro's office. Nobody goes in but the girls. Nothing comes out that isn't in boxes. Right? I don't want one item missing, do you understand?"

The man didn't say a word. Instead he just disappeared into the general manager's office, closing the door behind him very, very carefully. Then, gradually, Danny Lehman started to return to normal. He'd had enough of confrontations to last him not only for the rest of the day, but for the rest of the year.

Forty-five minutes later the offices of De Niro and Salgo had already been totally sacked and packed. The head of the casino security force and the tough cookie who was De Niro's secretary jointly presented him with the key which, they explained, would open the storage room on the basement level adjacent to the underground garage, Storage Room G. By this time Danny had seen more keys in one day than in any other day of his life, and was sick of the sight of them. Nonetheless, he accepted this last one graciously and then addressed a few more words to the pair standing in front of him.

"First, I want to thank you. Second, I want to tell you that you're both fired. I'm going to have paychecks ready for you within the next half hour. I'm going to include in that final paycheck an extra month's salary in both of your cases. I'll arrange that you can pick them up at the cashier's counter downstairs. Now I want you both to pack up your things, and I want you and those things out of this building in no more than one hour from now."

Danny then turned his back on them and walked into the general manager's office, which was now bare of everything but a desk, a chair, a coffee table, a couple of sofas, and some empty bookshelves. He picked up the phone and asked to be connected with the assistant manager in charge of room reservations. Once again he said, "This is Daniel Lehman." Then: "I'm not sure if you know who I am."

"Yes, sir, I do," the man on the other end replied. "Yes, sir."

"Tell me, what is the best suite in the house?"

"It's Suite 1515, sir, in the tower."

"I want it to be ready for occupancy within thirty minutes."

"Sir, I'm afraid it's already occupied by a very important client who is, I might add, a close friend of Mr. De Niro."

The word had apparently not reached the reservations desk. "Well, kick him out. I want all of his stuff out, and the room ready to go within thirty minutes. Do you understand?"

"Yes, sir."

About forty minutes later, Danny was lying in the bathtub

in Suite 1515 of the Raffles Casino and Hotel, singing loudly and badly. He never got too far with any given song, since, after just a few bars, he remembered the tune but not the words. No matter, he told himself. Danny Lehman had beaten them. Danny Lehman had arrived. Danny Lehman was happier than he had ever been at any time in his life. Because he now knew that it was just a matter of time before he got this money machine under control; just a matter of time before he would start to produce profit upon profit, starting a process which, he knew, would make him richer than all the people in the Lehman family combined, back through generation after generation to the old country and, beyond that, back no doubt to Palestine.

Atlantic City! For months now he had barely had time to think about Atlantic City. And it would still be a while before that project could be moved front and center. First he had to clean up Raffles. But after that: imagine what Atlantic City could be once again! No, not *could* be, *would* be. An instant money machine on a scale that even Henry Price could not imagine, because the Prices of this world had run out of dreams. But not Danny Lehman. Now that the nightmare was over he would have time to indulge in daydreams again. Because he'd won.

◆   ◆

At eight o'clock sharp, bathed and shaved but dressed in his normal style—a pair of brown slacks and a yellow sport shirt—Danny Lehman was back at exactly the same table in the same bar as he had been just a month previously. Now as then, he sat in splendid isolation, since RESERVED signs had been put on all the tables surrounding his. But this time they had been placed there at his instruction, placed there by a maître d' whose demeanor exuded a humility, a docility, and a subservience so pronounced that Danny decided then and there not to fire the guy; at least, not to fire him that day.

Just a few minutes later, William Smith arrived. He greeted Danny with a big open smile and a very firm and enthusiastic handshake. "I'm off duty, Danny, and if you don't mind I'd like a double vodka martini with a twist and tell them to forget the vermouth."

Danny decided to follow suit and told the maître d' to get two of them. The man arrived back with the drinks so quickly that it was surprising that the speed of his movement didn't result in a meltdown of the ice cubes. Smith downed most of his in one go, flashed another big grin at Danny, and said, "Are you ready?"

Danny answered, "I'm ready."

"All right. Here we go. From the beginning." He paused. "De Niro and Salgo have been skimming this place for years. That's hardly an unknown practice in this town, but the word was out that these two guys were going at it in a very hot and heavy fashion. You're obviously brand-new at this business and you probably were not aware of the talk that had been going on around here. It proved, if anything, to *understate* the situation. We've now counted the cash we found in those safety-deposit drawers in the counting room, and just so you'd know, I wrote down the exact figure on a piece of paper."

He now pulled it out of his pocket. "Here are the grand totals: De Niro, five hundred and sixty-nine thousand four hundred dollars. Salgo: four hundred and eighty-seven thousand six hundred."

"I'll be damned!" Danny exclaimed. Which required a bit of acting. In actual fact, the numbers added up almost exactly, based upon what had been coming through Philadelphi via Amaretto up until just over a month ago.

Smith continued, "Now, we have no way of ascertaining over what period of time these two men accumulated these amounts. But our suspicion is that it was probably over no more than a couple of months. So you can see, Danny, this has been a very big operation indeed. From what we have been able to determine, it started out on a really small-time level, almost on the level of petty theft when compared to the amounts we are talking about today. It seems that initially these two geniuses were raiding the slots, stealing silver quarters and silver dollars and peddling them through some fencing operation that we still haven't been able to track down. There's a big missing link there. But that's getting off the main track."

Let's hope it stays on that sidetrack, Danny thought.

"Okay," Smith continued, "what they figured, and I guess correctly so, was that the controls in the counting room were simply too good and complete for any large-scale theft to escape the attention of the accountants around here. That changed, and I'll come to that in a minute. But anyway, in the beginning it was silver."

"How?" asked Danny. "How did they do it?"

"Well," The FBI man answered, "now that you own and run this place, it's the sort of thing you'd better bone up on." He continued, "As is usual with so many of these types of crime, it was so simple that it's hard to believe it worked, in retrospect. The answer lies with the people that constructed this place six or maybe seven years ago. The counting room was originally built in exactly the same position and in the same way as it stands today: nothing more, really, than a big concrete bunker slapped against the rear wall of the casino building. What somebody did—and God knows who authorized it, but somehow it must have arisen from a demand of the building inspectors, trying to conform with the building code in regard to precautions against fire—was put a fire door in the damn counting room, a door that led directly onto the parking lot at the back of the building. Now, you probably don't know this, but there is an absolute rule laid down by the Casino Control Commission of the state of Nevada from the very beginning that there can be only one method of access, one door and one door only, leading into, or, more pertinently, leading from the counting room of any casino in this state."

Smith was right. Danny had no idea that there was such a rule. In fact, he was totally ignorant of all the rules, not just this one. Out of this realization, which just now began to dawn on him, an idea started to develop. But for the moment he let Smith continue.

"What I suspect happened is that the guys who owned this place knew about that door. I don't mean De Niro and Salgo, and I don't want to mention any other names here because I don't think they're going to come up in this investigation. As you must know, the guys that owned this place, the guys you bought out, were men of not exactly impeccable credentials. What I think happened is this: one, or more, or all of them saw the opportunity, the latent opportunity one could say, inherent in having a fire door in such a wonderful place. And instead of having the construction corrected, they had the thing plastered over, or at least more or less plastered over. Now, what I suspect further is this: those guys never used that door, but at some time, and our guess is it must have been two or three years ago, either Salgo or De Niro, we'll never know how, found out about this door and started to use it."

Smith stopped there for a moment to finish his second drink, but then continued, "I've already mentioned that we think they started out on a truly penny-ante basis."

"How?" interjected Danny.

"Well, what happens in this casino where the slots are concerned is that they're emptied out every twenty-four hours. What they do is wheel those carts—those metal carts you saw in the counting room—from machine to machine, and dump the coins into the metal boxes, which are then chained back to the cart so nobody can snatch one and run. This is done at four in the morning, because that's when the level of activity in casinos, not just this casino but every casino in general, is at its lowest ebb. You can understand why you'd want to do it at this time because it'd be rather embarrassing if too many people were there to witness the amounts of their money that the casino was carting off to its counting room. There's nobody in there counting at that time in the morning. The first shift of women counters, the ones we saw in there, arrives at eight o'clock in the morning and works until two, and the second shift works until eight o'clock in the evening, and that's when at least the counting activity in the counting room is shut down for the day. Okay so far?"

Danny nodded.

"What De Niro and Salgo would do is this: they'd go into the counting room, since they've got all the keys and combinations, and they'd unchain some or maybe all of the boxes. Then one of them would cart the metal containers out through that fire door and dump some of the contents onto a pick-up truck or whatever was parked back there, no doubt leaving a sufficient quantity in each box in order not to raise any suspicions. Then the stuff would be moved off and, as I said, would later be sold through some fencing operation, probably out of state. While one guy was raiding the counting room, no doubt the other was in the other section of that place where the credit files and everything are stored, creating a diversion, or in any case distracting the attention of anybody that was still in there at four o'clock in the morning."

Smith emptied his glass and then continued. "After all, whoever was there at that time of night would probably be as low as you can get on the organizational chart of this place, and there, 'visiting' him, would be one of the two guys who run the place. Who's going to even consider blowing the whistle on any suspected hanky-panky under those conditions. Right?"

The more Danny listened to the FBI man, the more that idea was growing in his mind.

"Anyway, I'm sure that whether it was De Niro or Salgo who was doing his thing in the counting area, he worked in the dark, so it was really not very difficult to pull off, even on a regular basis. You've seen the physical setup there. But there's a limit to how much you can do with this sort of operation, because to a substantial degree the movement of coinage inside a casino is circular, if you know what I mean. The customers buy their rolls of coins from the cashiers; they then lose them to the slot machines; the slot machines are emptied out every twenty-four hours and the drop is brought to the counting room. It's counted, and the coinage is recycled back to the cashier. So, as I said, there is a limit to the amount of money those two guys were able to take out of circulation, and I think they reached this same conclusion at a fairly early stage. But what was still available to them was the fire door, provided that somehow they could find a way to circumvent the auditing system in regard to the really *big* money floating around in this casino, namely, that generated at the gaming tables.

"Am I going on too long?" Smith asked Danny.

"Hell no," Danny replied. He really liked this guy Smith. And he could tell that the feeling was mutual. And, despite himself, Danny was fascinated with the story Smith was telling. It was, in fact, the mirror image, seen from the inside, of the same story he'd been following for years from the outside, through the periodic visits of Amaretto and Sarnoff to Philadelphia. *Déjà vu* was the phrase they now used for that, Danny thought.

"Okay. Now, I don't know if you noticed, but there is really no way to get around the controls in the counting room. The women in there count out loud and this count is picked up by a microphone and recorded. Likewise there is a camera, in fact three video cameras, trained on that room. Again, all the activities are put on tape. Well, the audio and video tapes go to the auditing department and are used to verify the results recorded in the ledger accounts. Right? So how do you get around it?"

The question had perhaps been asked in a rhetorical manner, but the answer suddenly flashed into Danny Lehman's mind and he blurted it out. "Downey! Rupert Downey!"

"Precisely," said the FBI man. "Once they had the chief auditor in cahoots there was no problem. All they had to do was change the numbers in the ledger that they received from the counting room, and at the same time lose or make inaudible and unviewable those portions of the tapes covering the period of time for which the count was to be falsified. Once they had 'fixed' the exact amount, out the fire door it went. This time no truck was necessary. Either De Niro or Salgo would just fill up his pockets at five o'clock in the morning and walk out the back, and that would be it. One would disappear into the night while the other was putting up the smoke screen. So when it got to this point, the operation was literally unstoppable. When you've got collusion between the managers of a place like this and the auditors, then the sky's the limit. And De Niro and Salgo, I guess, were determined to aim for just that: the sky."

"But why were they stupid enough to keep that amount of the skim in the safety-deposit boxes?" Danny asked. "Why didn't it go out the fire door immediately?"

"Well, that's the sixty-four-thousand-dollar question. And we don't know the answer. All we do know is we got a tip from somebody, and we don't know who it was. It was by telephone. It might have been somebody inside the counting room or somebody out on the casino floor, or even somebody upstairs in the executive office. Or maybe somebody on the outside who's got somebody on the inside. But it certainly came from a person who was out to get the asses of De Niro and Salgo and their buddies. The informant tipped us off to the effect that there was a lot of money accumulating in some safety-deposit boxes in this casino. The same source gave us some—I might as well be honest—most of the information about who was involved and what they had been doing. But back to your question. Why did they keep that amount of money here?" Smith stopped for a minute, and then said, "What follows, you understand, is pure conjecture.

"Okay. What we suspect is this: that the fencing operation De Niro and Salgo had been using for some years folded. Or perhaps they had a falling out with the people who had been fencing for them. But in any case, the pipeline got plugged all of a sudden. Got stopped up. They nevertheless kept stealing, but they were probably afraid of keeping the steady accumulation of new cash around at home or elsewhere. I mean, there are lots of crooked people around who might have stolen it!" Smith laughed at his own joke, and Danny joined him.

"Furthermore, I think these guys had developed such a

degree of overconfidence that they didn't think twice about keeping this amount of money in their safety-deposit boxes until they found a new pipeline to a new fence, or unplugged the old pipeline to the old fence."

"But how can you prove it?" Danny asked. "They claim they won it all at the baccarat tables."

"All right. I told you it was just conjecture. And between you, me, and the fence post, we can't prove a damn thing. Because, let's face it, they must have been spreading lots of joy among the baccarat dealers."

Danny privately thought that they were the next ones who had to go.

"But not to worry. We brought the IRS into this case just about an hour and a half ago. We explained our problem to them and they confirmed to us, in fact absolutely guaranteed, that the Treasury Department is going to go after these guys for criminal tax evasion, which no doubt extends over a period of years and obviously amounts to millions of dollars. As you know, the IRS doesn't normally have to prove anything. They'll just take the amounts that we found, over a million, extrapolate back, and come up with some number that is so staggering that it would mean that these guys would go to the slammer for the next twenty years at least. They'd probably get off a lot easier if they had just killed a few people. Or maybe I should say if they had killed a few more people, going on the rumors that have been floating around these two guys for years now. But all this we're hardly going to admit. As far as they will ever know, we've got the goods on them for grand theft on a monumental scale. Our story is that after we've arranged for them to get fifty years, then we'll turn them over to the guys from the IRS, who will get them for another twenty years. In the end, no doubt, we'll make a deal, as much as I hate deals. But we won't have the evidence we would need to take it all the way in a court."

"What about Downey?" Danny asked.

"Well, that's different. I kind of think that we should make a deal with him right now if he turns state's evidence. Wouldn't you agree?''

"Yes," said Danny. "For sure. I feel sorry for that guy."

"Don't," replied the FBI man. "He's a crook just like the rest of them. So that's it, Danny. . . ."

"Wow," said Danny. "Would you care for another drink?"

"With pleasure."

When the second round came, the two men just sat there in silence for a while, savoring the moment. Then the FBI man said, "When I was walking in, one of the fellows told me that you fired the head of security here. Is that right?"

"That's right," Danny replied. "In fact I want to talk to you about that." His idea was now a certainty.

"No need to talk to me about that, Danny. I have no doubt that if there was anybody else here who was also involved in this operation it must have been him. We'll get to him in due course."

"I didn't mean that at all, Bill."

"No? Sorry."

"I might as well be rather blunt. I'd like you to take over as the head security man of this place. I'd pay you very well and you'd have a completely free hand. I'd ask you to do just one thing initially and that would be to clean this place out from the top to the bottom. Anybody, and I mean anybody, with anything in his background, with any funny connections, in fact with anything suspicious at all, I would want out! Because what I intend to do is to start this place all over again with an absolutely clean slate, and then I intend to keep it that way. You know there's this myth going around that the very nature of the gaming business dictates that casinos can only be run successfully by crooks: gambling, prostitution, crime, criminal operators . . . they all go together, in fact, have to go together. Well, I'm going to prove them wrong.

"Secondly," Danny continued, "I want you to teach me the rules of the game here, and how everybody tries to break them. Do you want to help me?"

Bill Smith did not hesitate. "You're goddamn right I do, Danny. Give me three months. I've got a few things to take care of first, not the least of which are De Niro and Salgo."

"Three months. You've got it. Are you willing to shake on it?"

"Yessir, Mr. Lehman."

"Okay, Bill, thanks, and we'll be in touch. In fact I'll have a contract drawn up this week."

"You're on," said Smith as he left.

After Danny had lost sight of Smith he suddenly realized they hadn't exchanged one word about salary. And the more he thought about it, the better he liked it. Smith trusted him to be fair, to be correct. When his luck had turned, it had turned completely!

It was now approaching nine o'clock on that evening of June 9, 1969. And hookers' corner was starting to fill up with hookers. In contrast to the last time Danny had sat there, however, not one of them cast even a passing glance in his direction. But there was no apparent change in the level of activity on the casino floor. It was then, exactly then, that Danny decided what had to be done.

He started to raise his right hand to beckon the maître d', but it had barely risen six inches above the table before the man was already scurrying in his direction. "Bring me a phone!" Danny ordered. And the maître d' not only brought a phone, plugged it in, and placed the instrument in front of Danny, but also picked up the receiver and placed it gently in Danny's right hand, which was still six inches above the table. Then the man retreated, facing Danny in a position of obeisance that would have been more fitting in Riyadh at the king's palace than it was in Las Vegas at hookers' corner. But it was a scene that further confirmed the validity of the idea that was starting to grow at an almost explosive rate in Danny's mind.

It took the hotel operator ten minutes to finally track down Danny's hotshot Hollywood agent. All that Danny said to Mort Granville was that he'd appreciate it if he'd get on the next plane to Vegas. Granville agreed immediately. The next call was to his Philadelphia lawyer, to whom he gave exactly the same instructions.

He had just replaced the receiver for the second time when somebody patted him very gently on the head from behind. He tilted back to see who had dared pull such a stunt. And when he did, all he could see, at least initially, was a white sweater filled with two of the most shapely breasts that Danny Lehman had ever observed, at least from that angle. But then he saw above that marvelous sweater an even more marvelous face. A black face.

"Sandra Lee!" he exclaimed. "You're exactly the right thing at the right time."

The woman leaned down and kissed him first on the forehead and then firmly on his lips. "Your place or mine?" she asked.

"Mine," Danny replied immediately.

"And what about," she continued, "you know, about that problem you were having the last time we saw each other?"

"Solved, baby," he replied. Then he remembered that those two goons with blue suits were probably still on the payroll. But that, he thought, could certainly wait. After all, a man had to set priorities in life.

"So where's your place?" the woman asked.

"I'll show you right now."

He got up and took her hand, and again, as last time, this incongruous pair began to work their way across the casino floor. But this time, instead of heading for the exit, they headed for the elevator bank in the hotel tower. As Danny walked beside this woman he was, if anything, more impressed than before by the regal manner in which she carried herself, looking first left, then right, upon the motley crowd, more than a few of whom stopped to gaze. Now, as before, her style of dress could not have been more simple: a white cashmere sweater and a narrow black skirt, slit almost up to the hip, but worn in a way that was highly suggestive, yet at the same time perfectly discreet.

When Danny led the way into Suite 1515, the first thing that Sandra Lee did was emit a low whistle. "Honey," she said, "I knew you had class, but I didn't know you had *this* much class." She then turned to him and nearly smothered him with a bear hug into which she put every ounce of pressure she could muster from her long, lithe black body. "Sweetie, I think what you need now is a good piece of ass."

With that she stepped back and pulled off her white sweater with one swift movement. She wore no brassiere because it was obvious that what had been beneath that white sweater needed no support. She then opened four buttons at the side of the black slit skirt and, again with a swift, sure movement, stepped out of it. It was now apparent that Sandra Lee was not in the habit of wearing any underwear at all.

"Now I am going to do you, Danny," she said. With the same sureness she soon had Danny undressed and on the bed, where she proceeded first to arouse him with a tongue that seemed never to stop, and then to fuck him with a style, imagination, and vigor that left Danny Lehman as drained as he had ever been in his life. And when she rolled off him for the last time, she said, "Honey, let's eat."

Lying on her stomach with her long legs and high heels in the air, heels that for some reason she had decided to put back on, she picked up the phone and ordered a complete menu that began with Iranian caviar, moved on to onion soup, and ended up with roast pheasant. When she was done she put her hand over the phone, and asked, "Danny, what do you drink?"

"It's got to be champagne, the best they've got. This is my big night. The first that I've had in many a month. So get the best."

She returned to the phone and said, "Send up a bottle of Cristal. No, make that *two* bottles of Cristal. Two bottles and two ice buckets and listen, honky, I want that up here within five minutes or it's going to be your ass!" And she slammed the phone down and laughed like hell.

Danny loved it. He started to giggle, which got her going even more, and when they both finally got over it, he said, "Sandra Lee, I like you. I like your style. I think we're going to get along just fine." Then: "We'd better get some clothes on before that champagne arrives."

"Fuck the clothes," said Sandra Lee. "All we are going to do is turn the thermostat up, and if somebody doesn't like it, screw him."

The champagne arrived in less than five minutes, and it was Sandra Lee who went to the door to let the waiter in, still simply clad in gold earrings and high heels. But Danny couldn't do it. He was from Philadelphia, after all. So he waited in the bathroom until the delivery had been completed.

"Come on out, you chickenshit," called Sandra. He did, and when the food arrived they both went through exactly the same performance. Then the two of them spent the next couple of hours eating and drinking and fooling around a little bit, then drinking and eating some more. Around midnight Sandra Lee suddenly made a proposal. "Danny," she said, "I don't want you to get bored. I've got a girlfriend, in fact she's my roommate, and her name is Shirley-Anne. Why don't I call her up and ask her to come over and we'll make a little party of it?"

She did, and later the two black hookers and Danny Lehman were all sitting in the bathtub in Suite 1515 of the Raffles Hotel and Casino. They were drinking from their fourth bottle of Cristal champagne, which had been delivered by a waiter who wouldn't have exchanged jobs with anybody that night, especially when he entered the room with the third and fourth bottles, as well as two additional ice buckets, to find not only one tall black naked woman there, but now a second, dressed in exactly the same way, except for the fact that she wore neither earrings nor shoes.

"Girls!" Danny suddenly asked. "If you owned this place, what would you do with it?"

"For one thing," Shirley-Anne said, "I'd sure as hell put in a much bigger bathtub. Man, I'm wedged in here so I can hardly breathe, much less move."

Danny reached down and put his hand between her thighs, explaining that he was going to try to unwedge her, a move that got all of them giggling again. But then Danny said, "You're right, you're absolutely right. What else?"

"Are you asking *me* again?" Shirley-Anne said.

"Yes."

"Mirrors. You've got to have mirrors all over the place, don't you. It's more fun that way."

Again Danny said, "You're right."

"What they should do," the black woman continued, "is make this look like a very elegant cathouse. That's what the guys want when they come here, isn't it? And so do their wives if they get taken along. But they don't want to come to a cheap whorehouse like this goddamn place. It's got to be like an expensive French or Italian whorehouse. Aren't I right, Sandra Lee?"

"You're absolutely right, honey, but why are you asking all this?" Sandra asked Danny, as she gently began to massage him below the sudsy waves.

"Because I think I probably forgot to tell you girls something. I bought this place a couple of months ago."

"You bought what, honey?" asked Shirley-Anne.

"This place, Shirley-Anne. Raffles. The casino. The hotel. Everything. It's mine now."

Sandra Lee suspended her massage, extracted her hands from the suds and put them on both sides of Danny's head, pulled him even closer, and said, "Danny boy, I knew it, I just knew it; you're something special!"

The phone rang, and when Danny dripped his way across the room to pick it up it was his Hollywood agent reporting in. Six hours later it was Sandra Lee who took the phone when it rang again, since Danny Lehman was out cold, having been totally saturated with both alcohol and sex. She told the man who claimed he was Mr. Lehman's lawyer from Philadelphia the same thing that Danny had told his Hollywood agent, namely, "Mr. Lehman's busy right now. Call him this afternoon, but not too early this afternoon." In fact, all three of them slept until 3 P.M.

◆   ◆

The next day Danny Lehman was installed behind De Niro's desk and was explaining to Benjamin Shea and Mort Granville, the two charter members of his board of directors, everything—almost everything, since he left out Sandra Lee and Shirley-Anne—that had happened during the previous forty-eight hours. When finished, he expressed his conclusions very succinctly. "What all this proves is that this casino operation has always been profitable in the past. But when you look around, you can't help but wonder for how long. If things around here are run in the future the same way as they've been in the past, this place is going to go broke fast, even with De Niro and Salgo gone. It's quite obvious that Raffles is rapidly losing its clientele. Who wants to gamble in a place with lousy service in a building where the paint's peeling? This casino's got to be rebuilt from the ground up, and that's exactly what we're going to do. I'm going to close this place down at the end of the week. I'm going to fire most of the employees. I'm going to hire the best goddamn architect and the best goddamn interior decorator that I can find. And we're going to redo it from scratch."

Neither Granville or Shea wanted to say a word, it seemed, so Danny continued. "And I'm going to do something else. I'm going to change the name. I've been thinking about it for some time now, and what we need is a name that expresses class. We want this place to be a place that people are going to fight to get into, because it's going to be, you know, like a private club, not for everybody, not for the masses, but something exclusive, a place with a reputation like a Maserati or a Rolls-Royce, and the name's got to express all of that.

"Last night I spent six hours looking around this town. I was in and out of every big casino here. And the only place that even comes close to what I'm talking about is Caesars Palace. The place was packed! So I asked around why. It's all due to a guy by the name of Cliff Perlman. Everybody says he's a genius. He created Caesars Palace and in the process established new standards for the entire casino industry. So why not copy success? I'm going Caesars and Perlman one further.

"What we are going to call this place is"—here he paused for dramatic effect—"the Palace. That's what we're going to call it. *The* Palace. Because it's going to be *the* gambling palace, not just of Vegas and not just of the United States, but the gambling palace of the whole goddamn world. Our place, our Palace, is going to be to the gambling business what Buckingham Palace is to royalty: top of the league. In one word, it will have *class!*"

10

Exactly one week later, the Raffles Hotel and Casino closed its doors for good. True to his word, Danny fired almost two thirds of the casino's employees outright, and the rest of them were told that when the casino reopened, possibly, just possibly, they might be kept on.

Danny then decided that what he needed most urgently were two things: an architect and more money. The architect he found almost immediately. He had heard about a man up in San Francisco who was a friend of the Shah's, a confidant of the Kennedys, an architect to the people who ran the world, a man who built elite structures for the elite. Danny flew up to San Francisco to meet him and immediately told him what he wanted: a palace in the desert of Nevada, a palace to top all palaces. Not, he stressed, just a copy of Caesars Palace, but something on a much higher plane: a place that *reeked* of splendor and exclusivity.

Danny confessed that he didn't know exactly how to go about it, but he'd heard that this man knew about desert palaces because of the work he did in places like Iran and Saudi Arabia. It was true. Then Danny mumbled something about structures he had read of in such places as Persepolis, Jidda, and Riyadh. The architect knew of them, but he wondered if it would be altogether appropriate, and he tried to put it as discreetly as possible, to use as models structures that had been essentially inspired by the Muslim faith. Well, that stopped Danny for a while, since he hadn't really thought about it from that angle. His inspiration had come from the *National Geographic* magazine, which his mother subscribed to for him. As the conversation progressed over a fish lunch at the Washington Square Bar and Grill in the North Beach part of town, the architect gradually brought him around to the concept that what he really wanted was a palace, yes, but a modern palace filled with exclusivity *and* decadence. He suggested that though Danny had been right in searching for the correct motif in antiquity, he had probably gone a thousand or so miles too far east. The best model for Danny's purpose in all of the ancient world was still that of Rome, he said. Cliff Perlman had been right when he had used the specifically Roman motif for Caesars.

Then he continued, "But what you need is *real* fountains and baths, *real* marble and *real* statues. Not kitsch like they've got at all the other places in Vegas. Let me make a suggestion, Mr. Lehman. Go to Rome, go to Pompeii, visit Herculaneum, tour Sicily and look at Syracuse and Agrigento, and you will see what I mean." Then he added, "But to make it work we'll have to add neon lights and air conditioning and wall-to-wall carpeting."

Danny was already convinced. Would the Shah hire a dummy? Then came the real question. How much would it take to re-create out of the old casino the ultimate Roman palace with neon lights that the architect envisioned? "Twenty-five million dollars minimum, but my guess is it will cost forty million before we're through" was the answer.

"How much would you need to start?" Danny asked.

"Twenty million, Mr. Lehman. In the form of an irrevocable letter of credit. After all, you're not exactly General Motors." It would be a very hot day in San Francisco, the architect had already concluded, before he ever saw that kind of money from this guy.

◆   ◆

The first thing Danny did when he got back to Las Vegas was to go to a phone booth and put a call through to Fort Lauderdale. Meyers was not in. Then Danny had a better thought. Why give those guys any big ideas that he, Danny Lehman, really needed them?

He went back to the hotel, called his attorney, and instructed Shea to put the American Coin, Metals, and Currency Exchange on the market. He told him he wanted a quick sale. He realized that this would not allow him to get the maximum price, but he was now convinced that he was on the right track in Las Vegas and that time was of paramount importance. If he ever went back to Fort Lauderdale for more money it would be for some really big money—Atlantic City money.

Between now, Las Vegas, and then, Atlantic City, he had to prove himself, on his own. A letter of intent was signed ten days later. A New York conglomerate agreed to buy Danny out for a total consideration of twenty-two million dollars cash, a price well above anything Danny had expected to get. After paying off the loan he had taken out in Philadelphia in order to finance the initial cash down payment on Raffles, he would still be left with a hefty bank balance. Armed with that, or at least the documented promise of a hefty bank balance when the transaction was completed, Danny managed to negotiate a twenty-million-dollar revolving line of credit with a Los Angeles bank, or at least his lawyer from Philadelphia did. Days later, Danny was issued a letter of credit for the same amount.

When his bank in Southern California called the architect's bank in Northern California informing them of this fact, the temperature in San Francisco was ninety-two degrees, the hottest day of the year. The architect was impressed! A Shah this Lehman fellow wasn't, but to produce both the twenty million and the hottest day of the year that quickly: not bad.

The reconstruction of the casino and hotel began on July 15, 1969. Three days later, Danny Lehman, who by this time had reached a stage of total exhaustion, decided to take some time off. He would go to Rome as the architect had suggested. And while he was in Europe, he would also look around at some of the casinos in that part of the world. For Danny had a feeling in his gut that he might find something, maybe the missing elements in that grand design which he sought, just as only five years earlier he had discovered the entire foreign exchange business, something which, after all, had just netted him twenty-two million dollars!

◆   ◆

When three days later the man in charge of the reception desk at Claridge's looked up to find a short, pudgy American wearing a wine-colored sport jacket over a pink shirt with white tropical trousers above brown loafers, and beside him a six-foot-two-inch negress clad in a green dress with a décolletage that ended well down into her midriff, he knew that somebody had made a terrible, terrible mistake.

"Your name, sir?" he asked.

"Lehman. Daniel Lehman," the American answered.

"One moment, sir," the man at Claridge's replied. He went into the office to check the list of those clients' names deemed acceptable for lodging by the management of Claridge's, because he knew that nobody, but nobody, could have been granted a reservation if his name wasn't recorded there. And how did one get one's name registered in the beginning? Only upon the recommendation of somebody who was already there, naturally.

Lehman? Lehman? Of course. There was the obvious reason for this unfortunate mistake. Someone had surely mistaken this man for one of the partners of that prestigious firm of Lehman Brothers in New York. George Ball had been at the hotel just last week. Or was it a mistake? One could never be absolutely sure where Americans were concerned. The assistant manager, clad in tailcoat and striped trousers, went back to check. "You are, sir, I believe, from Lehman Brothers, the investment bank in New York, are you not?"

"No. I'm from Las Vegas. But it was a bank, my bank in Los Angeles, which promised to fix me up with a room in London. I borrowed twenty million dollars from them, and so they got me a hotel room here."

It was now decision time for Claridge's: to kick out a couple who looked like a pimp and his whore and risk a confrontation that might never end, causing discomfort for the other guests who, as they came and went, were already casting curious glances in the direction of this unique pair—or to check them in as quickly as possible and hope that they would stay in the room and out of sight during their stay. The assistant manager decided on the latter course of action, and within minutes Danny Lehman and Sandra Lee were in a suite on the third floor, one that was going to set Danny back two hundred pounds a night, but he didn't know that yet. It was a curious mélange of Victoriana and art deco that is uniquely Claridge's.

"Do you like it, honey?" Sandra Lee asked.

"It looks awfully goddamn old-fashioned to me," Danny replied. "It reminds me of something my grandma might have liked. Man," he added, "the British are sure behind the times."

Sandra Lee, of course, checked out the bathroom next, but she was delighted at what she found. "Come here, Danny. Look at this stuff. My God, the fixtures must be fifty years old. And look at the old mirrors, the marble! It's beautiful. I love it!"

"Beautiful? Are you kidding? It looks a bit seedy to me. I'll tell you something. I was afraid to ask that guy downstairs how much the room would be, but boy, if they charge me more than fifty bucks a night for this, I'm sure as hell going to complain."

They took a nap, but by six-thirty that evening Danny was ready to go. "Honey, I'm going to call the front desk," he said, "and ask them for the names of the best casinos in this town."

"You don't have to do that," Sandra Lee answered.

"Why not?" he asked.

"Because I know where they gamble here. Since they don't have casinos, they do it at clubs. I not only know the names of some clubs, I even know where they are. The best one is Crockford's. It's off Pall Mall. The other one is Les Ambassadeurs. It's a restaurant and nightclub, with gambling thrown in. That's where the very upper class go to have their fun. It's opposite the Hilton."

While she was speaking, Lehman watched her, his eyes growing ever wider with amazement. When she was done, he expressed just that. "How in the world do you know these things?"

"Honey, what you mean is how in the world could a black hooker from Las Vegas know anything about London, England. Isn't that right?" It was right and they both knew it. "Well, Danny, if you want the truth I'll tell you," Sandra Lee said. "I've got a client from London. He always stays at the Sands. He is in the insurance business, with Lloyd's, if I remember correctly, which also happens to be here in London, my dear. About a year ago he did very well at the Sands and I sat beside him the entire time. The next day we both flew to London. He put me up in the Savoy and every night we dined exquisitely and danced and gambled, and that's how this big black hooker from Las Vegas knows a few things she probably shouldn't."

"Ahh, come on, cut it out, Sandra Lee. I didn't mean it that way."

"I know you didn't mean it that way, but that's the way it came out," she answered.

"I'm sorry. What matters is that we know where we want to go tonight. What was the name of the first one again?"

"Crockford's. But it's not that simple. You can't just go there. Like I told you, in London these casinos are private clubs. You have to be a member to get in, or"—and she paused—"you have to know a member who will vouch for you."

"And do you think your friend would do that?"

Sandra Lee smiled. "For me, yes. For you?" She spread her hands, but then burst out in a giggle. "Of course he would, Danny! I'll call him, or at least I'll try to call him. I don't think he's in London that much of the time. But who knows, we might be lucky. I'll give it a try, okay?"

She went to her immense black purse and pulled out a very thick, well-thumbed notebook, which Danny eyed with some awe and more than a little discomfort.

She read his mind immediately. "Danny, this isn't just where I keep my telephone numbers. I keep everything in here that I want to remember: recipes, restaurants, the names of records I want to buy. All sorts of things worth remembering for a lady," she said, letting just a little mocking smile cross her face.

Danny liked the answer. In fact, he liked the answer a lot because it seemed honest and rather sensible. He liked girls who played it straight.

"I've got two numbers," Sandra Lee went on. "One here in London and one in a place somewhere outside called Amersham. I think I'll try the second first." She did, and she got the man she sought immediately. They chatted back and forth for a number of minutes, and when she hung up, she said, "Danny, we have to get dressed in a rather spiffy way. My friend—his name's Douglas Penn—will be picking us up 'within the hour,' as he put it. Now I'm going to put on something very black, and I would suggest—if you don't mind, Danny, I would strongly recommend—that you put on that blue suit of yours with a white shirt and a striped tie. You know, the one I gave you just before we left."

"The tie? Are you sure?"

"The tie, Danny."

Douglas Penn looked exactly the way an upper-class Englishman should look. He was tall. He was thin. He carried an umbrella and he wore a waistcoat. He had red cheeks and he spoke with both a slight stutter and a slight lisp. He drove a Bentley aggressively, so he covered the distance between Claridge's and Crockford's in less than a quarter of an hour.

Crockford's was housed in a huge, rather unassuming stone mansion, and though a club, it was hardly in the class of an Athenaeum or a White's. There was, however, a substantial overlap of their membership lists. That meant that one could have lunch in the hushed dining room of the Athenaeum with one's banker, bishop, or member of Parliament, and then have dinner in the very lively atmosphere of Crockford's, perhaps with a chum from one's years at Oxford, or more probably with one's mistress, or with someone one hoped might become one's mistress.

At Crockford's they greeted Douglas Penn like an old friend and not an eyebrow was even slightly raised when he introduced Sandra Lee and Danny as his guests. A quick look around downstairs indicated why: the posh restaurant there was crowded with a mixture of types and races. Here a Chinese person, there a Pakistani. At one table a group of young Arabs in full desert dress, there a colored gentleman in tribal robes dining with a young girl who looked as if she was playing truant from her English boarding school. It was London at its best.

"Now," said the host, "should we eat here and then go upstairs, or should we go upstairs and eat later at Les Ambassadeurs?" Mr. Penn looked at Sandra Lee and Sandra Lee looked at Danny Lehman.

"After the flight, neither of us is particularly hungry right now," Danny said, "so let's go upstairs if that's where the gaming takes place."

"So be it," replied Douglas Penn.

They walked up the stairs and entered a foyer in which a number of people were gathered in front of what looked like a cloakroom, which, upon second glance, was hardly its function anymore, since there was a man behind the ledge who was accepting not overcoats but money. To be more precise, money was being handed over from the outside and chips, very large chips by American standards, were being returned from the inside. It was not just the size of the chips that was in stark contrast to what Danny was now used to in American casinos; the crowd stood in equal contrast. For almost everybody up there, from the money changer in the "cloakroom" to the "host" who now greeted them, and especially Douglas Penn, in an especially effusive manner to the "guests": all were dressed in a manner which, at least from an American point of view, one would consider as formal. Most of the men were in black tie, and more than a couple of winged collars; the women, for the most part, wore full-length gowns, enhanced by what appeared to be extraordinarily expensive jewelry. Crockford's no more resembled a Vegas casino than did Claridge's a Holiday Inn. In fact, Douglas Penn had explained to them in the car, no one, but no one, referred to an establishment such as Crockford's as a casino, it was a club with gaming privileges, which prompted Sandra Lee to say to Danny, "I told you so."

The upstairs itself was divided up into three large rooms, each the size of very large living rooms, and a fourth, smaller one, which functioned as the bar area. Two of the gaming rooms were devoted to roulette and what appeared to be bridge games, of all things, and the other room to a game that was quite obviously the one that was providing all of the action. There were three tables in that central gaming room and all of them were going full blast. In fact, around each table, which seated a dozen people, there was a full row of spectators standing behind the seated players, sometimes two deep.

"What's going on there?" Danny asked as he stood in the foyer, still observing the scene from a distance.

"Chemmy," answered Douglas Penn.

"What?" Danny asked.

"Chemmy, my dear chap. That's the word we English use for chemin de fer. Foreign languages have never been our strength, you know," explained Douglas Penn.

Danny still looked puzzled, since his French was not weak, it was nonexistent.

"You call it baccarat," continued the Englishman; "that's from *baccara,* which is also French, named for the town of Baccarat, where the crystal is made.

"I see," said Danny, who didn't know anything about crystal either, and who still understood very little about gaming. To save him any embarrassment, Sandra Lee intervened. "You know, Danny," she said, "they tried to make it the big game at the Sands. It was Toni Manzoni who introduced it there. He's one of our local Cubans. It was a bust when they started it, and it's been nothing else ever since. It's too complicated."

"My dear," the Englishman said, "you are half right. I've been to the Sands." He was too discreet to mention with whom. "And I agree fully with what you just said about it being a bust there. But I can't understand it. Chemmy is definitely not complicated. You Americans don't know what you're missing. Baccarat, chemmy, chemin de fer, I don't care what you call it: it's the most exciting game in the entire world. Come watch and listen for a while. You'll agree with me, Mr. Lehman, I assure you."

All three of them moved up to the middle table. It was a long table shaped like two horseshoes facing each other, or to some people's eyes it was shaped more like a kidney. In the indenture sat a dealer. Around him, to his right and left, sat nine players. They seemed to be starting a new game because there was an immense number of cards on the table—312, or six decks to be precise—upon which the dealer was focusing his attention. They were shuffled and then stacked and then shuffled again and then stacked again, and put into a wooden container shaped somewhat like a shoe, which is exactly what they call it in the United States. In Europe they cling to the old terminology and call it the *sabot*, which means exactly the same thing. And it is the path of movement of this sabot, which travels slowly and counterclockwise as it is passed from player to player till the game is over, which gave rise to the original name of this game, namely, *chemin de fer*, the French term for a railroad. It was a railroad that could provide a very fast ride to both the promised land and to disaster, as subsequent events soon demonstrated.

At the table in front of them, with the six decks of cards now in the shoe, the dealer looked around at his audience of nine people and asked, "Ladies and gentlemen, what am I bid?" Somebody called out a hundred pounds, the next a hundred and fifty, a third, a woman, three hundred; and then a swarthy fat man, who seemed to be of Egyptian origin, bid one thousand pounds with an air of finality that immediately proved justified. For it stopped the bidding. "I don't follow," Danny now whispered to his English host.

"In chemin de fer the person who bids the highest initial bet gets to act as the 'bank.' He can continue to act as the bank as long as he continues to win. When he loses, or decides he's had enough, he passes the shoe to the person on his right, who becomes the new banker. The cards in one shoe are dealt down to the last seven or eight cards. Then they shuffle up again, and the bank is auctioned off anew. Remember: the holder of the shoe is always the banker. And all the other players can only bet against him. So now it will be everybody against that Middle Eastern gentleman."

The Egyptian who had been willing to put up the largest amount of money in order to act as the bank now pushed ten chips, denominated at one hundred pounds each, into the center of the table. The dealer counted them and then pushed the sabot to the Egyptian, now officially the banker, and the action began. It was amazingly quick and amazingly simple. The Egyptian drew one card from the box and slid it across the table to the croupier. The card remained there face down.

Then the Egyptian drew a second card and, without looking at it, slid it partially under the shoe from whence it had come. Then he drew out a third card and again slid it across the table to the croupier, who again let it lie there face down. Finally a fourth card was extracted from the shoe and likewise partially shoved under the wooden sabot immediately in front of the banker.

The croupier, still not turning them over, took the two cards in front of him in the middle of the table and pushed them over to a woman seated immediately to his left. She took them in hand, looked at them carefully, and then put them back down on the table, again face down, and passed them back to the croupier. Danny Lehman, who was getting impatient, leaned over to his English companion and asked in a hoarse whisper, "What was that all about?"

The Englishman leaned back and told him, "It was about nothing, absolutely nothing. It is just courtesy, a formality. This game is a pure matter of luck. No skill at all is involved, absolutely none. That woman has no more influence on the outcome of this game than the man in the moon, or for that matter that Middle Eastern type or even the croupier. It's simple luck where all of them are concerned."

Danny Lehman grunted, and it did not seem to be a grunt of approval. The pointless ritual completed, the croupier now took the players' two cards and turned them face up. They were a king of spades and a five of hearts. The king of spades equaled zero and the five of hearts its face value.

"*Cinq,*" he called out. He then nodded to the Egyptian, who drew his two cards from beneath the shoe and turned them over. They were a three of diamonds and a five of clubs: a total of eight.

"*Huit,*" the croupier called out and added, "*La petite.*" The next move was that the croupier quickly proceeded to rake all the players' chips from in front of each of them and added them to the banker's pile in the middle of the table. He then slid the four used cards through two slots, the banker's cards through a slot on his right and the players' two cards, or those that had been dealt to him in his role as agent for the other players, through a slot on his left. This done, he nodded to the Egyptian.

Danny Lehman again whispered, this time rather loudly, to the tall, slim Englishman beside him, "What happened?"

The Englishman raised his hand just slightly, indicating that Danny had somewhat exceeded the tolerated level of decibels, and answered, "The banker, the Egyptian, had a five and a three, which equals eight. The croupier, acting for the rest of the players, drew a king and a five. In this game, all the face cards count zero, so the players' king and the five added up to five. The whole point of this game, in fact the one and only point, is that he who gets closest to nine wins." He continued, "As you saw, the Egyptian got closer with his hand, totaling eight, than did the players with their hand, totaling five. So he won."

"That's it?" Danny asked.

"Almost. If after the initial two cards are dealt neither hand totals either eight or nine, then a third and final card is dealt to both, according to a fixed set of rules. The hand with three cards with a total count closest to nine then wins, regardless of the total, unless they tie. That's the whole game. The only other thing that you have to know is that no hand can ever total more than nine. For instance, if a hand contains two cards, say a six and a five, which would otherwise total eleven, in chemin de fer and baccarat you drop the first digit so that the eleven becomes a one; or a fourteen becomes a four. A three-card total of twenty, for example, ends up as a zero, as far from nine as you will ever get. It's the simplest game in the world. It's also the most exciting game in the world. Now just watch!"

The Egyptian dealt a card to the croupier, one to himself, the second card to the croupier, the second card to himself. The croupier pushed his two cards across the table, this time to an elderly, very distinguished-looking man dressed in tails, who in turn gave them but a cursory glance and pushed them back to the croupier, who then turned them face up. This time the players' cards added up to seven; the Egyptian's cards again added up to eight.

"*Huit,*" the croupier again called out, and again added, "*La petite.*"

"What did he say?" Danny asked.

"Eight, verifying that the banker has won again. And '*la petite,*' indicating that it was a 'natural,' meaning that the banker's hand wins automatically. No more cards need to be dealt."

There was no doubt that the Englishman's interpretation of events was correct, since the croupier now raked in all the chips that had been bet in front of the eight other people sitting around the table, and once again pushed them onto the pile in the center of the table. The Egyptian then proceeded to win three further times in succession, each time with a natural. Each time the bank money in the middle of the chemin de fer table again doubled: from four thousand pounds to eight thousand, then sixteen thousand, and finally thirty-two thousand pounds sterling. The murmur around the table had increased to a quick chatter when the pile had grown to sixteen thousand. But now at thirty-two it had grown to a sharp hum, very muffled, to be sure. Although the two other chemin de fer tables in the room were still going full blast, neither had a single spectator around them any longer. All attention was focused on the center table, and on the Egyptian who just sat there saying not a word, sipping from a glass of champagne once in a while, and puffing incessantly on a dark, ugly-smelling cigar, a habit that seemed to have been lifelong from the stained appearance of his teeth.

But now he stopped both the sipping and puffing and just sat there in an almost comatose condition. All eyes around the table and from the rows behind were on him, and quickly the room became almost totally silent. Even the players at the other tables seemed to sense that something very big was about to happen. Finally the banker moved to reach into his pocket and pull out what looked like a card.

Danny could not contain himself any longer. He again leaned over and asked, "Now what the hell's he doing?"

The Englishman replied, "He's probably got a system. Maybe he's checking the odds. By God, if I were he I would get up and leave!"

The Egyptian did not get up and leave. He nodded at the croupier and then the croupier nodded at the players around the table, soliciting once again their bets against the thirty-two-thousand-pound bank. But suddenly the center of action shifted. A young man in Arab headdress, who had thus far not even participated in any of the action, now spoke out in a very soft but highly distinctive voice. "Banco," he called.

Danny had watched him on different occasions out of the corner of his eye because, although the man had a large pile of chips in front of him, all one-hundred-pound chips, he had just sat there, betting nothing, drinking nothing . . . just chain-smoking Gauloise Bleu cigarettes. Now his single word galvanized the room to the point of evoking gasps from a couple of ladies in the audience, no doubt more theatrical than real, but certainly adding to the fun. Again Danny nudged his English companion and asked, this time with rising excitement in his voice, "What's happening?"

"That Arab chap has just preempted everybody. Any player can do that anytime by calling out 'Banco.' He's pre-empted on all other bets that the other players could have made against the Egyptian. Now it's the Arab against the Egyptian. It's going to be the bank's thirty-two thousand pounds there on the table against the thirty-two thousand pounds which that Arab chap is now going to have to come up with. Watch and you'll see what I mean about this game. Just keep quiet and watch!"

The Arab motioned to the croupier; he, in turn, motioned to another official of the gaming room, who walked over to stand behind the Arab and then leaned down to listen to what the Arab wanted. The man nodded, went back to the foyer, and returned immediately bearing a large number of chips of a quite different color. He handed them to the Arab. When the Arab pushed thirty-two of them onto the center of the table, it was obvious to all of the participants and spectators that they were denominated at one thousand British pounds each—the equivalent of two thousand four hundred dollars each. This meant that the grand total that was now in the pot at the center of that chemin de fer table at Crockford's in London on that warm summer night was the equivalent of almost one hundred and fifty-four thousand dollars. Or calculated in yet another way, one that was quite appropriate considering that there were now only two men, both from the Middle East, involved, the pot was the equivalent of sixty-six thousand, six hundred barrels of oil. Remember, this was 1969.

By now one could notice just the slightest hint of perspiration on the brow of the Egyptian as he began to deal the cards from the shoe. When the cards were turned face up, the player's hand consisted of a three of hearts and an ace of diamonds. In chemin de fer an ace counts as one, so the player's hand totaled four. He signaled that he needed a third card. In front of the banker there were a five of clubs and a king of hearts. His total was five. First the player had to draw.

The perspiration on the Egyptian's brow was becoming a full-fledged sweat, and in fact a couple of drops of water now began to roll from his forehead to his cheek and then dripped down onto the green baize that covered the table. He drew a card and shoved it across to the dealer, who immediately turned it up. It was a four. A four of hearts. The room was now completely silent. All action at the other tables had completely ceased.

The banker had the option of drawing a third card or not. He hesitated and then drew the next card. The last card. The bank's card. Slowly the Egyptian turned it up. It was another five of clubs. Zero. The bank had lost. Finally and massively.

The dealer, having taken 5 percent of the chips from the pot for the house, swiftly pushed the remaining chips over to the soft-spoken Arab on his left. The room was by now totally still as everybody watched the Egyptian. He was still sweating profusely. But his color had not changed, his hands were still steady as he lit up a new cigar, and his voice was firm as he ordered more champagne. He obviously intended to stay. "Bravo," said Mr. Penn. And he was seconded by at least half a dozen other voices. This Egyptian knew how to play the game.

The object of their admiration now passed the shoe containing the cards to an old dowager sitting to his right. Then he requested more chips, this time in a loud voice. He wanted thirty-two thousand pounds' worth. He then stared at the young Arab, and his gaze remained fixed on him for a full half minute.

Suddenly Danny Lehman had had enough. "You're right," he said to Penn. "You're absolutely right. There is no game in the world that even comes close to this. Let's go."

The next stop was Les Ambassadeurs, where they dined and danced, and then went upstairs to the gaming tables of that establishment. It was the same thing all over again: roulette and chemin de fer, but, as at Crockford's, it was the chemin de fer tables that attracted all the attention, and it was at these tables that tens of thousands of pounds were changing hands every ten or fifteen minutes. It was there that Danny again noticed that at the end of play the croupier was always extracting a certain number of chips and placing them in front of himself.

"Why's that?" he asked Douglas Penn.

"Oh," the Englishman said, "the house always collects 5 percent from all winning banks."

Danny made a mental note of that. True, it needed even further simplification to go over really big in the States. But there was now not even the slightest doubt in Danny's mind that chemin de fer—or baccarat—and The Palace were meant for each other. Just one day in Europe and he'd already found what he wanted. When they got back to Claridge's, Danny told Sandra Lee that he'd had enough of London. There was no way that their visit there could possibly be further improved upon. It was time to celebrate, and leave.

So he picked up the phone and ordered a bottle of Bollinger champagne and caviar for two. Then he called British European Airways and booked two seats on the 10 A.M. flight from Heathrow to Frankfurt. His next stop was going to be Baden-Baden. If he had learned this much from the lazy Limeys in just twenty-four hours, imagine what he might pick up from the industrious Krauts!

11

They collected a mercedes from Hertz at Rhein-Main Airport, picked up the main north–south Autobahn which runs along the border of the airport proper, and headed in the direction of Freiburg im Breisgau and ultimately the Swiss border. About two hours later they came to the Baden-Baden exit and, within minutes, found themselves in one of the true garden spots in all of Germany.

One enters Baden-Baden via the Lichtentaler Allee, the Champs-Élysées of southern Germany. It is a broad boulevard lined with huge trees and gardens full of rhododendrons, azaleas, roses, and zinnias. The lush vegetation is more Mediterranean than Central European. And it was, in fact, a Roman emperor, Caracalla, who gave the town its first big start. However, it was the warm springs, not the vegetation, that attracted the health-conscious Caracalla, who apparently spent most of his time building spas. The Emperor also set the tone and established the social standards for visitors to Baden-Baden. The more plebeian types were definitely not welcome. It was to become and remain a playground for the very rich for almost two millennia, for Romanovs and maharajahs, Habsburgs and various Princes of Wales. The final centerpiece was added in the middle of the nineteenth century with the reconstruction of the Kurhaus, the oldest casino in Greater Germany, in the style of Versailles. There, until World War I, one played with chips of gold and silver.

The girl behind the Hertz desk, no doubt inwardly amused at the embarrassment she would inevitably cause these singularly common Amis, had recommended Brenner's Parkhotel on the Lichtentaler Allee. As they drove the Mercedes up to the entrance of this palatial establishment, which was surrounded by immense, impeccably kept grounds dotted with statuary and ponds, to any outside observer it would have been obvious that this pair was even more out of place there than they had been at Claridge's. But apparently quite oblivious to the grandeur of it all, the two of them simply walked into the place, blithely in Danny's case and gamely where Sandra Lee was concerned, and requested the best accommodations the house had to offer. They got what they wanted: a superbly appointed suite of vast proportions, overlooking a huge swimming pool area that had been built as an almost perfect replica of a Roman bath.

It's all starting to come together, Danny thought as he opened the french doors leading from their bedroom to the terrace overlooking the magnificent pool and grounds below.

Sandra Lee had gone down to the lobby just to look around. So Danny decided to stay out on the terrace for a while, taking in the late afternoon sun. About half an hour later Sandra Lee was back and joined him out on the terrace.

"Fabulous, isn't it, Danny?" she said.

"Yeah," Danny admitted, and then went on, "but you know something? This is the first time I've ever been in Germany. I never wanted to come here. Not that I ever thought of it that much, but I'm a Jew. And this may sound stupid, but you know the girl at that Hertz counter in Frankfurt, and the people downstairs, somehow I had this feeling they knew I was a Jew. And they knew they had to be nice to me, or at least appear to be nice to me. But I'm just as sure they don't like us any better now than they did thirty or forty years ago."

"Ah, come on, honey," said Sandra Lee, "that may be true of some of the old guys around here, but I don't believe for a minute that any of the young ones feel anything but shame for what their parents did to your people. Look, it's the same way with us blacks, except it wasn't the Germans that were doing us in, it was the Americans. But that's also changing. You can't forget that stuff, but please don't let it get you down. We are here in a beautiful hotel in a beautiful town. So loosen up. Let's enjoy ourselves. We're on vacation."

"What've you got there?" Danny then asked, pointing at a small paper bag in her hands.

"A book," she answered.

"You read German?"

"No, dummy, they've got a whole section of English paperbacks at the newspaper stand in the lobby."

"So what did you get?"

"One I had to read in college, and now I am going to read it again. Not because I have to, but because I want to."

"College, what college?"

"Mills College."

"Where's that?"

"In California. In the Oakland hills."

"What kind of college is it?"

"A women's college. Liberal arts. Very fancy."
Silence.

"Now, honey, you were wondering what a black hooker was doing at a fancy women's college, weren't you?"

"Ha ha, and you were the one who told me not to be paranoid about the past," Danny replied. "Look, I'll forget about what the Germans did to me if you forget about whoever it was that did whatever they did to you, okay?"

162

"I'll tell you what I was doing at Mills College: I was there on a scholarship. It's a very liberal school. They believe it is good for the upper-class girls to mingle with, you know, with us."

"So what happened?"

"I left, after my sophomore year. I couldn't stand it. I don't mean I couldn't stand the school. It was wonderful. But I didn't have a nickel to my name. I had a scholarship and I still had to wait on tables. I just couldn't put up with the condescending attitude of those prissy young bitches."

"So what did you do?"

"I went back to Los Angeles to try to make some money."

"Well, you've been making money, a lot of it. So quit bitching."

"I'm not bitching, Danny. I am trying to explain."

"Well, then stop explaining if it bothers you. What did your parents have to say when you left this college?"

"I don't have parents. Most black girls don't have parents. They have *a* parent. My mother was sick, absolutely sick."

"What does she say now?"

"We haven't talked in three years." Then she continued, "What about your mother? Is she proud now that her son owns a casino in Las Vegas?"

"My mother is seventy-five years old. She knows absolutely nothing about either Las Vegas or casinos. She lives in the past but is perfectly content with her life."

"What do your brothers and sisters think?" she then asked.

"I don't have any brothers and sisters."

"That's too bad."

"Maybe. On the other hand, I like to be alone. I've enjoyed being alone since I was a kid."

"Now, come on. Didn't you have any friends?"

"Of course I had friends. In fact, the best time of my whole life was when my mother used to take me to Atlantic City from Philadelphia. My grandmother lived down there and she had a house right on the beach. There was a whole gang of kids, local kids. From the time school was out in June until it started again on Labor Day we had the best time. Most of them were Italians. Italians know how to have fun. You know, it's difficult for us Jews to have fun, then and now. I think we simply think too much."

"Ah, come on, Danny, don't let this place get you down. I've got an idea. Let's go down and have a drink. That'll pep us both up."

"Nah, I don't feel like it. I think I'll just sit out here on the terrace for a while longer. Why don't you go and read your book?" He did and she did. About an hour later it was starting to get a bit chilly on the terrace, so Danny finally came inside. Sandra Lee was lying on the bed, reading.

"What's the book called?" Danny asked.

"The Gambler."

"Who wrote it?"

"Dostoyevsky. You know, he also wrote *Crime and Punishment.*"

"Why're you reading that stuff?"

"Because I remembered that he'd written this short novel about casinos. It all took place in Germany, so since we're here I thought I'd reread it to really get into this place. Probably other people do the same. That's why they've got it here."

"What's it all about?"

"It's about a Russian. He's a compulsive gambler."

"What the hell was a Russian doing gambling here in Germany?"

"It was about a hundred years ago, Danny."

"So what's so interesting about a compulsive gambler anyway? Most gamblers are compulsive."

"I guess that's the point. This guy destroyed himself. He landed in jail. He lost his girlfriend and ended up absolutely ruined."

"So?"

"So? So? I mean it has something to do with what you're

doing, doesn't it? Let's face it, Danny, whether your mother knows about it or not, you're now in a business that destroys people."

"That's no concern of mine," Danny said, now getting just a little bit peeved and showing it. Sandra Lee decided not to answer. She just raised her eyebrows and went back to Dostoyevsky.

Then she looked up again. "What about the slots? You know who plays those: the cabdrivers, the waitresses, women who work in factories. They come to Vegas for one night and blow everything they've saved for the last couple of months. I mean, is that terrific?"

"Nobody forces them. When they come to Vegas, they have a good time. That's it. A casino isn't there to make anybody rich, especially the cleaning lady or the cabdriver. A casino, at least an American casino, a Las Vegas casino, *my* casino when I've got it fixed up the way I'm going to have it fixed up, is there to provide illusions; to invite dreams, to allow these people to get away from their crummy cabs and from the crummy factories and their goddamn lives and come over to a palace in the desert. That's all I'm going to give them: a palace in the desert. And so they spend five hundred or a thousand dollars. They'll get their money's worth. Because for twenty-four hours or forty-eight hours they'll live a life of class. In my casino I won't care who the hell they are. Everybody who works in my place is going to treat everybody, even the worst bum, in a classy way. And that goes for a Mexican bum, a Chinese bum, or a white or black bum. Now tell me what the hell is wrong with that?"

"And what if these bums end up like this guy in this novel, this Russian, and wreck their whole lives?" Sandra asked.

"Forget about your book which is a hundred years old and come back to reality. Look, a guy like that what's his name . . ."

"Alexis."

"Okay, this guy Alexis, he's a compulsive type, I'm sure. If he hadn't ruined himself gambling, he'd have ruined himself drinking, or, these days, ruined himself on drugs."

Sandra put the book down and was silent for a moment, but then said, "You may be right Danny. But that doesn't—"

"Of couse I'm right. Look, I don't want to hear this moral crap. I mean, what about the stock market, for God's sake? Or what about the commodities markets? What about all those people who are trying to make a fast buck in real estate? That's gambling! The only difference is you can't compute the odds when you do that. In a casino you take five minutes out and ask around and you can find out exactly what you're getting into. Strictly mathematics."

"You think that waitress is going to sit down and read a book on the mathematical odds before she sits down in front of a slot machine?" She added, "Shit!"

"Look, Sandra Lee, I'm telling you that I'm not interested in waitresses and I'm not really interested in slot machines. That's why we're here in Europe. I've got this idea that the real profit doesn't lie there like most people seem to think. I've got the very definite impression that the way to get really rich in this business is to concentrate on the rich. Leave the little fish to the other guys. I'm after rich guys. So who should care if they end up like bums? But let's stop this, for Chrissake; let's get dressed and see how they do it here in Baden-Baden.

"I don't expect to see too many Russians, do you?" he added with a mocking smile.

"What's that supposed to mean?" Sandra asked.

"Well, I don't like these deep discussions, but I'll tell you something: I'll bet there are about twenty million Russians that would give their left arm to be able to get out of that country and come over here and lose their life's savings on a weekend. Look, that's what free enterprise is all about: the excitement of being able to make a lot of money or lose a lot of money. Now, if you're a Russian, like what's his name again?"

"Alexis."

"Alexis. I'll bet if he lived in this century and spent his life driving a tractor in Siberia, this guy Dostoyevsky sure as hell wouldn't have written a book about him, you know what I mean?"

◆   ◆

The casino in Baden-Baden was a bust. Danny sensed that from the moment he stepped into the building. It resembled a mausoleum, not a place where you had fun. From the marble floor to the heavy curtains to the hush that pervaded the place, there was nothing that Danny Lehman found even remotely attractive, including the people, most of whom looked like visiting farmers. They actually had to pay admission to get in! And once they got in they walked around the place in the same way they had been walking around other tourist attractions earlier that day.

The only tables that had even a semblance of any action were the roulette tables, but when Danny checked out the stakes, he saw that the average bet was five marks! Less than two dollars!

"Where are all the princes?" he asked Sandra Lee as they walked through the place. She was, as usual, drawing the envious glances of everyone from the croupiers to the visiting farmers. "Back in Russia driving tractors," Sandra replied.

They went back to the Brenner's Parkhotel and had some Wildschwein with Preiselbeeren and Spätzle and a good bottle of Mosel wine, ending the meal with a delicious piece of Schwarzwälder Kirschtorte and finally a bottle of Sekt, the German version of champagne. By ten-thirty they were up in their suite.

"Say," Danny said as they undressed. "I'm going to stay up and take a look at that book of yours, okay?"

"Go ahead," Sandra said.

At two o'clock in the morning, when he had finished reading the book, Danny leaned over and shook Sandra Lee, who had been in a deep sleep for hours.

"What do you want, Danny?" she asked, suspecting that she knew full well what he wanted.

"Nothing. Just wanted to tell you that, as usual, you helped again to make this little trip worthwhile. Baden-Baden, the way it is now, stinks. But the way it was—that's something else. What I'm going to do is find the right people, and when I find the right people I'm going to give them exactly what this Russian guy Alexis was looking for: excitement, high stakes, a feeling of being part of the elite. Class. I'm going to get the same types to come to Vegas. He's got a count in that book. He's got those upper-class Englishmen. He's got a general. He's got them from all over the world. They all came here because the stakes were high and the atmosphere was right: not lousy five-mark chips, but gold florins. Not fucking nobodies, but aristocracy!"

"To hell with it. Go to sleep." She leaned over and gave him a pat, and then they both went to sleep.

The next day they caught a Lufthansa direct flight from Frankfurt to Beirut. The plane was barely off the ground when Danny ordered a bottle of champagne, real champagne, to celebrate having left German soil. Danny had concluded that there was nothing Jews could learn from Germans except that they should stay away from one another.

12

There was a message waiting for Danny when he checked into the St. George Hotel in Beirut. It could hardly have been more brief. "Call me at 702 555 9977 after 7 P.M. my time." And it was signed "William Smith."

Sandra Lee watched his face as he read it. "Not good, huh?"

"What time is it in Vegas?" Danny asked.

"It must be either nine or ten hours behind this place. So my guess would be around two or three in the morning." Which definitely put it after 7 P.M.

The first thing Danny did in their room after the bellhop had left was to get the telephone operator. This being Beirut, he figured that it might take a day or two to get through to the States. But Lebanon, at least in 1969, though in the Middle East was not really of the Arab world. It thought of itself as the Switzerland of the eastern Mediterranean, and its phone system lived up to that forced comparison. The head of the Las Vegas office of the FBI was on the line in less than thirty seconds.

"It's Danny Lehman. What's wrong?"

"Oh," and there was relief in Smith's voice. "Look, I found out where you were going to be from your attorney, Mr. Shea. I hope you don't mind."

"Of course not."

"I'll make it brief. De Niro and Salgo got out on bail the day you left town. Just like we expected."

"Yes," Danny said.

"Well, the day after that, two guys got killed. Their names were Amaretto and Sarnoff. Joseph Amaretto and Sam Sarnoff."

The missing link, as the FBI man had put it, was no longer missing.

Danny turned ashen. Sandra Lee was watching him and moved in his direction, but then changed her mind and disappeared into the bathroom.

"Ever hear of them?" Smith then asked.

"No," answered Danny, and there was a quiver even in this single syllable.

"I didn't think so. We're pretty sure we know why they got killed. And who did it: the who being Roberto Salgo."

"Shit," Danny said, with a fervor that came from being scared.

"Yeah, I know how you must feel. They were no doubt fencing for De Niro and Salgo. They worked out of Reno. Our office there has known about them for years, but never really came up with anything that we could nail them for. Which, frankly, didn't really bother anybody that much, since they were small-time. At least, we thought they were small-time."

Danny just listened.

"De Niro and Salgo no doubt figured out that somebody must have informed on them, and then they came up with these two guys."

"Were they?" Danny asked. "Working with you, I mean?"

"No. We had an informer. I told you: the one who phoned us. But it couldn't have been either of the dead ones. It was an insider. A casino insider. We'll never know who, I'm sure."

I'll bet Fort Lauderdale knows who, Danny thought. But he said, "Then how do you know it was Salgo who killed them?"

"We don't know for sure. But Salgo was seen with both of them at a bar in Reno two nights before their bodies were fished out of the Truckee River. The bodies were kind of mutilated, I should add."

"So is Salgo back in jail?" Danny asked.

"No. No way we could do that. Zero evidence. Which brings me to why I called you, Danny."

He does know something, Danny thought. I'm going to be the evidence.

"I want you to be very careful," Smith then said. "Salgo hasn't been seen since. If I could find out where you were with one phone call, I'm sure he and De Niro can do the same. They still know everybody at the casino."

"Why me?"

"You got them arrested. You ruined them; you and whoever informed on them in the first place. They think they took care of the latter up in Reno, and got away with it. Now if they try to get you over there in the Middle East, there's not a damn thing we can do about it either. There's nothing we can do to prevent it, and, if they succeed, there's no way we could go after them. Lebanon is, to put it mildly, a bit outside our jurisdiction. We could contact their law enforcement people through the embassy. But I'm sure those Arab cops are not going to care too much if some visiting Jew thinks he's in danger of getting knocked off by some American mob types, if you know what I mean."

"Yeah," Danny replied. First surrounded by Germans; now Arabs. Then he said, "Is it going to really be much different when I get back? If they've been crazy enough to go this far already, what's to stop them from staying crazy enough to get me, no matter how high the risk?"

This time it was the voice on the other end of the phone that remained silent. Then: "Not much, Danny. Even when I'm working for you, that's one thing I couldn't guarantee. There are lots of bad ones in this town, hell, that's obvious, but very few who act irrationally. It's part of the business. Right? But anybody who goes out and kills just like that . . . well, it's scary. So be careful."

"I will. And thanks. I'll be back in a couple of days."

When he hung up, Sandra Lee was coming out of the bathroom, wrapped in a huge white towel. "Come on, honey. I've got a bath ready. It'll relax you, and I'd say, by the look of you, that you need relaxing bad. So maybe we shall have to do more than just soap your back." She did more. Afterward he told her every detail of his conversation with William Smith. Then she gave him some valuable advice.

◆   ◆

It was around six in the evening when the phone rang, startling both of them. Incoming phone calls in Beirut could hardly bring anything other than more bad news. "I hope this is Danny Lehman," the man on the other end said, in a voice that seemed to combine a variety of accents.

"It is."

"This is Eduardo Cordoba. I run the Casino de Liban and I just heard that you had arrived in Beirut. Welcome!"

"Who told you?" The question was, of course, not unrelated to what he had just heard from the FBI. Any unsolicited phone call would now have to be regarded with suspicion.

"That's part of our business, isn't it? But to put your mind at rest, it was the concierge at the hotel. You apparently inquired about our casino; he then automatically let our people know. For a small gratuity, of course. But please accept my apology if I'm intruding."

"No, no problem," said Danny, for it was true. He had asked the concierge while waiting to be checked in.

"When your name came up, I, of course, recognized it. We heard about your taking over Raffles. I thought I'd like to welcome you to the fold by inviting you for dinner this evening. On the off chance you might accept, I've already taken the liberty of instructing one of our limousines to pick you up at the St. George at eight. I do hope you'll come."

"With pleasure. I'm with somebody—"

"I know," interrupted Cordoba. "The concierge also told me. We are all waiting to meet her!"

It was a Rolls that pulled up at eight on the dot in front of the St. George, and there were whiskey and gin beside the ice bucket on the bar in the back, ready to go. Plus a dozen canapés topped with black caviar. There was a bottle of almost frozen Stolichnaya to go with that.

The drive from Beirut to the Casino de Liban takes about an hour, and it is a spectacular one. The farther north one goes, the wilder the coast becomes, as the rocks rise above the sea. It is comparable to Highway 1 in California, especially where it runs high above the Pacific between the Russian River and Mendocino. The car was air-conditioned, but Sandra Lee asked the driver to turn it off. She wanted to take in the Mediterranean evening air. It was still very warm but, amazingly, not humid.

The casino itself is perched on a huge rock formation overlooking the sea. It stands in splendid isolation. And it is a splendid structure, combining the Middle Eastern with the French, as is only proper in a former colony, or mandate, of France. Eduardo Cordoba was waiting for them at the entrance, which led to a huge high-ceilinged hall. The gaming facilities were off to the left; the dining and entertainment rooms, to the right. There were tons of marble and acres of rugs and carpeting. "First dinner, then the show, if you agree. I know that to invite you Americans to watch our show sounds, let us say, presumptuous. But I am willing to take that risk, if you agree."

His English was near-perfect, but hints of both French and Spanish accents were discernible. His was an attractive voice, and Cordoba was an attractive man. He seemed to be in his mid-forties. He was at least six feet tall, but probably no more than 160 pounds, a lean man whose most striking features were his black eyes and his hair of the same color. He was dressed in an old-fashioned pinstripe suit, and his tie could have been one of a Guards regiment. His manner was that of a good-natured sport, but his eyes, to which one was drawn, were coolly appraising. Sandra Lee had been attracted to him from the very beginning, and it was she who now took Cordoba's arm as well as Danny's and said, "Honey, we will do anything you say. Provided . . ."

"Provided what?"

"That we get some very cold Dom Pérignon very quickly."

The champagne was cold, the dinner superb and served with an elegance that matched anything that Lasserre, or Grand Vefour, or Tour d'Argent could have offered that year in Paris.

The show that followed was absolutely spectacular. It was set in an amphitheater that had more Greek than Roman influence. The show opened with a belly dancer and a flaming sword–swallowing act, which did not bode well for what might be coming later. But then, all of a sudden, the performance took on a dramatic nature. The stage tripled, at least, in size and now there were elephants and camels and Arabian horses apparently coming from all sides. Then, from above, immense chandeliers, golden chandeliers, three of them, began to descend, and from each chandelier hung six nude women, all completely coated with gold. When they reached stage level, there was suddenly the sound of rushing water, while at the same time short walls, obviously hydraulically powered, began to rise, sealing off what once had been the aisles in front of the semicircle of seats nearest the stage. Soon filled with rapidly

flowing water, they became canals along which floated an entire procession of golden swans, each bearing two golden
showgirls, plus an assortment of monkeys, parrots, and other
exotic birds.

"This is the damnedest thing I've ever seen!" Danny exclaimed.

If Beirut was the playground of the rich of the Middle East,
where one could do what one chose without religious or social
restraints, then the Casino de Liban was the centerpiece of that
playground. It offered to the elite of the Middle East what many
of them wanted: European women, American cocktails, and, as
Danny soon discovered, English-style gaming where huge
sums of money could be won and lost without anybody really
taking any notice—and all this in an atmosphere of French
congeniality combined with the Middle Eastern penchant for
extravagance.

Again, of course, it was at the baccarat tables where all the
action was to be found. Even the stakes on the tables at
Crockford's paled by comparison. As if reading Danny's mind,
their host started very quietly to identify some of the men in the
room. "That man over there, for example, the very dark one
with the pointy beard. He is a famous Saudi Arabian prince. He
comes to Beirut every three months and spends every single
evening of his stay here at the casino. Those four women behind him at the chemin de fer table change every time. But they
are usually Danish.

"The man next to him," Cordoba continued, "the old one,
he is Turkish. He owns mines there. He comes once a year and
stays only three days, usually. His sons come more often. The
man on his right, he's also a prince, but an Italian prince,
which, you must appreciate, is not quite up to Saudi Arabian
standards. But nevertheless . . . His wife is Brazilian. She is
still eating, most probably; she is very fat. But she also owns an
awful lot of coffee in Brazil, as the old song used to say."

"Which song?" asked Sandra Lee.

"You're too young for that," Danny answered, and then

asked Cordoba, "Where on earth did you learn the words to that song?"

"I went to a prep school in Santa Barbara, California. During the war. My father was in the diplomatic corps of his country and was stationed in Washington for much of the time when I was a child. Then he was posted to Berlin. They left me in the States. Our country, by the way, was Argentina."

"What is your country now?" Danny asked.

"None," Cordoba answered. And then continued, "That man next to the Italian is from Israel. He is with, or works for, the Israeli Aircraft Corporation. He's an arms dealer. Works out of Beirut. Comes here once a week. Usually drops a couple of thousand pounds each visit. He's a terrible gambler. Also a very unpleasant man."

"That surprises me," Danny said.

"What surprises you?"

"That an Israeli would be allowed to live here, much less deal in arms."

"Beirut is an open city. Lebanon is the most liberal of countries, the most multireligious of societies. We have a million Muslims, to be sure, but also many hundreds of thousands of Christians, many tens of thousands of Jews; we have Copts and Jehovah's Witnesses and Old Catholics and—everything else. Even Israelis who deal in arms.

"Everybody is safe here—even the most rich. You will notice we have no armed guards here in the casino," Cordoba continued. "You will see no military people on the roads, because we essentially have no army. And the only police you will see are directing traffic. It is like Switzerland."

"You love it here," said Sandra Lee.

"Yes. Very much. And now it is all going to collapse. Nasser is losing control. There will be war, and the Palestinians will take over this country. And it will all be over. Very quickly." Then abruptly: "I expect you both must be very tired after your trip. If you agree, I will ask your driver to take you back to Beirut."

Both agreed spontaneously and, in Sandra Lee's case, with an audible sigh of relief. As he walked them to the Rolls, Cordoba made two remarks. "Danny, I would like to have a further word with you in private and about private matters. My private matters. Could we do that over lunch tomorrow?"

"With pleasure. Why not at my hotel? Say twelve-thirty?"

"Done. One other thing. A man has been asking for you for the past two days. In Beirut. And twice here at the casino. We don't know his name. He was obviously accustomed to dealing in cash only. Dollars. He brought a bundle with him to the casino. But he had bad luck the first night and ran out. He then had some funds wired to us. They came from a bank in the Cayman Islands, but he's American. I thought you might like to know this. Perhaps we might be of some help in the matter if, that is, it is a matter that should be dealt with. In any case, I have told the people at the St. George to keep an eye open. They will take good care of you."

And then he disappeared back into the Casino de Liban. Sandra Lee, who had heard it all, said nothing. But she held Danny's hand during the entire drive back. She knew that both had the same name in mind.

Salgo.

◆   ◆

The St. George Hotel had a wonderful outside terrace at the rear, looking out onto the beach below and the Mediterranean beyond that. The crowd was always a mixture of local Beirut society, almost exclusively Christian, and tourists who for the most part were French or Arab, and mainly from the financial elite of their respective countries. The women were also of two classes: the wives of local men, on average in their forties and overdressed; and mistresses, on average in their twenties and underdressed (in terms of quantity, not quality)—these being the companions of the foreign visitors. Race and color went truly unnoticed; jewelry, shoes, décolletage did not.

As it was a weekday, the ratio of women to men at lunch that day on the terrace was at least three to one. In fact, Danny Lehman and Eduardo Cordoba were the only men lunching together. The St. George was a place reserved for pleasure, not business.

"It doesn't exactly look like this country is anywhere near falling apart," Danny said, once their Campari and sodas had arrived.

"It's because you can't see them from here," Cordoba answered.

"Can't see what?"

"The camps. The Palestinians. Hundreds of thousands. Hussein kicked them out of Jordan; no other Arab country wanted a gang of troublemakers like they are. So they've invaded Lebanon. And this country is neither willing nor able to do anything about it. One of these days they are going to try to take us over." And Cordoba's hand traced a semicircle, taking in the elegant high-rise buildings that rose up from the sea: some banks, some hotels, some condominiums.

"That is why I asked if we could lunch together," Cordoba continued.

Danny waited.

"Let me tell you more about myself. I spent my childhood and youth in your country. But then, after the war, we lived in Paris. My father was at the embassy; my mother became, as usual, involved with the local society. I ended up enrolled at the Sorbonne, but, as so often happens, I spent my time with the girls, drinking, even dabbling in drugs, although that was a long time ago, and gambling. I don't know how much you know about the Paris gambling scene, but if I may explain very briefly?"

"Please do."

"It could hardly be a greater contrast to what you are used to in Las Vegas. Your casinos there have always been wide open for everybody. In France that was not true for many decades. Laws were designed to protect the workers, the poor, from the 'vice of vices.' Casinos were banned within a sixty-mile perimeter of Paris. The only exceptions to this ban were the 'big' games—baccarat and chemin de fer—which were tolerated, *provided* they were strictly limited to private clubs, clubs which were reserved for the elite, those who could 'afford to lose.' Where the clubs are concerned, or *cercles,* as they are known in France, gaming was supposed to be only a small sideline, a way of providing a diversion from the main purposes of these establishments. The main purpose of, for instance, the Cercle Anglais, is to foster Franco-Anglo cooperation. The Aviation Club supposedly only has members who are pilots. All this, of course, amounts to hypocrisy on a scale of grandness that only the French could manage. To sum up briefly, there are sixteen such *cercles.* The most luxurious of these is the Cercle Concorde on the Champs-Élysées; the place where the big action has always taken place was, and still is, the Cercle Haussmann. I began working there in, well, let's think now, it must have been in 1952. My parents could not understand it. But in 1952 Paris was still a very drab place, especially for someone who considered himself a hot-blooded Latino. I found the excitement I thought I needed at the Cercle Haussmann, and spent the next eight years there. Then I moved to London. Not Crockford's, but nevertheless a place that came close."

"Why?"

"Why did I leave Paris?"

"Yes."

"Well, in 1960 new ownership moved in at the Cercle Haussmann. One referred to them, one still refers to them, though never in their presence, as the 'Corsicans.' They took over one club after the other. Their names are wonderful: Raffali, Benedetti, Mondolini, Francisci. Believe it or not, they all come from a few villages in southern Corsica situated within half a dozen kilometers of each other. They wanted their own people in. But don't misunderstand. I left in friendship. In fact, I consider some of the members of those families to be among the best friends that I have in this world. We help each other solve problems. You know? In this business, problems tend to move from casino to casino, from country to country, and, today, from continent to continent. Some of us long ago recognized that mutual protection was necessary. The Corsicans are good at that sort of thing."

"And after London?"

"I spent a couple of years at Monte Carlo, and then came here to manage the Casino de Liban. I've been here five years now."

"And you're ready to move again," Danny then said.

"Yes. Could you use me?"

"As you said yourself, Vegas is another world."

"Ah yes. But it is in a world which has gotten very, very small. The high rollers don't care where they gamble. In your desert or beside my beach: it's the same to them, as long as they can find the action they seek, and as long as they are treated in the manner in which they expect to be treated. And justifiably so. They find action here or in London and also at the Cercle Haussmann; they also find the same standard of service in all of these places. Why go elsewhere? They don't and they won't, unless they can find bigger action and better treatment somewhere else. Provide it in Las Vegas, and they will come to you. More than that: you provide what they want and *I* will bring them to you."

"Define action."

"Simple. One word. Baccarat."

"That's it?"

"That's it."

"What else?"

"Credit. For the big ones, almost unlimited credit. And no interest on these credits. Ever. No matter how long they remain outstanding."

"What else?"

"Attention. Someone who will fulfill their every whim when they are in town. Someone who gets to know them so well that they anticipate that whim before it even becomes one.

Suites, girls, booze, airplanes, yachts, limos, everything. All on comp. No questions asked. Just no drugs. That's out."

"Baccarat. Credit. Comp," Danny said.

"That's it. Plus having somebody who knows who they are."

"And that's you."

"That's right."

"How many are there?"

"A couple of thousand really big ones. But the number keeps growing. Especially in places like Hong Kong and Mexico where nobody pays taxes."

"They can support a casino?" Danny asked, skeptically.

"Yes."

"How many of that couple of thousand do you know?"

"Five, maybe six hundred."

Danny did some calculations in his head, paused, and asked, "How soon would you be ready to move?"

"You give me thirty days' notice; I'll be there on day thirty-one."

Danny Lehman had already made up his mind. He knew it was impetuous: he had barely met the man. But . . . "One more very important question."

Cordoba interrupted. "I know what you're coming to: am I clean enough to be licensed in Nevada?"

"Beyond any shadow of any doubt clean enough?" Danny asked.

"Yes," Cordoba said.

"You're absolutely sure?"

"Absolutely."

"Then you've got a deal. Consider that thirty days as starting from right now. Unless you want to talk about, you know, what kind of a deal . . ."

Again Cordoba interrupted. "Not necessary."

"Shall we order lunch?" Danny then asked.

"Frankly, Danny, I'm not hungry. My stomach's nervous. This was important for me. Do you mind if I pass?"

"Of course not. I think I'll just stay here a while longer. I'll call you before we leave. You're sure, now, about our deal?"

"I'm sure." Cordoba extended his hand to Danny and left.

Danny sat down and was pondering whether or not to order lunch when Sandra Lee appeared at the doorway leading from the hotel to the terrace. She had a frantic look on her face, and when she spotted Danny and came to his table, the look did not change.

"What's wrong?" Danny asked, rising to meet her.

"Sit down," she hissed. "He's at the bar. I was sitting inside having tea when he walked in. He saw me. And he saw me recognize him."

"When?"

"About ten minutes ago. I was petrified. Did I make a mistake coming out here? He must know I came to find you." She was trembling. "I'm afraid, Danny. If he kills you, he's going to have to kill me."

"Where do you get this kill stuff? That's George Raft movie crap."

Now her expression changed, from one of fear to one of fear mixed with anger, maybe even including a tinge of hatred. "Look, don't give me any of that innocent act. Your business is nasty. It's run by nasty people. Don't try and con me. Con yourself if you want. But not me. Now what the fuck are you going to do? Salgo is not here to visit the Roman ruins up in those hills back there."

"I'll go talk to him," Danny said.

"Don't!" And this was said with vehemence. "Stay away from him. I told you that yesterday. Don't go near him and don't let him get near you."

"How? He's not going to disappear."

"Talk to Cordoba. He came here for a job, didn't he?"

"How did . . . ?"

"It was kind of obvious. He wants out of this place, and you're the solution."

"You're right. I hired him."

"So call him. Explain the problem. Now!"

"I can't now. He's on his way back up to the casino."

"So call him in an hour. In the meantime, let's get out of here." She left ten Lebanese pounds on the table and almost dragged him to the stairs that led down to the path to the beach. Instead of going toward the water, she left the path and moved across the lawn to the broad boulevard that bordered the Mediterrean, crossed it, and headed into a street that led up a hill. Danny literally trotted behind her. Two blocks and five minutes later they were inside the Inter-Continental Hotel. In contrast to the St. George, this hotel was completely new and totally modern. Its lobby could have been that of the Inter-Continental in Geneva, Frankfurt, or Hong Kong. The only clue to where it was located was that in the middle of this lobby there was a swarthy man in Turkish costume pushing around a cart bearing an immense brass urn from which he drew coffee when any of the guests he offered it to chose to partake of the "freebie," which many did, since freebies of any kind are, to put it mildly, an extreme rarity in the Middle East. Sandra Lee was one of the takers. With a tiny cup in each hand she then carefully moved to the rear and installed herself on a sofa that offered a view of the entire sweep of the lobby, including the front entrance. Danny had no choice but to join her. He reached for one of the coffees.

"First get a newspaper," she said, "a French one. Then coffee."

He got a two-day-old *Figaro* from the newsstand and re-joined her.

"You read and sip that godawful coffee. I'll watch. At three on the dot you call that Cordoba."

At three he did just that, and the conversation was brief. "That man you said was looking for me—he's at the bar of the St. George, or at least was there an hour ago. His name is Salgo. He's trouble."

Cordoba said nothing.

"Perhaps very big trouble," Danny then added.

"Big enough trouble to affect your casino operation?"

"Yes."

"Were you seen together?"

"No. Sandra Lee saw him. That's all."

"Where are you?"

"At one of the phones in the lobby of the Inter-Continental."

"Give me the number."

Danny did.

"Stay there." No more than seven minutes later the same phone rang. "He booked himself into the St. George at noon," the Argentinian said. "Let me suggest something. First, is Sandra Lee with you?"

"Yes."

"Good. In fact, perfect. Go to the front desk of the Inter-Continental in about five minutes. I'll have a room arranged for you in my name. Sign nothing. Just pick up the key and go to the room."

"And what will happen?"

"Some friends of my Corsican friends happen to be in town. *Pieds-noirs*. This place reminds them of their old homeland, Algeria, although some of them miss the violence. So I'll actually be doing them a favor. All right?"

Danny said nothing. Then: "All right."

◆   ◆

Roberto Salgo was sitting at the bar of the St. George Hotel, waiting. Now that he was so close, he was getting increasingly impatient, even though it had been only days ago, after he had cut off Amaretto's fourth finger as De Niro held him down on the floor of the RV in the desert outside of Reno, that the man had screamed out Danny Lehman's name. Salgo had gotten so goddamn mad that he had taken the meat cleaver and, in one go, taken off what was left of the entire hand. In the end, when they had dumped both Amaretto and Sarnoff in the river, there were a lot of pieces missing from both bodies.

"Gimme another one," he said to the bartender. It was his third bourbon and soda. It had been so easy to track down the son of a bitch. Raffles had always referred its clients to the same travel agent in Vegas, because he and De Niro had always gotten a kickback on all the commissions, and free trips to anywhere on earth they wanted to go. Lehman had used the same agency, and now he knew why. The black bitch! She no doubt took *her* clients there, and for the same reasons.

The owner had delivered him a printout of their itinerary in the parking lot of the Dunes. From the look of intense fear on the man's face, there was no chance that he would ever talk about it, or even dream of doing so. The dilemma now was the black bitch: she'd recognized him. He was sure of it. But had she been stupid enough to tell Lehman? After all, she must know what would happen to her if she did. Christ, she'd been working Raffles long enough to know all about him and De Niro. Right?

"I'll be right back," he told the bartender and walked to the bank of house telephones in the main lobby of the hotel. He picked up one of the phones and asked, "Could I please have Mr. Lehman's room?" He let it ring three times and hung up. Still there, he said to himself, so she must not have told him.

"Check," he said. He paid it in dollars and, forgetting about any change, hurried back into the lobby and out of the door.

"Do you need a car, sir?" The man stood beside a Cadillac limousine, with dark windows. He had seen at least a dozen such cars in Beirut. The Arabs obviously went for them.

"Naw," Salgo said, and then he changed his mind. "How much for a grand tour of Beirut?" He looked at his watch. "Say for about two hours, and then back here." That should be about the time that Lehman and the black bitch would be back in their room, getting a little rest before preparing to go out for dinner. He knew the habits of tourists.

"A hundred dollars," came the answer.

"You got it," Salgo replied.

"If you sit up front beside me," the chauffeur said, "I can better explain Beirut to you. That is," he added with a show of humility, "if you don't mind."

"Hell no," said Salgo, although he never would have agreed to do it in Vegas where they knew him.

The dark-complexioned man, who spoke with a French accent like so many natives of the Lebanon, which, after all, had been a French protectorate for many years, opened the car's front door for Salgo, closed it behind him, and then scurried around to the other side and into the driver's seat. He moved the car swiftly away from the curb.

"I suggest we first go north for a bit," the driver said. "The coast is very beautiful. Like parts of the Riviera in France. Or," he added, "Corsica, where my family comes from."

Five minutes after they had left the outskirts of Beirut, following the same route that Danny Lehman and Sandra Lee had taken the night before on their trip to the Casino de Liban, the smoked-glass panel that separated the backseat from the driver's compartment slid back very slowly, very quietly. Just as slowly and just as quietly two gloved hands appeared, holding a length of piano wire.

With one swift move, the wire was looped over Salgo's head and then drawn across his throat. It took a good fifteen seconds before the violent thrashing of Salgo's body ceased. Then the garroting of Roberto Salgo, Corsican-style, was completed.

Ten minutes later, the driver turned left off the main road and followed a dirt track that took them to the very edge of a cliff overlooking the rocks below and the Mediterranean beyond. When the Cadillac came to a halt, two swarthy men emerged from the backseat. They expertly removed all the clothing from Salgo's body and put it in the trunk of the car. Then they threw the body onto the rocks seventy feet below. The incoming tide was already beginning to cover them.

The driver, who had been watching all this while smoking a cigarette, looked at his watch. *"Cinq heures,"* he said. *"Alors, allons donc avoir un apéritif."* He drove the Cadillac back to the main road and headed north once again. At the next village they pulled over in front of a primitive sidewalk café. While they talked about soccer, two of the Corsicans had Pernod; the third, a Perrier.

◆   ◆

At eight o'clock the next morning the phone rang in Room 1137 of the Beiruit Inter-Continental Hotel and Sandra Lee took it. "It's Eduardo Cordoba."

"Yes?"

"You can go back to your suite at the St. George. Maybe wait an hour or so, so it looks like you're returning from a morning stroll."

"But—"

"Nothing more needs to be discussed. Tell Danny that I will cable him in Las Vegas when my travel plans firm up. In the meantime, *bon voyage* to both of you."

"Thanks, Eduardo."

"You're welcome."

◆   ◆

Beirut Airport lies south of the city on a flat, dusty strip of land between the mountains and the sea. Sandra Lee and Danny boarded Middle East Airlines Flight 11 to Rome at noon the next day.

By one o'clock Danny had finished his third vodka martini and ordered a fourth.

"Is that smart?" Sandra Lee asked.

"No," Danny answered.

"What's bothering you?" she asked.

"Salgo."

"Why? He's gone."

"And those two other guys, too."

"Those two in Reno?"

"Yes."

"So what? From what Smith told you, Salgo killed them.

Now with a little help from your new pal, Eduardo, we . . ."
She decided not to finish that sentence.

Sandra Lee remained silent for a few minutes, and then said, "What about De Niro?"

"What about . . . De Niro."

"He's still alive and well."

"Maybe. You leave him to me." Now he was getting a bit aggressive.

Sandra Lee decided to shut up for a while. In the meantime, Danny was into his fourth martini. "You're wondering why I'm doing all this, aren't you?" he suddenly asked.

"Maybe. I mean, is it worth getting killed for, for Chrissake?" she answered.

"Maybe it is," he replied. After a few seconds: "Didn't expect that answer, did you. Hah!" Then: "Look, I've got one hell of an idea. The biggest. The best ever!"

"Sure. Baccarrat and *real* Roman baths," Sandra Lee said.

"Naw, not that. Well, that, too. But I mean the *biggie*. Atlantic City. You know what I'm going to do? I'm going to single-handed—is it handed or handedly?—never mind; I'm going to remake Atlantic City into the hottest place on the East Coast. I loved Atlantic City when I was a kid. Did I ever tell you that?"

"Yes, Danny."

"I used to stay with my grandma. We used to go to the Boardwalk and to the shows on the steel pier. It was great. The best times I ever had in my life." He picked up his glass again. "Then the politicians moved in and cleaned the place up, as they say. They wrecked it is what they did. It's a terrible place now. Well, not much longer. I'm going to put it back on the map. And nobody's going to stop me. Not the mob, not the bankers, not the pussyfoots like my lawyer. I'm going to build the best goddamn casino in the world in Atlantic City before I'm done. And then if you want to have a good time in Atlantic City again, you'll be able to. You want to drink, make a fast buck—come to Atlantic City. Do you know that there are over fifty million people living within a two hundred-and-fifty-mile radius of Atlantic City? With nothing, nothing to do? Did you know that? No, you didn't, did you? Now let me tell you something else. I told all this to a guy called Henry Price. You never heard of him, did you?"

"No, Danny."

"Well, he's a hotshot on Wall Street. Really big. Understand?"

"Yes, Danny."

"Well, I told him just what I'm telling you now. And do you know what the bastard did? He looked at his watch, and threw me out!"

"The son of a bitch!"

"Exactly. And now, Sandra Lee, we are going to show them all: Henry Price, the fucking mob, everybody, that two little nobodies, like you and me, can . . ." He rambled on. And on.

Over Brindisi he fell asleep.

Two hours later they landed at Fiumicino Airport. Sandra Lee reminded Danny that the original purpose of the trip had been to take a look at the Roman ruins.

"Fuck the Roman ruins," he said, and they went over to the Pan Am counter and booked a direct flight to Los Angeles. So Danny saw no Roman ruins. In fact, later he never even remembered seeing the airport at Rome.

◆   ◆

The next day Danny had the worst hangover of his life. But when he entered the casino grounds of the "old Raffles place," as it was now known, the first person he asked for was the architect. For the next three hours they pored over the plans. Danny sensed immediately that the architect had come up with exactly what he wanted. The Palace would be a place where even Caracalla, or whatever the hell his name was, would have been more than happy to take a bath. But Danny was sure that guy Alexis would feel right at home.

But there was still a loose end: De Niro. Danny decided

that it was one that, if left alone, could prove fatal to everything that was now falling into place. So early the next morning he summoned William Smith, who still had a while to go with the FBI. It was Smith who directly broached the subject the moment after he arrived in the executive offices of the casino, about the only part of the entire building complex that was not under renovation. "Did Salgo turn up in Beirut?"

"Not that I know of," Danny answered.

"Funny," said the FBI man. "We managed to track him as far as Rome, and it seemed like a cinch that he was headed directly for you."

Danny shrugged and then asked, "What about De Niro?"

"He's still in town. Never left town, from what we know. Still living in his house on the north side with the wife and kids, playing the good citizen."

That clinched it for Danny. Try as he might, he simply could not get Lebanon and Salgo out of his mind. For the truth was that he, Danny Lehman, had killed that man. No matter how indirectly, no matter how much it was justified, no matter that it was in self-defense, he had had a man killed. That simply could not be allowed to be the first step down a road that would lead God knows where. Because he knew what would happen to De Niro if he was brought to trial for murder the way things stood now. They would sentence him to life, and in seven years he would be out on parole. And out to get Danny, just like Salgo had been. Except that now De Niro would be doubly motivated. And then what? He'd have to send for the Corsicans again. And after De Niro, who would be next? Not only that: what if he got caught? Danny had to admit it: *that* was what ultimately scared him most. It was one thing to take such a fucking risk five thousand miles away in Beirut where nobody seemed to care. But here in America, if you were caught for being involved in murder, they punished you, no matter what. He wasn't even sure if they had a death penalty in the state of Nevada. Danny then concluded, The very fact that whether they do or not is now important to me is nuts! Uh uh; no more of that stuff.

So he said, "Look, I've been thinking, especially after your phone call in Beirut. I've got nothing, absolutely nothing, to gain from pushing this thing any further. You know?"

"I agree."

"What if we decided not to press charges?"

"You mean the casino?"

"Yes. Could he be convinced to leave town then? For good?"

Smith puzzled on that one for a full minute. "Probably, except for the IRS; they press their own charges."

"What if they get paid what they think is due them?"

"That's something else."

"How much is involved?"

"Well, we've talked about it vaguely. They're thinking in terms of a couple of million dollars' unreported income. So you'd have the tax on that, the penalties, and the interest on both. Million bucks would probably do it. If it was a cash settlement."

"A million for De Niro, or for both De Niro and Salgo?"

"Million each."

"Would they drop the criminal charges if they got the million? Got it now?"

"My guess is that the IRS would drop both the criminal and the civil charges. Such good shape they're not in. De Niro and Salgo are going to keep claiming that they won all that stashed cash at the baccarat table and that it all happened very recently. With those two guys dead up in Reno, it's going to be damned hard to prove otherwise. That goes for us as well as the IRS. So they might be very happy indeed to get their money and close this thing out. If they do, we'll follow."

"Can you find out. Today?"

Smith did. And the IRS confirmed what he had thought right down the line. De Niro was picked up at seven that evening in one of the casino's limousines and brought back to the gaming establishment he had run for most of its existence. Danny and his attorney, a very uneasy Benjamin Shea, were waiting for him in his old office on the third floor. Danny indulged in no small talk whatsoever. He told De Niro he wanted to make him an offer and that it was not open to any discussion. All he wanted to hear was a simple yes or no.

The offer: first, the casino would withdraw its complaint against him and would urge the federal authorities to drop their charges against him. Second, the casino would grant him a one-million-dollar "golden handshake," provided the funds were put into an escrow account and designated for tax settlement purposes, this to take place with the full knowledge and agreement of the IRS. If the authorities all agreed—and there was no reason to believe they would not; in fact, quite the contrary—that would allow him, De Niro, to leave town permanently, in fact to leave Nevada permanently, to begin establishing a new life for himself and his family. This new life would have nothing to do with the gaming industry in any form, in any place, or at any time. These were the only conditions. But they were absolute conditions.

Yes or no?

De Niro knew that being persecuted by the IRS was about the worst thing that could happen to any American. It was, after all, the IRS, not the FBI, who had finally put Al Capone away. They had no fucking regard whatsoever for the rights of anybody, especially those they had designated as targets, for reasons that had nothing at all to do with taxes. They would, he knew, hound him publicly, to make an example of him, to keep him in a state of perpetual fear. The IRS would haunt him literally to the end of his life. So he accepted the offer with an alacrity that appeared genuine beyond any doubt.

Salgo's name never came up, but De Niro obviously knew. For after the deal had been struck, as he turned to leave what had been his office at the casino for the final time, the emotion reflected in his last glance back at Danny Lehman was one of neither thanks nor fear. It was one of deep, black rage. He left town exactly thirty days later.

The loose end had been taken care of.

Or had it?

◆   ◆

Danny Lehman wasted no further thoughts on De Niro. As usual, he became completely absorbed in the project at hand, the conversion of Raffles into the casino of his dreams.

Two years later, in the summer of 1971, The Palace opened its doors, and became an instant smash success. During the next five years, Danny Lehman spent almost all of his waking hours, day and night, inside The Palace, making sure that it stayed that way. He lived alone in the penthouse, which had been especially designed to suit his tastes, so the less said about it the better. For a while Sandra Lee shared it with him, until one day, for the first time in their relationship, she asked a favor of him.

"Danny," she said over breakfast one February morning in 1972, "you don't even know I'm here anymore."

"What was that?" he'd replied, since he was absorbed, as usual, in poring over the latest computer printouts of the casino's financial results.

"Now listen to me for a change, Danny! Would you do something for me?"

"Sure. Name it." He had stopped reading.

"I want you to let me work the baccarat tables. I want to become a dealer."

"Why?"

"I want to be independent."

"You don't have to do that, honey. You know you can have anything you want. Just ask."

"I don't want to have to ask. I want to do it on my own again. Okay? Believe me, Danny, I won't let you down." He'd never heard her plead like this before. It was embarrassing.

"Sure," he said. "Look, I'll talk to the dealers right away. When do you want to start?"

"Tomorrow."

A month later she moved out, explaining that it was never good when an employee fucked the boss on a regular basis. She didn't let Danny down. Quite the opposite. She was soon so good, so attentive to the regulars, that the high rollers started to come from as far away as London and Hong Kong to spend an evening at Sandra Lee's baccarat table at The Palace.

During the rest of the decade, they drifted even further apart, as Danny Lehman's attention increasingly turned to Atlantic City. By the last year of Jimmy Carter's term in the White House, Danny's new casino on the East Coast was fully operational. As a result, Atlantic City was in the process of being reborn.

By 1980 Danny Lehman had finally arrived.

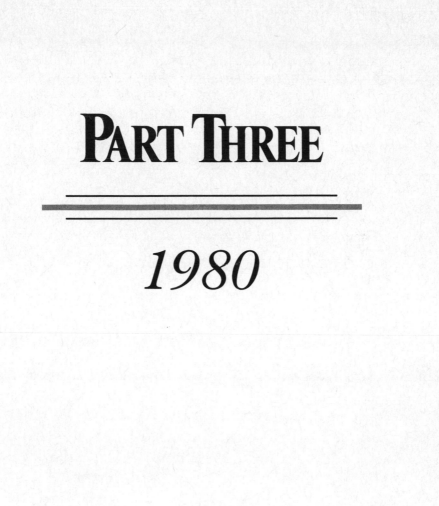

PART THREE

1980

13

Nineteen eighty was not the best of years for the United States, but nobody noticed that in Las Vegas. Good times in Vegas are strictly defined by the number of overnights at the hotels and the aggregate win of the casinos. Both were up in 1980. So who cared about the rest of the country? For The Palace it was truly a banner year; the casino was in the process of grossing over $500 million, netting $35 million, returning, therefore, $1.75 profit on each of the 20 million outstanding shares that The Palace Inc., the parent holding company, had issued when it had gone public in 1973, allowing Danny to pay off the note and consolidate The Palace's financial situation, at least temporarily. Its shares were quoted on the New York Stock Exchange.

So who cared about hostages in Iran or inflation at home? If anybody in the entire state of Nevada did, it was not Danny Lehman. If there had been any problem at all for

Danny during most of the years of the Carter presidency it had been the fact that Wall Street and the investors of America had for four years steadfastly refused to recognize The Palace for what it was, a real money machine. So its shares had sold—even in the best of times when the Dow was flirting with 1000—at only nine times earnings. As late as 1977 you could have bought into The Palace for a lousy two dollars a share, meaning that Danny Lehman, who owned exactly 10,312,000 shares of The Palace Inc., could claim a personal net worth of only $20 million. In retrospect, he probably would have been worth a lot more if he'd stuck to the currency and bullion business, especially since silver was heading from $4 toward $40 an ounce and gold from $150 to $800—while The Palace's stock seemed stuck forever at $2.

But all that changed in 1979. The Palace Inc.'s common had suddenly soared to a peak of $35.25 on August 17 of that year, putting the price at over one hundred times earnings, meaning that Danny's stake in his own company was suddenly worth just over one third of a billion dollars. Wall Street had come around, suddenly and massively, for two reasons: it finally recognized the past, and it anticipated the future.

The past was represented by the magnificent casino-hotel-entertainment complex, the world's largest, which Danny had produced out of the decay of the old Raffles operation in Las Vegas. His formula had worked magic: the Roman motif, baccarat, unlimited interest-free credit for the high rollers, and the high rollers themselves. The Palace had become their place, just as Eddie Cordoba had promised that summer day in Beirut in 1969. And just as Danny had thought from the moment he had been caught up in the action at Crockford's in London, it was that American version of chemin de fer—baccarat—which had attracted them to the American desert. Amazingly, it was also the baccarat tables that would contribute well over 60 percent of the total casino win that The Palace would chalk up in 1980. To be sure, The Palace had become one of the biggest credit institutions in Nevada in the process: it now had as much as $30 to $40 million in outstanding "advances" to its clients at peak times. As in Saudi Arabia, no interest was ever charged. But as with Bank of America in San Francisco, write-offs due to defaults were infrequent, averaging less than 2 percent. Eddie Cordoba saw to that, sometimes with the help of his "Corsicans," but occasions requiring their intervention were extremely rare, in any case a lot rarer than outside skeptics would ever have believed, Why? Because the high rollers all wanted to be able to come back to the new Mecca of the gaming world someday, so they paid their debts voluntarily.

But what really made the difference, what set The Palace apart from every other casino in Vegas and, for that matter, in the world, was the attention its management lavished on its clientele, and the entertainment it offered clientele and visitor alike. Visitors now came from all corners of the world to have fun, for The Palace had the reputation of being Hollywood, Disneyland, Wimbledon, the Hockenheimerring, and Madison Square Garden all in one. It was the Casino de Liban magnified, not tenfold, but one hundredfold.

That was the past and the present. The future that beckoned, a future that had just begun, was Atlantic City. The high rollers and baccarat might have been Eddie Cordoba's contribution; Hollywood, Disneyland, and Madison Square Garden might have been Mort Granville's; but Atlantic City was solely and exclusively Danny Lehman's baby, from the moment of conception to that of realization. The hotel-casino on the Atlantic coast had now been in full operation only six months. What had startled Wall Street was that it was not only already in the black; it was ten million dollars in the black! Result: high rollers and baccarat in Vegas, plus instant massive profits in Atlantic City, added up to a success unparalleled in the history of the gaming industry.

And all this was only the beginning. Danny Lehman had just announced his decision to build a second hotel-casino in Atlantic City, and the land had already been acquired. Not that he was standing still in the West: Danny had simultaneously

embarked upon a program that would double the capacity of
The Palace in Vegas by 1982. He had just bought a hotel-casino
complex on the south shore of Lake Tahoe. What Howard
Hughes had tried to accomplish in the 1960s, and miserably
failed to do, Danny Lehman had attempted in the 1970s, and
was now achieving to a degree that was truly astounding. He
seemed virtually unstoppable.

Even Henry Price could no longer ignore the phenomenon. For the details of Danny's success were highlighted in a
report by Mercier Frères, the international investment bank, a
copy of which Price had taken with him from New York to his
Virginia farm one summer weekend in 1980. As usual, Natalie
Simmons was at the farm with him. "If he is a latter-day Howard Hughes, then why does he need you?" she asked.

Henry Price, who had been reading parts of the report
aloud, took his time to answer. "Because something must be in
the process of going wrong."

"How do you know?" she then asked.

"Look at the share price. Thirty-five dollars last summer.
Twenty dollars now. Short interest growing steadily. Nobody
on the street seems to know why. I suspect that Mr. Lehman
might, however."

"But why is the firm trying to bring you into this?"

"Because it seems that he asked for me."

"You mean Lehman?"

"Yes."

"He sounds like a very grubby little man."

"He is, my dear. I've actually met him. Don't you remember my telling you about him? Maybe not. It was eleven years
ago. In any case, 'grubby' was precisely the word I remember
using to describe him then."

Natalie was now watching Price closely. "And in spite of
that, you are seriously considering it." Her attitude on the
matter was made quite clear from the tone of her voice. Her
tone was not by any means a harsh one, it just conveyed a
certain anxiety, which she then reinforced with the following

words. "I suggest, Henry, that before you really get involved, even in the slightest, you consider your own reputation. After all, it is not just this man Lehman. It must be everybody in the gambling business. All grubby, and no doubt all crooked, if you ask me."

"Gaming," he said.

"What?" she asked.

"Gaming," he repeated. "They prefer that you call it gaming, not gambling. It sounds better."

"How silly."

"Perhaps. But just remember that every two-bit stockbroker in New York now calls himself an investment banker."

"Ah yes, but the difference is that you *are* an investment banker. And bankers and gamblers should not be seen in public together, in my opinion."

"Maybe. Maybe not. One must be realistic. Let me read what our hotshot analysts at Mercier Frères have to say." He picked up the report again. " 'We estimate,' they say, 'that domestic legalized gambling is running at an annual rate approaching thirty billion dollars per annum, and that the inclusion of the underground gambling pool, principally sports betting, raises the ante to a ninety-to-one-hundred-billion-dollar market annually, a size sufficient to rank gambling as'—and now listen carefully, my dear—'as the *third largest United States industry.*' "

"Amazing. You've made your point."

"Quite."

"But it's not really that, is it?"

"No, probably not."

"You want to shock a few people."

"Maybe."

"You want to do whatever has to be done to bail this grubby Lehman fellow out of his overextension or whatever, and then be able to say to everybody back in New York or over in Paris, 'See, Henry Price can still pull off his stuff wherever he wants, whenever he wants, if he chooses to.' "

"Perhaps."

"And assuming you do, what will you really have proven?"

"Nothing, *really,*" and the emphasis showed that he was becoming rather peeved.

"Let's face it," he then went on, "what happens to Mr. Lehman and The Palace will hardly alter the course of Western civilization, will it? It's meaningless."

"Then why do it?"

"To use that tired phrase yet again: because it is there, my dear."

"May I come along?"

Price now smiled for the first time during the conversation. "Aha," he said, "you see! You too!" She almost never accompanied him on business trips. She said it was because she had to keep her eye on her Georgetown art gallery. But he knew that it was also because, despite the fact that they had been "going together" for more than a decade now, she didn't like it when strangers gave her a certain look when it was explained that, no, she wasn't Mrs. Henry Price, but rather Miss Natalie Simmons.

"All right, I'll admit it. I *am* curious. But also I don't want you to have to stay in that ghastly place all by yourself, assuming you are really going to go there."

"Of course I'm going, so start packing. Two medium-sized suitcases should do."

"You mean now?"

"Well, yes. Lehman is sending his Lear to pick us up at four at National Airport."

"You told him already that I would be coming along?"

"In fact, I did."

"You know something? I like that. After twelve years you're still full of surprises." Natalie came over and gave him a peck on the cheek and then disappeared up the stairs.

◆　　　◆

That evening, a Lincoln was waiting for them on the tarmac about a dozen yards from where the plane parked at McCarran Airport in Las Vegas. The driver, not a regular driver but one of Bill Smith's security staff, first settled them in the backseat and then retrieved the suitcases from the airplane and put them in the trunk.

"Welcome back to Las Vegas, sir," the driver said as they started to pull through the steel-mesh gates onto the public roadway.

"Well," said Price, "when you say 'back' in my case you are referring to a long way back. The one and only time I have been here was in 1962, and then only for a few hours."

"And you, ma'am?" the driver asked.

"Never been here at all," was her answer, stated in a tone of voice that indicated that conversing with drivers was not one of her regular habits. So he decided to shut up and drive.

If one could pick an ideal moment to arrive in Las Vegas for the first time, it would have to be at early dusk during late summer. The heat still shimmers on the desert floor, but beyond, and not that far beyond, lie the mountains, which take on a blue, then a purple hue as the evening light becomes increasingly diffuse. Then, suddenly, one is confronted with the Strip: and the dramatic natural scene fades behind the lights of the Sands, the Dunes, MGM Grand, the Silver Slipper, the Flamingo, the Golden Nugget, Caesars Palace. The casinos seem to get brighter and brighter, and closer and closer together, the farther down the Strip one goes. And they blot out totally any lingering romantic thoughts about deserts or mountains.

"My God," she said. "Look at this!"

"Isn't it something?" he said. "I had no idea what had become of the place. I seem to remember it as a small town. This makes the bright lights of Broadway look like a joke."

A mile later they halted at the brightest intersection of all: on one corner the MGM, on another corner the Flamingo, on the third corner Caesars Palace. But on the fourth corner, bathed in a greenish-bluish light emanating from the enormous fountain on the grounds in front of it, stood The Palace. The sign that said so rose fifty feet high, broadcasting the name with a light display of such intensity that the sunglasses Natalie still had on were not even out of place.

"My God," she said again. "I thought I'd already seen everything, but this place beats them all."

The driver decided to break his long silence. "That's what everybody says about The Palace, ma'am. There's only one of them, isn't there?" It was obviously the prime object of local pride. The Palace was to citizens of Las Vegas what the Golden Gate Bridge was to the San Franciscans: not just a tourist attraction but a monument, a true landmark.

When they pulled up under the immense cantilevered canopy, two doormen rushed up to the car, obviously having been alerted to the fact that one of Bill Smith's men would be bringing in a VIP of the highest order. Inside the casino-hotel proper there was a scene that verged almost on bedlam. The casino floor was jammed; the noise level, fifty decibels plus. All tables, every single one, were going full blast. And the area that contained this manic scene seemed to be measured not in square feet but acres! Though it seemed hardly possible, what one's senses were confronted with inside The Palace was even more astounding than the neon dazzle one faced outside.

Like two people who had just landed on the moon, and whose initial reflex was to take it all in quickly lest it should suddenly disappear, neither Henry Price nor Natalie Simmons said a word as they proceeded the fifty-odd yards from the entrance to the hotel reception area. The minute the doorman mentioned Price's name at the desk, an assistant manager came out to greet them personally and, waiving all formalities, asked them just to follow him; their accommodations, it seemed, were ready and waiting.

The accommodations were one of the four suites on the twentieth floor of the new tower. They defied description. The living room was built around a fountain. Beyond the fountain was a statue, a replica of Michelangelo's *David*, not quite full-size, but at first glance it seemed so. To the right of *David* there was a circular bar equipped with everything one needed, including a live bartender. Beyond the bar was a dining room of immense proportions. The candles on both the buffet and the dining table, twenty-four in number, had all been lit. The entire main section was partially ringed above by a balcony. One reached the balcony via a semicircular staircase. Upstairs there was a suite within a suite, again one of extraordinary size reflecting equally extraordinary taste: a blending of ancient Baghdad and modern Anaheim. For instance, the sunken bath was made from mauve marble.

"Never in my entire life . . ."

"Darling, nor in mine."

When Natalie Simmons had finally taken it all in, she asked, "But do we have to stay here?"

"We hardly have a choice, but then why not? This probably will never happen to us again. Let's lie back and enjoy it."

"But it's absolutely outrageous," she said. "Look at the color scheme!"

They both shook their heads and began to giggle, while Natalie sat down on the circular bed, watching her image in the lighted mirrors above and around the bed. Then she opened the soft gauzelike curtains that encircled the bed, so as to be better able to reappraise the entire suite. The so-called color scheme was one of oranges upon greens upon purples upon pinks, but to call it garish would not only have been an understatement of vast proportions, it would also have been unfair. For as ludicrous as it all appeared when seen through the eyes of "civilized" Easterners, who were used to the best that New York or Paris or Rome had to offer, The Palace had not been built to please *them*. And it was Henry Price who said just that. "Darling, this is probably not for us to judge and most certainly not for us to put down. This man Lehman undoubtedly knew exactly what he was trying to accomplish here. I suspect he wanted to provide adults with the counterpart of Disneyland,

and I think he has succeeded to a remarkable degree. But I think I could use a fairly stiff drink.''

"Make that two," Natalie answered, "but first tell me: must we dress for the occasion or may we go as we are?" And again she giggled.

"I think we will chance it and go as we are."

At that moment he noticed that lights were blinking on the many telephones scattered around the place, so he picked one up, and when connected with the message desk was informed that Mr. Daniel Lehman was expecting them for dinner at nine o'clock in the Florentine Room, and that dress would be formal. After he had hung up he turned to Natalie and said, "Well, my answer was premature. I fear that, contrary to all expectations, we will indeed have to dress for the occasion. I do hope that you have brought some evening clothes along."

She had. "And what is the occasion, if I might ask?"

"We have been beckoned to dinner by Mr. Lehman."

"You mean you were invited via the message desk?" she asked.

"Well, my dear, this is a new and strange country. As Goethe once said, *Andere Länder, andere Sitten.*"

◆   ◆

The Florentine Room, The Palace's most elegant dining room, had nothing garish about it. It had been designed as a replica of the Tour d'Argent in Paris, and rather successfully. It suffered somewhat from the fact that the view was not of Nôtre Dame and the Île de France but rather of the parking lot and, beyond that, a vacant lot. But copies can hardly ever be expected to live up to the original totally. The ambience, at least one's initial impression of it, was not enhanced by there being three armed guards at the entrance to the restaurant, who, however, immediately let them pass when Henry told them his name. They had barely stepped into the room when a man about five foot six inches tall, slightly on the plump side, immaculately groomed and just as immaculately tailored, and smiling a smile that could only be described as very broad, open, and friendly, came up to them.

"You're Henry Price," he said. "We've met before, if you recall."

"I do recall," replied Price.

"You've done us all a great honor by coming over here today, and I hope everything so far has met with your approval. If there's anything, anything at all you need, just call my office. My secretary knows you're here. Tell her what you want and you'll get it."

"It's all fine. And we're delighted, simply delighted to be here," Price responded. He almost meant it, since Danny Lehman was obviously trying his damnedest to be a gracious host. So why not be agreeable? After all, there could well be a great deal of money to be made here.

Danny's eyes then turned to the young lady at Price's side; she appeared to be in her mid-thirties, in contrast to Price's sixtyish appearance. It was very apparent that Danny approved of what he saw, and he said as much. "Well now, Mr. Price, tell me who this young lady is, and before you do let me warn you: I like beautiful young women, but I promise you that she is going to be, beyond any doubt, the most beautiful woman in this room tonight. And we have some pretty good-looking women coming, I can tell you."

"Mr. Lehman," said Henry Price, "please meet my fiancée, Natalie Simmons."

Danny took Natalie's hand in both of his and gave her a smile which, if anything, was still broader and even more open. "If you don't mind, Mr. Price, I would be honored if your delightful fiancée would join me at my table this evening." Without waiting for an answer from Henry Price he turned to Natalie and asked, "Would that be all right with you, Miss Simmons?"

"I would be delighted," she answered, although her voice indicated more than a small measure of, not hesitancy particularly, but rather a feeling of having been taken back by the abruptness of it all. Danny Lehman noticed nothing, naturally, and once again took hold of her hand and said, "Come. I must introduce you to some of my friends who will be joining us for dinner this evening."

The first friend was Eddie Cordoba. Eddie, Danny explained, ran The Palace. Then he introduced an Englishman, Chapman by name. He was, he declared, the owner of the Lotus auto racing team. Next in line was an Argentinian, a friend of Cordoba's. And so it went on until finally she spotted Henry Price again across the room. She made her excuses to Danny and hurried to Price's side.

"What are all these people doing here?" she immediately asked.

"It seems," he began, "that there is going to be a Grand—"

At that moment there were three loud claps, and when everybody turned to see what was going on, it was Danny Lehman who clapped again. "It's time to eat, folks, so let's sit down."

Natalie had drawn the place of honor, to the right of Danny Lehman. To her right sat the Argentinian, a Mr. Reutemann according to the engraved place cards, who explained why all these people were gathered together. "You see," he said, "we are having a race here tomorrow and Mr. Lehman has put on this dinner party for some of the owners and drivers. I guess you could best describe it as a pre-race get-together."

"And where is this race going to take place? Out in the desert somewhere?" she asked.

"No," answered the Argentinian, "actually, in the parking lot."

"The parking lot?"

"Yes, the parking lot. You see, Mr. Lehman and Mr. Cordoba decided last year after having watched the Grand Prix in Monaco that it would be rather nice to have one here. They couldn't put it on, or at least couldn't *quite* put it on in the casino, so they decided to stage it in the parking lot."

"But how on earth can you convert a parking lot into a Grand Prix race circuit?"

"With about two million dollars," Reutemann answered.

"You must be kidding." She noticed he wasn't. "Then how does one park from now on?"

"No real problem. It's only taken them three days to put it up, and they tell me it will only take them one day to rip it down. So it won't cause that much difficulty. Now tell me, Natalie, isn't it, you are coming to watch me race tomorrow, aren't you?"

"Well no, no one has said anything."

Reutemann then leaned across her to address Danny Lehman. "Danny," he said, "Natalie insists upon coming to the races tomorrow. You've fixed her up, haven't you?"

Danny shook his head and said, "No, but it will be done right now." He waved a finger at Cordoba, who immediately sprang up from his seat to come over and, having received his instructions from Danny, left the room directly. He was back three minutes later. "Everything has been arranged, Miss Simmons," he said. "The necessary credentials will be delivered to your room in the morning. My suggestion is that you come to the track around eleven o'clock. That way you can choose the spot you want before the race starts."

The dinner was excellent by the standards of any country. Surprisingly, nobody drank much, even though the white was a superb product of France and the red, a 1978 Jordan, California's finest. At ten-thirty, Danny Lehman stood up to say that the party was over and in a courteous but strangely abrupt fashion said good night to everyone at the table, including Henry Price and Natalie, and disappeared from the room.

Henry immediately came over, put his arm around Natalie, and, leading her out of the dining room, asked, "Darling, shall we go down to the casino and try our hand at something?"

"Please, no," she said. "I hope you don't mind terribly if

we call it a night. I think I've had already too much of this sort of thing. Let's go upstairs and see if our personal barman can serve us a nightcap."

"I don't mind it at all," he answered with disappointment. "In fact, I've had quite enough myself."

They proceeded back to the tower suite, which was now *sans* bartender, so Henry did the honors. Within a short time they found themselves beneath the sheets, and also beneath the immense circular mirror that reflected their immense circular bed. The entire room was bathed in a pinkish sort of light from a hidden source, and presumably controlled by a switch that was also hidden, so well hidden that they were not able to find it. This did not bother Henry; he went to sleep right away. Natalie, however, could not; and, for one of the few times in her life, rather regretted that the man beside her was, well, not insisting. Even the light would not have bothered her, though she did prefer to do it in total darkness. She did not find the sight of *it* exactly aesthetic. But somehow, that night, she had the feeling even *that* would not have bothered her. How to explain it? She speculated for a few minutes and then came up with the answer: either it was the Latin machismo of the Argentinian racing driver, and she giggled, or, and this surprised her, perhaps it was the way Danny Lehman had insisted upon touching her from the moment he had first laid eyes upon her. All through dinner he had found a way to touch her that was apparently very casual, yet somehow intimate. Anyway, it was all rather ridiculous, she told herself, and after glancing up at the mirror one more time, she turned away from Henry and drifted off to sleep.

14

It began with what sounded like a series of staccato sputterings, which soon merged to become a growl, a growl that quickly became a howl, and then a roar.

"Henry. Are you awake?"

He was now. He looked at the bedside clock. It was 7:30 A.M. "What's going on?"

The roar was now interrupted, or perhaps a better word is "superseded," by a series of fierce, ear-splitting whines, now building up, now fading. And all this was permeating a room on the twentieth floor, one that was still bathed in a faint pink light, and that appeared to be as hermetically sealed off from the rest of the world as possible.

"It's outside," Price then said.

"Obviously," she replied, sliding out of bed. She first tried to find a way to open the curtains, and God knows what else that stood between her and the window, but failed completely. Nothing budged. So she ducked under them, slid open the window, and looked down upon a huge racetrack, surrounded by a series of grandstands and bleachers, around which at least twenty racing cars were pounding at what she could tell were ferocious speeds, even from that height.

"Henry!" she screamed. "You've *got* to see this."

Henry, sleeping naked as usual, crawled under the curtains and was soon crouched at her side, equally stunned by the scene below. "Christ!" he exclaimed. A few minutes later his arm went around her shoulder and then descended almost immediately to her left breast. His right hand came around to take her right hand and pull it to his erection. She had no choice but to grasp it. Then, reluctantly, her hand began to move, first slowly, but then more quickly and intently.

"Honey," he said, "let's go back to bed." But she just kept going.

"There," she said soothingly, looking neither at him nor, most definitely, at *it*.

Then chimes sounded, so loudly that one could hear them above the bedlam outside. She pulled her hand away immediately and, flushed with embarrassment, ducked back out into the bedroom. "It's somebody at the door," he said, reemerging behind her.

"I'll see," she said. "You're not dressed." She put on a rather flimsy robe over her nightgown and quickly descended the semicircular marble staircase to the living quarters below. When she opened the door, it was to find Danny Lehman, tennis shorts and all, waiting there with a huge package in his arms. Again he greeted her with a huge grin. "Gee," he said, "I hope I'm not too early."

"No, not at all. I mean, with all that noise out there we've been up for at least an hour," she lied.

"Look," he then continued, "I've got your credentials for the race and I brought something else I thought you might like. May I?"

"Of course," and she led him into the living room.

212

"Here," he said. He proceeded to rip open the package and hand her two huge jackets, obviously his and her versions of the same design.

"What are they?" she asked.

"Racing jackets. Special editions. We only made twenty of them."

They looked like the sort of thing that Canadian hockey players wear when trudging through the snow to and from the rinks in northern Quebec. Made of shiny satiny material, they were predominantly fire-engine red, with black stripes down the arms and, in huge letters on the back: THE PALACE GRAND PRIX—1980.

"Try it on," he said. As she struggled to do so, both her robe and nightgown ended up in a total state of disarray.

"Let me help you," said Danny, ever the gentleman. After he had gotten her arms through the sleeves, he then fumbled for the zipper, finally had it, and then drew it slowly up to her chest, giving her breasts a very careful examination in the process. By the time he was finished, he and Natalie were only inches apart. He stepped back and, without looking down at her body again, said, "It looks terrific on you."

He then reached into his pocket and handed over two badges. "Put them on the jackets. They'll get you anywhere you want to go down there. I look forward to seeing you around eleven." And he turned and left.

Trembling, Natalie slipped out of the jacket and slowly went back upstairs where Henry Price was waiting, still naked. "What was that all about?"

"Tickets for the race. And jackets for both of us to wear that are simply unbelievable," she answered.

"Let me go down and look," he said.

"In a minute, Henry," she replied, reaching down again to grasp what remained of his erection. When she took it into her mouth a couple of minutes later he was so surprised and excited that he came within ten seconds.

She'd never done *that* before. Maybe there was really something to Las Vegas.

◆　　　◆

At eleven o'clock precisely, Natalie Simmons and Henry Price entered the elevator at the top of the Palace tower. It was a slow descent to the main lobby, stopping on almost every single floor until the elevator had reached its weight limit. The routine was the same at each stop. The gaming *cum* race fans, almost all men, stepped in, looked at Natalie Simmons and then at her racing jacket, and either silently, or in three cases out loud, said the only thing one could say, "Wow!" The guy that got in on the sixteenth floor went one further. "I'll give you five hundred dollars for those jackets." When neither Natalie nor Henry said a word, he added, "Each!"

Rather than get into any detailed conversation, Henry just shrugged and said, "Sorry," and the elevator, this time, headed straight for the lobby. When they had managed to push their way through the crowd in the casino, already going full blast at 11:00 A.M., and out of the back entrance of the casino, normally used for access to the parking lot, they were met by a deafening silence, and by a very long line leading to the gates of the track itself. Almost immediately a scantily clad young lady came up to them, gave the badges with their blue ribbons and the letter *A* a quick glance, and asked Henry where they would like to go.

"What are the choices?"

"That badge, sir, will get you anywhere you want to go."

"Where are the cars?" Natalie asked.

"In the pits for the final checks."

"We couldn't actually go there, could we?" Natalie asked.

"Anywhere you want" was the answer.

And a few moments later they found themselves on the narrow strip separating the track itself from the pits, in which the crews were really doing nothing but standing around having their pictures taken. They had been there but a minute when Eddie Cordoba appeared, as if by magic. "Mr. Lehman asked me to see to your needs," he said. "May I suggest that we say hello to a couple of the drivers, and then, perhaps, have a glass of champagne?"

They went first to the Saudi Leyland cars to meet Alan Jones and Carlos Reutemann, who vaguely remembered having met Natalie the evening before, then over to the Brabham piloted by Nelson Piquet and, at the insistence of Henry Price, who seemed to know more about this than either Natalie or Cordoba expected, to meet Mario Andretti, who was driving an Alfa Romeo 179C for the Marlboro team. "We've served as advisers to Alfa in Milan on various occasions," Price said. "In fact, it was upon our advice that they got back into racing."

They ended up in a roped-off area at the finishing line, where a boxed lunch consisting of Iranian caviar, gravadlax, and cold pheasant awaited them. Men in white jackets were serving Moët ct Chandon to counteract the heat, which had already built up to the low nineties.

When the race began, it was the incredible noise that impressed the pair from the East Coast the most. But then, to stand behind a thin concrete wall, only waist-high, and be able to lean over and look at a Grand Prix racing car coming at you at a couple of hundred miles an hour and passing within feet of where you were standing—and then another and another—was, as Henry put it, "a rather uncommon experience." It was over in two hours, and it seemed that nobody really cared who had won, except for Reutemann and his team, who were the victors. Within thirty minutes of the end of the race the crowd of fifty thousand was gone and a crew of construction men were already starting to rip down the grandstand.

Inside the casino the unreality of it all was further magnified. For on passing through the doors one left the heat, the blinding sun, the dust, and the rowdy atmosphere of the sports crowd behind, and suddenly found oneself in a perfectly controlled, sixty-eight-degree, dust-free atmosphere: a world of dimmed lights and the muffled background noise of coins jingling in slot machines. It seemed that almost every one of the fifty thousand who had attended the race, paying up to two hundred and fifty dollars for the privilege, was now in the casino.

"That man Lehman is a genius," Price said as he and Natalie walked from the "new" casino, which had been built at the base of the new tower, to the vast area of the "old" casino. They passed thousands of people whose main aim in life now appeared to be to find a place at a blackjack or craps table and come out a winner, just as Reutemann had in the Grand Prix.

When they finally got back to their suite, it was to find the message lights on the telephones blinking. Again it was an invitation to dinner. Another black-tie event, but this time in Palm Springs. Someone would pick them up at their room at six-thirty. It was optional, the message said, whether they packed an overnight bag or not. There would be accommodations waiting for them in Palm Springs; a plane would be available if they chose to come back to Vegas.

The same man who had picked them up on their flight in from Virginia knocked at the tower suite at exactly six-thirty to escort them to the limo. The plane that awaited them at McCarran was a Gulfstream II, the luxury liner of the private airways. It was the ever present Eddie Cordoba who was standing next to the plane greeting the dozen or so passengers, all of whom seemed to arrive at the same time. They were all strangers, so Natalie and Price kept to themselves. The trip from Vegas to Palm Springs was short and dramatic: straight up over the sharply rising mountains, and then straight down into the valley that contains the Springs. On the ground it was a repeat of the scene in Las Vegas: a fleet of black limos ready to transport each of the couples in air-conditioned privacy.

The common destination was the Palace-owned villa, built in the Spanish style, surrounded by plush green lawns, and overlooking the obligatory immense pool area. It was there that they were all gathering, and it was there that Daniel Lehman was greeting the guests as they came down the steps.

Naturally a mariachi band was playing in the background, and just as naturally a team of Mexican waiters was circulating with the champagne and caviar canapés. To say that Danny greeted everyone effusively would not even begin to describe the enthusiasm he displayed. What did not seem particularly appropriate was that the guests did not seem worthy of such effusiveness. To Price and Natalie they looked like rich retired butchers with call girls. The men had paunches in every case, and all smoked cigars. Two thirds of the women were blond.

Before Price and Natalie had quite made it up the receiving line to Danny, the Mexican band burst out in a very brassy version of "For He's a Jolly Good Fellow," and Danny went up to the microphone in front of the band, accompanied by someone who looked like the president of the butchers' union: an old guy, probably over seventy, but one who was still built like a linebacker, and who had a blonde in her mid-twenties in tow.

"Listen, everybody," yelled Danny, once the band had stopped. "This evening's a surprise party for a man you all know and love, and who deserves this celebration more than anybody I know in the whole world. He's a man who is all heart. Somebody once asked me who I'd turn to if everything went bad, when everybody was walking away. You know what I mean?" A lot of nods. They knew, it seemed.

"Well, the man I named was this man." And Danny put his arm around the seventy-year-old hulk, and squeezed. "So let's hear it for Mannie!" he then yelled.

He got the applause he wanted. Mannie just stood there grinning, and then held up his arms in the boxer's victory salute to himself when the band this time began a Tijuana version of "Happy Days Are Here Again." Then a parade of waiters suddenly began to come down the steps from the main house, each carrying a tray of delicacies that would make up the buffet dinner. The crowd of about fifty murmured appreciatively and began consuming exotic pâtés, marinated seafood, and clams cooked in a superb combination of herbs, in a manner that indicated that they were quite used to this sort of cuisine.

"Casting pearls to the swine, wouldn't you say?" These words came from Eddie Cordoba, who stood behind Natalie as she helped herself to some clams. She was too startled to reply. "No offense, I hope," Cordoba continued. "It's just that sometimes it's hard to stomach even for me."

"Who are they?" Natalie asked.

"Shall we all sit together and perhaps I can explain," replied Cordoba.

The three of them found a table as far away from both the buffet and the band as possible. No sooner had they sat down than a waiter filled their glasses with a 1978 Meursault-Perrières. He returned a few minutes later to ask which red wine they preferred, a 1969 Lafite or a 1959 Latour.

At the tables around them, the wine seemed to disappear almost as soon as it was poured, and the volume of noise began to build up in spite of the fact that they were outdoors. A few of the blondes let out a screech now and then. Then Danny Lehman suddenly materialized at their table. He spoke directly to Natalie. "Great that you could come. How do you like the place?"

"It's marvelous," she answered.

"I hope you're staying the night. You've got the best of the bungalows if you do."

"No. It's so kind of you, Mr. Lehman. But we decided to go back. This is all still a bit . . . strenuous for us, isn't it, Henry?"

"It is. Overwhelming."

"Henry," Danny then said, abruptly. "Could we talk tomorrow morning?"

"Sure."

"Eight o'clock all right? My office?"

"Perfect."

Danny was then accosted by one of the blondes, who almost dragged him off toward the area where the dancing had begun. He looked back at Natalie with a feigned look of help-lessness and disappeared.

"He's something, isn't he?" commented Cordoba.

"Yes," answered both Natalie and Henry. And both meant it in their own ways.

"Now," said Natalie, "tell me. What is this place?"

"This"—and Cordoba's hand traced a circular move-ment—"is one of the establishments which The Palace keeps for its clientele, those high rollers you've no doubt heard about. There's this place, then there are the yachts, plural, one in Miami, the other in San Diego. Then there's La Costa, and the place in the Catskills, and . . . well, there's one more, but I forget where it is at the moment."

"But who are these clients?"

"From all over the world, usually. But not tonight. This is strictly for our American friends."

"And who is the guest of honor?"

"Mannie?" he replied, and then just sat there thinking for a minute. "Mannie has been coming to The Palace once a month since I've been here, and that's been, what, ten or eleven years now? Anyway, Mannie used to just come to shoot craps. After all, that's what all his friends play, or at least used to play. And a lot of his friends are here tonight. Mannie was originally from St. Louis, I might add. Now he's from"—and again his hand traced a semicircle—"from these places.

"Anyway," Cordoba continued, "about ten years ago I introduced him to baccarat. He loved it from the first minute, because it's fast, very fast, and you don't have to think. Ever." Cordoba leaned over the table toward Henry Price and said, now in a much softer voice, "Mannie drops an average of two million dollars a year at the baccarat table. Every year except this year. So far this year it has been almost five million. He dropped a very, very big packet a month ago, the weekend of the big fight."

"Amazing," said Price.

"Let me suggest something," Cordoba said. "Hang

around for another fifteen minutes. Then I'll have your car ready to take you back to the airport. Our plane is there waiting. It'll start to get a bit crude soon. And I think you might . . ." His voice tailed off. "Now please excuse me," he said.

Cordoba walked over to the dance floor and retrieved Danny from the blonde; then both of them disappeared in the crowd. Five minutes later Danny once again climbed onto the podium and the band stopped. "Mannie," he said. "Where are you?"

Mannie staggered onto the dance floor, supported by two young ladies. "Mannie, I've got something I want to show you, and everybody! Follow me!"

Danny stepped back down to the dance floor, pushed one of the ladies aside, and, taking Mannie's left arm, led him, and the parade of guests that followed, up the stairs toward the main house. But rather than entering, he followed the walk around the left side, leading toward the circular driveway that wound around the inevitable fountain, as spectacularly lit as the neon "signature" of The Palace. He shooed the second blonde away, and then, with Mannie, walked up to a red Rolls-Royce Corniche that stood alone in all its splendor. Danny raised his hands to silence the crowd and then spoke in a voice that all could hear. "Mannie, happy birthday!" And he handed over the keys of the Corniche to the huge, aging man. Tears began to stream down the man's face. He embraced Danny, then Cordoba, who stood beside Lehman, then a third man, who Henry later found out was one of Danny's partners, a man by the name of Granville. Then all four stood together with their arms around one another. The ceremony ended almost as quickly as it had begun. The crowd moved back toward the dance floor, lured by the sound of the Mexican band. Cordoba found Price and Natalie and took them over to a waiting limo. "How did you like the last little scene?" he asked, as he opened the back door of the car for Natalie.

"Truly touching," she answered.

"Glad you thought so. Mannie will lose the cost of that Corniche in less than two hours tomorrow night." He grinned and closed the car door behind them.

Half an hour later they were airborne, heading back to Vegas, alone in the Palace's Gulfstream with the two pilots. "Henry," Natalie said, once they were up.

"Yes."

"Maybe it's better we go back to Virginia."

"Yes, I know what you mean."

"Let's. We can get our things at the hotel and just leave."

"No. That wouldn't be fair."

They sat in silence for the next five minutes. "What kind of people are they?" Natalie asked. Henry Price just shrugged. But Natalie knew that he was bothered, perhaps deeply bothered, by what he had witnessed.

"Look," he then said, "Lehman knew it. You noticed he did not introduce us to a single one of those people."

"Thank God. But then why invite us?"

"Maybe to show that one client, the birthday boy, covered the total cost of today's extravaganza in the parking lot. Even with the Rolls thrown in. If that was Lehman's point, he made it, at least as far as I'm concerned. You could not help but wonder how much the other twenty-five guys who were there this evening 'contribute' on an annual basis."

Then he added, "Don't worry, dear. I'll be careful tomorrow. If I don't like what I hear, we'll be on a plane before lunch."

15

Henry Price rose at six-thirty the following morning. Without waking Natalie, he left their suite, went down to the pool, which was totally deserted, and did his normal fifty laps. Then it was back up to the suite, where he showered and shaved and, finally, put on what Natalie called his banker's uniform, a pin-striped suit. Ready for work, he headed for Danny Lehman's office.

When Price left the small elevator that led to the executive offices of The Palace it was exactly eight o'clock in the morning, and he was amazed to see the place bustling with people and activity. Somehow, he thought to himself, one had the idea that everybody in Las Vegas probably started work around ten, or even eleven. Danny Lehman was in the hallway waiting for him. As Price walked up, he looked at his watch. "Eight on the second. I knew it," he stated. "Now, do you want the grand tour, or would you prefer to get right down to business?"

"Business," Price replied.

Henry Price knew only too well the value of grand tours. Invariably when he visited a company for the first time, one that needed either his money or his services in obtaining some, the first day was totally wasted with a tour of their bricks and mortar and whatever was contained within: assembly lines where automobile firms were concerned; laboratories if the company was in either pharmaceuticals or electronics; vaults and computers if it was a financial institution. On one memorable morning, when visiting a rubber company, he had been forced to watch women dressed in white testing prophylactics mechanically. That was outside Tokyo. In London, the chairman of the board of an office machine company there had personally picked him up at Heathrow in the company Rolls, since the chauffeur had suddenly been taken ill. They needed a hundred million dollars, and it was thought that when Price saw their magnificent new plant on the outskirts of London, Mercier Frères would simply thrust the funds at them. For an hour and a half the chairman, Price, and the Rolls wandered around an industrial park somewhere off Western Avenue, searching, asking directions, searching again. They never found it. Needless to say, the company also found no lender in Henry Price.

"Then let's look at the numbers," Danny said.

Price immediately agreed. He was a numbers man.

"I'll bring in our financial man. His name's Matthew Kelly."

Kelly was a man of very few words and masses of computer printouts. He presented page after page of statistics to Price, and Price either nodded or grunted, after which they moved on to the next set of numbers, or he asked a very brief question and got an immediate equally brief answer. Danny had quietly left the room and returned after fifteen minutes bearing two cups of coffee. After having placed them on the coffee table in front of the two men, he left again. When he returned for the second time an hour later, Kelly was in the process of packing up and leaving.

"What're you doing, Kelly?" Danny asked, visibly disturbed.

"Mr. Price said he doesn't need me anymore, sir."

"At least for the moment, Mr. Kelly," Price added.

Danny's expression brightened. His financial man left.

"Well, what do you think?" he then asked Price.

"You need a quarter of a billion dollars rather badly."

"Exactly."

"You've overdone it a bit in Atlantic City, I'm afraid."

"For the moment, yes. In the long run, no."

"What's the long run in your business?"

"Two years."

Price laughed. He liked the answer. "Why are your banks not willing to carry you that long?"

Danny shrugged. "Because they're banks, I guess. They gave us a construction loan. The construction's done, with the usual overruns. They want their money back."

"That's all?"

"As far as we know."

"How's Atlantic City done thus far?"

"On schedule."

"I saw your projections. You expect to gross two hundred million this year, if I recall the number. And to net 10 percent. Right?"

"Yes."

"What will be the combined cash flow next year of both the Nevada and New Jersey operations?"

"Sixty million."

"Assuming you net twenty million in Atlantic City."

"Right."

"If I arranged to get you that quarter of a billion for five years at 15 percent it would be a rather tight squeeze, wouldn't it? If Atlantic City somehow didn't quite pan out, I mean."

"Yes, it would."

"That's no doubt why the banks want out," Price stated.

"Probably," Danny answered. "They're worried that Atlantic City won't be able to support all the new casinos that are about to open: Bally, Caesars, Boardwalk, Claridge, Playboy, Resorts International, Ramada Tropicana, Golden Nugget, Harrah's. It's a long list."

"Will it?"

"Of course it will. Just look at the demographics. There are fifty million people within a day's drive of Atlantic City. We've just scratched the surface so far."

"Some of the people on Wall Street seem to disagree."

Danny shrugged, but said nothing more.

"What kind of a kicker could you offer Mercier Frères' clients?" Price then asked.

"What would Mercier Frères expect?"

"We would want the notes to be fully convertible."

"That's pretty steep."

"You've got a risky situation here."

"At what price could the conversion rights be exercised?"

"That would take some study. I don't think that would present a big problem. Normally it's 20 percent above the market price at the time of the closing. We're investment bankers, not thieves."

"What else?"

"Immediate representation on your board."

"Who?"

"Probably me."

"Where will the money come from?"

"I'd put together a private placement syndicate. In Europe. We might put up some of the money ourselves. In any case, you would be dealing only with us. Nobody else. Now or later."

"How soon could you put such a deal together?" Danny then asked, this time after a substantial pause.

"Oh, say within two months."

"That quickly? What about the SEC?"

"No problem if we do it all outside of the United States, which is what I would intend to do. That, plus the fact that our name would be on it." Then Price added one further condition. "We would, of course, require absolutely and on a contractual basis that you would remain with this company as its chief executive officer during the entire term of this loan, Danny. We'd also expect that you continue to supervise Atlantic City. We know you personally got it off the ground; now we'd want you to keep it there."

"Henry, you couldn't keep me *away* from Atlantic City. That place was my idea and that casino is my baby." Then he added, "The conversion aspect bothers me. I've personally controlled The Palace absolutely from the very beginning. I don't believe in partners."

Price had an immediate answer: "If, and I emphasize the 'if,' we exercise the conversion rights it will only be because the combination of your leadership and our money has produced the success we both fully anticipate. After that we will be the most silent of silent partners. We exist to make money, not run casinos."

Danny nodded and asked, "Assuming we proceed with this, how long would it take you and your bank to come up with something concrete? You know, like a written proposal or whatever you call it."

"A week. How long would it then take you to get your board's approval?"

"A day," Danny answered.

And again Price laughed. It was hard to get one up on this guy.

"One more question," Danny then said. "How would we handle the publicity on this?"

"Easy. There will be none. When it's all done we'll place a couple of tombstones in the *Journal* and the *Times,* and after those bare-bones announcement ads, no further comment."

"Why? Are you still reticent to have your bank's name identified with our business?"

"Not at all. We always do it that way."

This time it was Danny who grinned. Both men then rose from behind the coffee table. "I'll need a little time to think about all this," Danny said.

"Take all the time you need. Now I have a request."

"Name it," Danny replied.

"Do you think I could get in a little tennis this morning?"

"Sure. Would you like to play with one of our pros?"

The thirty-two-year-old pro was waiting on the court half an hour later. An hour after that he was surprised to find out that his opponent on the other side of the net had actually caused him to work up a mild sweat. Some of those old geezers could really surprise you. And this guy was a fucking banker. No doubt that was why Danny Lehman had asked him the favor. On the other hand, it was comforting to know that even a rich bastard like Lehman had to kiss somebody's ass. The old guy must be *really* important, the pro concluded. So he let him win a game.

◆   ◆

Danny Lehman liked to walk when he had something important to think about. At nine o'clock in the morning The Palace was not exactly going full blast, so he could walk and think without having somebody stop him every dozen steps or so to say hello. By ten-fifteen he had made up his mind. So he took the elevator to the top of the tower and rang the bell outside the door of Price's suite. It was Natalie who opened it.

"I guess he's not back yet," Danny said, suddenly embarrassed.

"He's playing tennis," she said.

"I know."

"Come in for goodness' sake."

She was wearing white shorts and a white blouse, the top button of which was open. Danny looked at the desk and saw the stationery on top of it. "I'm interrupting," he said, ready to back right out of the room.

"Not at all. I was just catching up on a few letters. I need a break. I've got some coffee going in the kitchen. Would you join me? I'm sure Henry will be back soon. He's been gone for almost an hour already."

Damn good legs, too, he thought as he watched her disappear. And he also liked the short blond hair. She was quite a number. He sat down on the sofa, and when she returned with the coffee and leaned down to serve him, the second button on her blouse popped open. I'd better be careful, he thought. If I even *think* about fucking her, that guy Price is bound to notice and I'll blow this whole deal. He crossed his legs and he could have sworn that she was watching his crotch while he did so. Settle down, boy, he told himself, and then said to her, "You're an artist."

"No, no. I own an art gallery, that's all. In Washington. I first worked there, then bought it. Are you interested in art?"

"Definitely. Especially statues. What kind of art are you into?"

"Women painters, predominantly. It's the specialty of my gallery."

"You mean like Grandma Moses?"

"Kind of, yes. Although actually most of our painters are early-twentieth-century Europeans. Names like Paula Moder-sohn-Becker, or Suzanne Valadon." She saw that he was drawing a complete blank. "Now let me warm up that coffee for you," she said, and this time when she reached down to pick up his cup the *third* button gave way and Danny was faced with a luscious white cleavage.

Exercising self-control that he never thought possible, Danny said, "Here, let me." He stood up, reached across the table, and, one by one, as she stood there frozen to the spot, did up her buttons.

"There," he said, hoping that the trembling of his hands had not been noticed.

As if by telepathy, Henry Price chose that moment to walk into the suite, and Natalie turned to greet him, cool as a cucumber.

"Darling, I've been discussing art with Mr. Lehman."

While showing me your breasts, Danny thought, though he said, "Yes, and very interesting it was, too. Now tell me, was Joey nice to you?"

"He was. In fact, he let me win a game," Price replied.

"Do you mind if I get back to business for a minute?" Danny then asked.

"Not at all," replied Price.

"I've decided to accept your two hundred and fifty million dollars at 15 percent for five years," Danny said. "I don't like the conversion aspect, but I can live with it."

Price moved forward to shake his hand as Natalie retreated to her desk on the other side of the room. "Wonderful," he said. "I'll get right on the phone to New York. Then I think you should send that man Kelly up here. We're going to need a lot of data and documents from you people. I think he's probably the best qualified to help us there." Then he added, "If you agree."

Of course Danny Lehman agreed. New York was also enthusiastic, to put it mildly; 15 percent interest with conversion rights at a premium of 20 percent of current market, which would mean twenty-four dollars a share, and all this from a company in solid financial shape and one with a brilliant future: Henry Price had outdone even himself. In fact, it was a situation of such merit that the partners of Mercier Frères unanimously agreed that they should keep part of the deal for their own investment account through either London or Paris. And it was obvious that they would have to ration the outside allocations of this one very carefully. The Europeans had *never* been given an exclusive shot at a thing like this. It was agreed that the whole matter should be left in the hands of Price: he should decide which European banks to invite in; he would make the presentations personally. And the sooner the better, before

Lehman, or somebody on his board, had second thoughts; i.e., before they decided to check out Kuhn, Loeb or Lazard or Salomon Brothers for a better deal.

It was with some reluctance that Price agreed to his role in all this: running around Europe putting together yet another syndicate was not his idea of semiretirement. But he could hardly turn it over to anybody else at this point. Furthermore, it could hardly take very long. Unless . . . unless there was something somewhere in The Palace's financial woodwork that had not yet come out; something that could explain why Danny Lehman was so immediately ready to accept the terms he had proposed. The only thing that made him uneasy came up when he was going over the draft of the "History" section of the private placement prospectus that The Palace's financial man, Kelly, was helping him put together.

"Now, you say here, 'In the spring of 1969 Mr. Lehman assumed control of the company, then known as Raffles Inc., and three months later took on the position of chairman and chief executive officer.' "

"Yes," Kelly answered.

"Why the delay between 'assuming control' and becoming CEO?" Price asked.

"I believe that there was some kind of a conflict between Mr. Lehman and the senior management."

"Was that before your time here?"

"No, I was here, but not privy to that type of information."

"I see. Well, I don't think anybody in either New York or Europe is going to raise the issue. But there's one they might raise, and it's not covered here."

"What's that, sir?"

"Where did Mr. Lehman get the financing allowing him to 'assume control'?"

"He sold his old business in Philadelphia."

"I know that. But where did the rest come from?"

"I don't know. I'll have to ask Mr. Lehman."

Price made a quick decision. Despite his reservations,

there simply was no sense in introducing some new element at this point, one that might for some reason upset Lehman, and thus rock the boat. This was a deal that every house on Wall Street would leap at. He intended to close it as quickly as possible. "Hold on. I'm not sure it's pertinent. After all, the SEC is not going to have to pass on this. Again, it was a long time ago and really has nothing whatsoever to do with the current status and outlook of the company."

"I agree. But still, I could ask Mr. Lehman and—"

"No. Let's drop the matter. Now, back to the balance sheet. Accounts receivable. They seem awfully high. What's the reason?"

Boring stuff, but it had to be done. And three days later it was done. And at five o'clock on the evening of that third day, Price announced to Natalie that they would be leaving the next morning for the East Coast, and then, almost immediately, for Zurich. Then they would return directly to Las Vegas, provided everything could be worked out.

"I'd rather not come with you, Henry," she told him. "As you know, the Swiss lost their charm for me a long time ago. Do you mind if I just stay here for a few days until you return?"

"Of course not. But why here? I would have thought you'd prefer going back to your Georgetown house."

"Actually no. I'm rather enjoying myself. I think I might even rent a car, visit the Hoover Dam, take a spin in the desert."

When he left for the airport at six-thirty the next morning he decided not to wake her. In fact, he sensed that she again needed what she called "space"; better he let her be for a few days. He would call her from New York.

16

Natalie didn't get up until noon. She had awoken around eight and realized that Price was gone. It didn't bother her; quite the opposite. She then decided to rent a car for the afternoon. She picked out a red Cadillac convertible. Why not, she told herself. When in Vegas . . . She went to the dam, took the guided tour, and returned to the hotel around five o'clock, hot and dusty.

While she was taking her bath in a replica of Pompeii's finest, she thought to herself, What I need now is the calming effects of a very dry martini, ice-cold and straight up. But what on earth does a lady alone wear during the cocktail hour at The Palace?

She decided on a little black dress, which would probably turn out to be the only little black dress at The Palace that evening, but what the hell, she thought as she emerged from the elevator into the vastness of the casino, I'm probably also the only one in this place that has ever heard of Paula Moder-sohn-Becker.

Next problem: which bar? The one that was supposed to resemble a circus tent? The one that was built like a Viking ship? The one featuring country and western? She kept walking and was suddenly confronted with a bar that was built like a bar, looked like a bar, and had no music. Furthermore, she noticed with comfort, it was also a bar in which there were a number of other women sitting by themselves. Good for them! She took a step up, passing through a gap in the brass railing that separated the area from the main casino floor, and was greeted by the maître d', who scrutinized her and, failing to recognize her, said nothing and just proceeded to show her to a table well off to the right. Then he seemed to decide that he should say *something*, so: "I thought you might want to watch some of the big action. It usually starts to happen right about now."

A girl in a slave costume took her order, and when the drink arrived, Natalie decided to just settle back, take in the scene, and wait for the promised action to take place.

Just minutes later, it was clear that whatever it was, was going to happen in the area immediately to her left, the one that was roped off from the rest of the casino. Four terribly elegant Chinese women had suddenly appeared and taken their places at the long gaming table that was situated not more than five yards from where Natalie was sitting, though one step down and separated from the bar area by the brass railing. Nonetheless, it was so close that Natalie could not help but hear that the foursome were chatting in Chinese, not English. Then one by one they were joined at the table by what looked like an Italian count, then by what must have been a New Jersey truck driver, followed by a housewife, no doubt from Orange County, and finally by two fellows, both in yellow pullovers, who, she thought, were definitely from San Francisco. As striking as the group was, it and its components were pushed into the background when the person who was obviously the croupier walked up to the table and took charge.

She was dressed in a classically cut suit, had perfectly styled hair, stood at least six foot tall, and was perhaps in her early thirties, but a woman who moved in the knowledge that she was the most attractive woman in the place, even a place that by now contained a couple of thousand people, maybe a third of whom were female.

She was soon joined by two men, both in black tie, one on her left and one on her right, and finally, in the background, yet another man, also in a tight-fitting tuxedo, who took his place in a high chair as if ready to begin judging a tennis match. But though surrounded by these three white men in formal attire, it was the black woman in her dark suit and white blouse who was in charge. She was the one who gave the signal that the game was to begin, handing three decks of cards to each of her colleagues and retaining two more decks. They then all began to shuffle their cards. After re-collecting the now shuffled eight decks, she stuffed them into a shoelike container. Then came a couple of mysterious moves. First the black woman inserted what looked like a plastic joker into the shoe. Then, for whatever ritual reason, she extracted the top card from the shoe and turned it over. It was the eight of diamonds. She put it into a slot in the table in front of her, and then extracted eight additional cards from the shoe and put them into the same slot. Finally, she handed the shoe to one of the Chinese women, the one sitting to her extreme right, who immediately proceeded to take over by starting to deal around the table. The dealing done, the bets were placed, some cards were turned over, and, to Natalie's surprise, it was all over within two minutes. The Chinese woman ended up with all the chips, and her three friends chirped with happiness. Seemingly unaffected, she simply returned to the shoe and started dealing the next game. The bets were again placed. The cards were turned up. The winner was declared by the black woman. Another two minutes. Natalie didn't have a clue what was going on, but it was clear that vast sums of money were changing hands at a speed that was simply appalling.

Two martinis and forty-five minutes later they changed dealers at the number one baccarat table at The Palace, and Sandra Lee headed for her regular early evening drink. She had noticed the white woman in the little black dress watching her almost incessantly while she had been calling the cards at her baccarat table and decided, for no real reason, to ask if she could join her. A lot of customers liked that: a baccarat caller was big stuff in any casino, and to be seen drinking with one brought the prestige of being considered a high roller, since nobody but high rollers ever dared go near the baccarat tables.

"Do you mind, honey?" she asked, looking down at a now somewhat flustered Natalie Simmons.

"No, not at all. In fact, how very nice of you. I guess you must have noticed: I've been admiring you from afar."

"Yes. Aside from all the staring, you have no idea how you stick out in this particular crowd." Her eyes moved around the bar area.

Now Natalie became truly disconcerted. She knew she didn't perhaps fit in her Givenchy and Charles Jourdan satin pumps, but . . . The black woman read her mind perfectly. "No," she said and then giggled, "it's not what you're wearing. It's just that it's obvious that you intend to keep it on, that's all."

Natalie still couldn't figure it out.

"This's known as hookers' corner, ma'am, and, for reasons known only to himself, Steve put you at the table which has the largest turnover in the place; I'd say probably twenty tricks a night at two hundred dollars a pop originate right from where you're sitting."

Instead of blushing, or showing even the slightest annoyance, the white lady in the little black Givenchy seemed to love it. "Oh boy," she said, "wait until I tell this to Henry Price!" She thought of that first night they had been together in Paris. How she had, well, not exactly hated to have to do it, but still

. . . and no doubt it had showed. Maybe it still did. A hooker she was not cut out to be. But to hell with it, there are other things in life. Right?

"You're sure I'm not bothering you?" her visitor then asked.

"No, please stay. I was just reminded of something."

"You just mentioned a man's name. You're thinking of him, I'll bet."

"Actually, yes."

"Now the question, the one that I'm immediately asked by men, is how did a nice girl . . ."

". . . like you end up in a place like this?" said Natalie.

"That's it. You want to know the answer?"

"Only if you want to tell it."

"Why not? It's got two parts. First, I got kicked out of Mills College. Second, I started sleeping with the boss."

Natalie never even raised an eyebrow. "That's actually rather commonplace. Except the bosses are usually stockbrokers or regional sales managers or congressmen. At least you must have picked a fun boss."

"Do you know Mr. Lehman?"

Natalie had not expected *that*. "Oh my," she said. "I really didn't mean to pry."

Sandra Lee leaned across and patted Natalie's hand. "It's not exactly a secret. Anyway, we don't really see that much of each other anymore. Haven't for years. Danny's been too busy, I guess. And during recent years he's been spending most of his time in Atlantic City. So I guess it's the old 'out of sight, out of mind' thing where I'm concerned." Sandra Lee paused and scanned the bar area. "Oh, oh, more company."

Natalie was about to turn to see what she meant, but what she meant was already at the table, totally overlooking Natalie and glaring down at the black woman. "I've told you at least a dozen times, Sandra Lee: I don't want you hanging around this bar on your breaks. It gives a bad impression. You're a dealer, not a goddamm hooker anymore."

"Fuck you, Cordoba," was her answer. Then, to Natalie: "Pardon the French, dear."

Eddie Cordoba looked as if he were going to have a fit right there and then. Then Natalie intervened. "How very nice to see you again, Mr. Cordoba. Do join us, if you have a minute. I'm afraid that Henry has left for a few days. I'm sure he would've wanted to thank you for all the arrangements you made for us the other night over at Palm Springs."

Cordoba was instantly transformed. He reached down to take Natalie's hand and now, ever the charming Argentinian, implanted a Continental kiss between her second and third knuckle. "How very, very nice to see you again, madame. This calls for a small celebration." A wave of his hand brought Steve, the maître d', who in turn brought an ice bucket containing a bottle of Dom Pérignon. He was followed by one of the slave girls bearing three glasses. When they were filled, Cordoba raised his, looked first, rather pointedly, down at Sandra Lee, and then at Natalie Simmons, and said, "Peace." And sat down.

Then, addressing Natalie, he said, "The word is that Mr. Price and Mr. Lehman are getting along very well."

"Yes, they are," Natalie replied. "But don't ask me anything about business. When they start talking in terms of hundreds of millions of dollars I'm way out of my depth." Turning to Sandra Lee, she said, "My major at Vassar was art history."

Cordoba interrupted. "Excuse me, I'm afraid I'm out of my depth now. And I've got some work to do. One of our most important clients from Mexico City is coming in shortly. Mr. Lehman insists that I personally take care of him. It's been a great pleasure to see you again." He rose from the table, bowed to Natalie, ignored Sandra Lee, and left.

Sandra Lee's eyes tracked him until he had disappeared into the crowd now milling around the periphery of the casino floor. Only then did she again address Natalie Simmons. "What's your fiancé do?"

"Henry's an investment banker."

"Yeah. It figures." She paused, and then continued, "I know it's absolutely none of my business but I'm still going to say it: be careful of that man, especially if your fiancé is going to bail Danny Lehman out."

"Bail him out? Of what?"

"Honey, everybody in this place knows that Danny Lehman made a big mistake going into Atlantic City the way he did. It's eaten up all his time during the past year, and word has it that it has also eaten up most of the profit from this place. That the money guys, the banks or whoever, have him by the balls and are getting panicky now that everybody and his brother is about to open up a casino in Atlantic City. What they're saying is that the six-month joy ride that Danny had as a result of his being the first one to open up there full-time is about to end with a big bang. So the greedy bastards have told Danny that either he gets them their money back or they move in and Danny moves out. Cordoba, who's been running this place almost single-handedly since Danny's been tied up in New Jersey, can hardly wait. He's got the furniture movers on standby. He probably measures Danny's office once a week to make sure his new rug will fit."

"I see," said Natalie.

"I hope so. When you told him your friend and Danny were talking about hundreds of millions of dollars the guy actually turned white. I'm not kidding. You yanked that new rug of his right out from under Danny's desk where he thought he'd be sitting in a few weeks."

"So what can he possibly do about it?"

"It's not just Cordoba. He's gotten to be very palsy with one of the guys, one of the key guys, on the board of directors of this place, one that used to be a Hollywood agent. You can imagine how loyal he is to Danny! He and Cordoba are cut from the same cloth, believe me. I see them huddled together right here, two or three times a month. Muttering among themselves and laughing. Not when Danny's in town, though."

"Have you told Mr. Lehman about all this?"

"Danny wouldn't listen if I did. First, he's been so successful it could be he now thinks he's invincible. Second, Mort Granville and Eddie Cordoba have conned him into believing they worship the very ground he walks on. But third, and no doubt least important, like I told you, we don't see each other in that way anymore." She paused and then said, "Let's face it, Danny Lehman's become a very big man. So he now associates with people like *you.*" For the first time there was bitterness in her voice.

"I don't understand," Natalie said.

"You will in a minute. What Cordoba said was absolutely right. Sandra Lee is nothing more than an ex-hooker: a *black* ex-hooker. Danny Lehman fixed me up as a dealer a long time ago. And since then we've gone our separate ways."

She looked at her watch. "Gotta get back to work. It's been nice chatting with you." She got up to leave.

"Please wait just a minute," Natalie said. "I've got an idea. Are you free for lunch tomorrow? We could eat up in my suite."

The black woman hesitated. "What time?"

"Is one o'clock all right? I'm in 2001. In the tower."

"I'll be there," Sandra Lee answered, all hesitancy now gone. "I can't believe it: the two of us girls at lunch. I think I might wear a hat." She giggled, leaned down and kissed Natalie on the cheek, and left.

◆　　　◆

At one o'clock the next day—but one o'clock in the morning—just five hours after Cordoba had strode away from hookers' corner, he also showed up for a rendezvous; one that was, however, strictly for the boys. And it took place out in the desert, a good hour's drive northeast from Las Vegas, in the Valley of Fire where the red stalagmites rise on either side from the desert floor. Wayne Newton has a ranch there where he breeds Arabian horses. Cordoba at his own place bred and trained polo ponies; when the Argentinian polo team finished their American tours they inevitably ended up at his home; so did many of their ponies.

Cordoba had left the gate open and all the lights on, inside and out. He filled a bucket with ice, and put the whiskey bottles and two glasses on top of the bar in the living room, where red stone, redwood, and Persian rugs, and paintings of horses, cattle, and vaqueros had been put together in a spectacular display of good taste that was singularly lacking in the city that Cordoba had left earlier that evening.

Mort Granville arrived ten minutes later, and as he walked through the door and into the living room he was all smiles and good cheer. "What's the good news, Eddie?"

"Let me pour you a drink first."

"Oh oh," Granville said. "I hope it's not that kind of news."

Cordoba handed him his drink and took a good shot out of his own glass. "It is. That fucking Lehman has pulled it off again. He's working out a deal with Mercier Frères, *Mercier-fucking-Frères,* worth at least a couple of hundred million!"

"You sure?"

"I got it straight from Mercier Frères' girlfriend no more than five hours ago."

"How soon?"

"It doesn't matter how soon. When the word gets out that an outfit like that is going partners with Lehman, forget everything. The banks will never play ball with us."

"What do I tell my principals?"

"Don't gimme that goddamn 'principals' crap one more time Granville. Just tell your pals that they can take their money somewhere else. The Palace deal is dead. *Morte.* Finished."

For the next couple of minutes the two men stood in silence at the bar. Then Granville broke the silence. "You know something. You've got to hand it to Lehman. That guy has more luck than anybody on the face of this earth. First he gets that casino handed to him as a gift—and handed to him by *me* of all people. And now, when it looks like he's going to blow it, in comes this fairy godfather from New York with a freight car full of money."

"Let's not totally revise history, my friend," said the Argentinian.

"What's that crack supposed to mean?"

"You handed him *nothing*. I've heard the story. You sold him a place that was run by a bunch of hoods who wouldn't even let him in the front door, a door, by the way, that was about to be closed permanently by the gaming commission, leaving Lehman out about sixty million dollars. I hope you never do *me* any favors like that."

The ex–Hollywood agent decided to let that pass. Cordoba spoke again. "Who blew the whistle anyway?"

"What whistle?" Granville's voice was now low and sullen.

"The whistle that put the FBI onto that counting-room scam."

"How should I know?"

"Just asking."

"You never 'just ask' anything, Eddie." Granville's interest had returned.

"It's just an idea that suddenly occurred to me."

"Ask Bill Smith. He was the chief honcho at the local FBI office at the time. If anybody knows, he should."

"Naw. Can't ask him. He still thinks Lehman can walk on water."

Granville was thinking. "I'll tell you who has no doubt been asking the same question every day during the past ten years or more."

"Who?"

"Lenny De Niro."

"Who's he?"

"One of those hoods you referred to. He and a guy called Roberto Salgo ran the old Raffles. They and the chief auditor, I think his name was Downey, yeah, Rupert Downey, had organized the scam."

Cordoba did not particularly like the drift this conversa-

241

tion was suddenly taking. Salgo was a name he had totally blotted from his memory. When you have a guy killed, it's not smart to even allow the subconscious to retain the fact. Be careful how you handle this, he thought. If Mort Granville gets even the faintest whiff of something, he's just stupid and greedy enough to start turning over the soil looking for pay dirt. The problem was that the process would lead him to Beirut and Lehman, which would be the end of Lehman, but unfortunately it would also lead to Eddie Cordoba and his Corsican friends. That approach was very definitely out!

"Is this De Niro fellow still around?" he asked.

"No. Left town eleven years ago and has never been sighted since. Somebody made his pal Salgo disappear permanently, and I guess he got the message. But if he's still alive, I'm sure there would be ways to find him. What do you think?"

"Naw. Look, Mort, I think it's getting too late to think. Let your people know that The Palace is out. Maybe they should be looking at the Aladdin. I hear things are pretty unsettled over there."

Ten minutes later Mort Granville's car disappeared into the darkness of the Valley of Fire. Eddie Cordoba turned off all the lights, except one by the fireplace. He got himself a new glass of whiskey, lit a cigar, and stretched out on an easy chair. He was still there an hour, two whiskeys, and another Partagas later. He was thinking.

◆   ◆

Sandra Lee was true to her word: she turned up in a white Swiss lace dress that was spectacular enough in itself, but it was the hat, flowers and all, that made her appear like a bridesmaid headed for a wedding or, in another time and place, a young lady on her way to tea at Claridge's, the Ritz, or the Hassler with some of her ex-classmates from Katherine Branson, Smith, or Mills who were also touring Europe. To Sandra Lee's delight, Natalie Simmons had taken the occasion equally seriously and, although *sans chapeau,* was dressed in a Chanel suit and was the picture of Fifth Avenue elegance abroad for the season. Natalie had arranged a menu that began with champagne and ended with more champagne, interrupted by onion soup, then oysters on the half shell, and then a selection of pâtés, accompanied by a baguette, cornichons, and mustards.

Sandra Lee completely dominated the conversation. She knew at least one juicy detail about the sex life of almost every major entertainer who had done his or her stuff at The Palace during the past decade. And she had no qualms about discussing even the most graphic details of what went on on a regular basis "in this very suite." It often involved "stuff that even I would not have dreamed of doing," Sandra Lee said, adding, "before I retired from that particular line of activity."

Natalie sat there and drank it all in—in absolute awe. At three-thirty the phone rang, and it immediately produced a look of disappointment on Sandra Lee's face.

"Don't worry, it's probably only Henry checking in," Natalie said.

It wasn't. It was Danny Lehman announcing that he was going to drop in for a few minutes, if she didn't mind.

"I'll just leave quick," said Sandra Lee when she heard the news, immediately reverting to Southern patois.

"Please don't. I'd prefer not to be alone."

Sandra Lee seemed to have trouble figuring that one out, but decided to stay put, at least until the next act got under way. When Danny Lehman walked into the suite behind Natalie he was chattering like a little kid, until he spotted Sandra Lee.

"Surprise, surprise!" she said. "Didn't expect to see the help up here, did you?"

Danny's face broke into a grin. "It figures," he said. "But how the hell did you two meet?"

Natalie intervened. "We were both working hookers' corner."

"Yeah," Sandra Lee continued, "and now we're both staying with this high roller that lives up here. Henry's his name."

For just a second Danny was caught off guard, then:

"Come on. Henry's in New York." He paused, thinking. "However . . ." He stopped. Then he continued, "I've got an idea. Now before either of you say no, please hear me out. Okay?"

They nodded.

"I came up here to say goodbye because I'm leaving for Atlantic City again in a couple of hours. We're going to have a board meeting there later this week. You probably know why, Natalie. Anyway, I just thought of something. Sandra Lee, you've never even seen our new place there, have you?"

"No."

"And, Natalie, I'm sure you haven't been there either."

"No, I haven't."

"Well, how about both of you paying us a visit. You could fly out tomorrow morning. It'll all be fixed from this end tonight. I'll have our people pick you up at the airport in Philadelphia." Then to Natalie before either could reply: "I'm sure that not only would Henry not mind, but I think he'd be interested in hearing about what you see there. And I'd love to show it to you. And to you, too, Sandra Lee," he added. "Be a bit like old days. That is, if you still . . ." And now Danny Lehman seemed to revert physically to the boy from Philadelphia, way out of his depth in a place like this in the company of women like these two.

"You're on," Sandra Lee said. "We'd love to go, wouldn't we, Natalie?"

"Absolutely," she answered, without the slightest hesitation. Her two visitors soon departed, so it was she alone who finished off the third bottle of Dom Pérignon, ending up rather tipsy. But she went to bed early, rose at dawn, packed, and even sang while she packed. "This is getting to be fun," she said to nobody in particular as she stepped out of The Palace's limo at the airport to check in for United's nine o'clock flight to Philadelphia. Sandra Lee, it seemed, had been even less affected by the indulgences of the previous day: she was already standing at the United counter waiting.

17

The Palace East limo driver was waiting at the baggage carrousel in Philadelphia. The drive south to Atlantic City, through what seemed to be endless growths of spindly pines, took almost two hours. When they got to the outskirts of Atlantic City and then the city itself, neither woman said a word for a while. Then: "Pretty crappy-looking place, if you ask me." It was Sandra Lee.

"Very crappy-looking place."

"Very crappy place in many ways," Sandra Lee continued.

"I thought you'd never been here?" Natalie asked.

"I haven't. But that doesn't mean I haven't heard a few things. After all, I'm in the business."

"So tell me a few things?"

"All right. Part 1: Introduction to New Jersey in general. The history of New Jersey is absolutely *studded* with congressmen and other elected, and appointed, officials who have been indicted, been convicted, and served time for every conceivable type of behavior, the least of which is accepting bribes."

"Really."

"Where have you been? Look, New Jersey is the black-belt constituency in the world for corruption."

"Are you sure?"

"Well, in the *developed* world. Christ, Natalie, you can hardly expect me to include the Congo or Nigeria. Those are run by black folk."

Natalie giggled. "How do you know all this?"

"Eddie Cordoba, I'm sorry to say. An endless fount of knowledge where what he terms 'Danny's folly' is concerned."

"Why folly?"

"He says that in the end they're going to get Danny. 'Crooks cannot abide other crooks invading their territory.'"

"He says that about Danny?"

"Of course. I've been sitting in hookers' corner when he said almost exactly the same thing to Danny's face. Cordoba knows he can get away with murder with Danny. So naturally he pushes it to the limit."

"And what did Danny say?"

"'We're the best of the best. They *need* us. We're the bellweather. We come and then everybody comes.' Can't you hear him?"

"And?"

"Danny's been right. So far."

"What could go wrong now?"

"I don't know, honey, but I've got a feeling . . ."

"But then why does he do it?"

"Greed. Simple greed. He's got everything already, hasn't he? But not in his mind. He always wants more."

"More what?"

"Money. Just plain money."

At that moment the limo pulled up in front of The Palace East. And within minutes the two women had been whisked out of the car and up in the elevator and installed in a large, though hardly elaborate by Vegas standards, two-bedroom suite. The place was full of flowers, there were two buckets of champagne; two boxes of Godiva chocolates; two bathrobes, with their names embroidered on them, of course; two bottles of Joy perfume; two . . .

"This is ridiculous," Natalie said. "Furthermore, it's a terrible waste!"

"Of course it is," Sandra Lee replied. "But, honey, this is vintage Danny Lehman. For him this"—and her hand went from the champagne buckets to the chocolate boxes to the godawful chintz sofas in the room—"this is what life is all about." She paused. "But he means well."

"You still like him a lot, don't you?"

The black woman nodded and disappeared into her bedroom.

◆   ◆

Their invitation to dinner was extended in the usual intimate Lehman manner: through the telephone message service. A section of the gourmet restaurant of The Palace East had been cordoned off, and screened. Natalie and Sandra Lee were apparently the only women who had been invited. Danny Lehman introduced them to the half a dozen men already gathered there as the fiancée of Henry Price of Mercier Frères and her traveling companion. The man from the governor's office and his sidekick, the man from the New Jersey Casino Control Commission, with his flunky, and two members of The Palace Inc. board, Mort Granville and Danny's old attorney from his Philadelphia days, Benjamin Shea, made up the rest of the guest list. Natalie was placed between Danny and the governor's man; Sandra Lee was apparently considered safe between Granville and Shea.

The talk over dinner was that of old cronies, men who knew one another well, allowing them to joke about balding heads, to gossip about mutual acquaintances not present, and to reminisce about the good old days in New Jersey and elsewhere. The only time it got serious was when Shea mentioned the subject of the casino's temporary license: not just that of The Palace East, but also those of Bally, Caesars, Resorts, and Playboy. The whole situation involving a "temporary" versus a "permanent" license had arisen because while on the one hand New Jersey wanted to get the economic benefits of the introduction of legalized gambling to the state as soon as possible, on the other hand it did not immediately have the legal machinery in place to process the licensing. So, in the interim, only temporary licenses could be granted, with no guarantee that a permanent one would automatically follow. Caesars got its temporary license on June 26, 1979; The Palace, Bally, and Playboy, six months later, on December 29 of that year.

"Together we've put almost a billion dollars into this town," Shea said to the commissioner, although the whole table was listening, "and you guys still have us operating on a temporary basis. This has got to change, and quickly. When we went into Vegas, the state of Nevada had us cleared, permanently, within sixty days with no fuss or muss. I know, since I handled it personally for Danny. In Nevada they understand that nobody can afford to put up that kind of money without security, permanent security."

"You've got security," the governor's man interjected.

"I don't see it," the attorney answered.

"You see it right at this table," the politician answered. The commissioner nodded his agreement.

"How about dessert?" Danny asked. After dessert he announced that he wanted to show the ladies around; he thought the rest of the guys might want to stay for cognac and cigars. The men all rose, Danny and the ladies left, and the boys got back to business.

Once they were safely out of the restaurant, Sandra Lee grabbed Natalie's arm. "See what I mean?" she said in a loud whisper. "How blatant can you get? And they don't even care who knows!"

Danny was plunging ahead of them, through a crowd that was much denser than either woman had ever seen at the sister casino in Las Vegas. But it was a different crowd: seedier by far. The women were fatter; the men were sloppier. The whole group had an unwashed quality to it, Natalie thought. Sandra Lee must have had precisely the same thought, since she glanced over at her new friend and pinched her nose between her thumb and index finger. Danny kept plunging through the crowd. They were now passing by bank after bank of slots. Again there was something new: people were actually *lined up* to get at them!

"Heh, not so fast!" Sandra Lee yelled at Danny.

He waited for them.

"Where are the baccarat tables?" she asked.

"Table," he answered.

"One table?" Sandra Lee asked in amazement.

Danny shrugged. "Different town. Come on." They finally emerged out of the back of the casino onto the Boardwalk. Danny stopped. The women joined him.

"Now look," he said. There must have been two hundred people lined up *outside* the casino, waiting to get in.

"Why?" Sandra Lee asked.

"Fire regulations," Danny replied. "We usually hit the limit around seven o'clock already. After that, nobody gets in until somebody goes out." He turned to Natalie. "That should interest Henry.

"Now let me show you one of the greatest places on earth," he said. His right arm encircled Sandra Lee's waist. He offered his left arm to Natalie Simmons. She took it, and down the Boardwalk the threesome strode. It was a marvelous evening: warm, almost hot; the moon was shining brightly; the Boardwalk was jammed with strollers. "I used to come here as a kid," Danny said, "every summer. With my mother. Atlantic City is an island, you know. The Indians used to call it *Absegami,* meaning 'little sea water.' The white people shortened it to Absecon Island. Then some guys from Philadelphia discovered this place and built a railroad between here and Philly. First they bought most of the island for ten dollars an acre. Changed the name to Atlantic City to hype it as a vacation place. Three years later they were selling the land at three hundred dollars an acre." They kept walking.

"I bought the property where the casino stands for two hundred thousand dollars seven years ago. Know what it's worth now? Just the land?" He answered his own question. "Ten million at least. Us boys from Philadelphia have done all right here in Atlantic City, and Atlantic City has been good to us.

"You know what Teddy Roosevelt once said about this place?" This question was, naturally, addressed to Natalie Simmons.

"No idea," she answered.

" 'A man would not be a good American citizen if he did not know of Atlantic City,' " stated Danny. "How do you like that? It was true then and now it's true again. And you know why?"

"Why, Danny?" answered Sandra Lee rather sarcastically. But Danny didn't notice; he was on a roll. "This may sound immodest," he began, causing Sandra Lee to roll her eyes, "but I brought the casino business here and it's going to change this town as much as that railroad did."

They kept walking and nobody spoke for a while. Then: "There's the pier," Danny said, pointing at the run-down pile of steel and concrete jutting out into the water. "John Philip Sousa played the pier every single summer for years until he died in 1932. Then Paul Whiteman took over. Everybody used to come here. One of things I remember best was going with my Mom to see Amos 'n' Andy when they did a show here the year of the World's Fair, 1939. That day they pulled in eighty-four thousand customers! Boy, it was really something." He kept on walking and talking. "Mom and I used to come down for Easter. Just for the weekend. Because there was nothing, anywhere on earth, like the Easter Parade on the Boardwalk in

250

Atlantic City. You know how many people would come? *Half a million!*"

"Then what happened?" Natalie Simmons asked.

"The do-gooders took over and ruined the place. Stopped the gambling, closed the brothels, banned the horse parlors. Since there was no fun here anymore, all of a sudden everybody stopped coming. After the war you could have bought the whole *town* for ten million dollars." He looked at his watch. "The show's starting in ten minutes. Want to go?"

When Liberace came onto the stage, it was apparent that fun had returned to Atlantic City, and with it so had the crowds. The place was jammed. The show was stunning. Nowhere—whether in Paris or London or Tokyo—was there entertainment that was even remotely comparable to that offered night after night by The Palace and The Palace East. Danny Lehman's philosophy in this regard had been clear from the beginning and he had never wavered from it since: get the best, and once you've got them, keep them, and don't ever worry about what it costs.

Of course, there was champagne again. Then more champagne when Liberace joined their table after the show, explaining that he had come over to say hello to "the boss," as he called him. At one o'clock the party started to break up. "I've got one more idea," Danny said as they left the nightclub. "A nightcap!"

The women, both a bit wobbly, thought that though it was not an original idea, it was a terrific one.

"My place or yours?" he asked.

"Yours," Sandra Lee answered. "The whole place is yours, isn't it?"

Danny's apartment embraced the entire top floor of the eastern wing of the hotel. When one walked directly out of the elevator into his living room there was no doubt that one had left the tacky atmosphere of the casino behind. Whoever the interior decorator had been, he or she had done a magnificent job. The air-conditioning had the place chilled down almost to shivering point. But no sooner had they entered than a butler appeared from nowhere to light the fire and take their orders for drinks.

"Absolutely the last one," Natalie said, moving on to scotch.

Danny installed himself on the immense sofa next to the fireplace and when the drinks arrived waved the women over. "Well, what shall we drink to?" he asked, as they settled on either side of him.

Natalie, with some difficulty now, answered, "To Atlantic City."

"The greatest place on earth," Danny Lehman echoed. Their glasses clinked in celebration. Then Sandra Lee added, "Just to show that we don't take you for granted, Mr. Atlantic City, I for one am going to give you a big good-night kiss." It was the chastest of kisses on the cheek.

"Well, I think I could manage that, too," Natalie said, and did so immediately.

Then the three of them settled back on the sofa in silence, Danny's arms now firmly around the waists of both women. Diana Ross's voice filled the room from the background hi-fi.

"Nice," Sandra Lee said.

"Yeah," said Danny, moving just a little, and cupped Sandra Lee's breast from behind with his left hand. The strap on Sandra Lee's gown seemed to have slipped. Danny drew Natalie closer and caressed her leg. After a few moments the entire front of Sandra Lee's gown was down, and the entire bottom of Natalie's skirt was up, and—

"No," she said. "I simply can't." And, just as abruptly, Natalie Simmons was on her feet.

"Honey, take it easy," Sandra Lee said, her gown now also back in place. "Everybody just got a little carried away."

"Time for us to go, lover boy," she said to Danny. "Thanks for the evening. And no hard feelings, okay?"

Natalie, who was in front of the elevator, waiting, turned to look back at him. "I'd like to second that, if I may."

Danny grinned. The elevator arrived. Five minutes later the two women were back in their suite, where both immediately went to their separate bedrooms. It was now well past two in the morning in Atlantic City and Natalie Simmons, if not smashed, was half-smashed.

Peculiarly, it was the sight of Sandra Lee's magnificent breasts that had set her off much more than had Danny's hand. "I wonder what would have happened if it had been Henry instead?" she giggled. But Henry Price doing it with two girls, one black and one white—that was about as remote a possibility as one could conjure up. So maybe she should have taken advantage of what might have been the one and only chance she would have to see if she . . . And then she pulled back. "No. Time to get away from this world. In fact, maybe it's what they mean when they say the 'nick of time.'"

She stopped talking to the room and went to sleep, her mind made up.

The next morning the two women went to the airport in Philadelphia together, neither of them referring to how the previous evening had ended. Sandra Lee got a plane back to Vegas; Natalie, one to Washington. They promised to keep in touch.

18

An hour after Natalie Simmons returned to her Georgetown town house, Henry Price called. "I was in Atlantic City," she blurted out the moment after he had said hello.

"And?"

"It's terribly tacky. But did you know they actually line up every evening outside the casino, desperate to get in?"

"No. Interesting. Anything else?"

"Henry, I don't know anything about finance or about the casino business, but I must say I overhead something at dinner there that did not sound especially good."

"What was that?"

"You apparently need a license from the state to operate a casino. Is that right?"

"It is."

"Well, the license that they have in New Jersey is just a temporary one. Did you know that?"

254

"Indeed we do know that. All the licenses granted thus far to casinos in Atlantic City are temporary."

"Well, isn't that dangerous? For you and your bank, I mean?"

"We've taken account of it."

"Oh well . . ."

"I'll be meeting with their board tomorrow. Then I think I'll be off to Europe right away, assuming everything works out, which I'm sure it will. Should be back in a week. Are you all right? You sound a little . . ."

"I'm fine, just fine, Henry. Just a little worried for you, that's all. You're sure you're doing the right thing with those people?"

"You mean Lehman?"

"Yes."

"I think so. Why? Is there something else?"

"No, nothing. It's just that they are all so . . ."

"So are a lot of people who run oil companies. And brokerage houses. And more than a few governments that we deal with south of the border. Lehman's certainly no better, but he's probably no worse. Anyway, we're big boys here. You worry about your art gallery, Natalie. I can handle the Danny Lehmans of this world. By the way, is everything all right with your business?"

"I haven't had time even to call in during the past few days, but I'll be going over right away tomorrow morning."

◆   ◆

Three days later Henry Price was in Zurich at the Baur au Lac. He was talking with Lothar Winterthur, one of the managing directors of the General Bank of Switzerland.

"I assume you or your people have gone through the financial data," Price asked.

"Looks fine, but tell me, why are we being so blessed?"

"SEC registration would take too long."

"How did you arrive at a share price of twenty-four dollars where the conversion rights are concerned?"

"It's not fixed yet. The actual conversion price will be set on the day of the closing and will be based upon the last trade in New York on that day. We'll take that price and add on a 20 percent premium."

"So you are assuming that that last trade will be twenty dollars. How can you know that now?"

Henry Price put up his right hand and pulled down some thin air. But then he added this: "I was hoping, Lothar, that you might help nudge it in that direction by a little judicious last-minute buying or selling in New York. Whatever's needed."

The Swiss banker smiled. He knew that such practices were illegal in America, though hardly in Switzerland. *"Ja,* we do that often. We call it *Kurspflege.* It should be no problem provided the whole deal comes together."

"Good. Our client and his board have already agreed to the deal. But they will expect a conversion price of at least twenty-four dollars, since that's the figure I've been using all along."

"I understand. Now tell me, Henry, who else do you expect to come in on this?"

"Why don't you just take a minute and read the memorandum. It covers everything you need to know, Lothar." He handed him a three-page document. Mercier Frères stood at the top of the list as the sole lead manager. The borrower was to be The Palace S.A. of Luxembourg. The guarantor was The Palace Inc. of Delaware. The amount was two hundred and fifty million dollars. The term was five years. The holders of the notes would have the right, at their option, to convert, starting one year from the day of closing, their note or any portion of its principal amount into fully paid-up, nonassessable shares of The Palace Inc. stock at the price to be set on the day of the closing.

The list of potential participants in the syndication, all of whom had already expressed firm interest, included two Dutch banks, one Belgian, three German, two French, and one Japanese bank's subsidiary in Luxembourg. That was where Price was headed next. But it was important to get the General Bank of Switzerland in first—firmly, and with a lot of money. Then everyone else would fall into place without hesitation.

"Is this man Lehman clean where the authorities are concerned?" the Swiss banker now wanted to know.

Henry Price shrugged. "A William Tell or Winkelried he's not. But he's an interesting fellow, even charismatic in his way. Knows the business better than anybody in the world; no doubt about *that.* And I think you will agree that this is what counts." Winterthur nodded. "But clean?" Price continued. "Nobody is clean in that world according to normal standards, Lothar. Certainly not if judged by Swiss moral standards."

Lothar Winterthur liked that. There were no moral standards on earth that could even remotely approach those of the Swiss, and this American knew it. That's why the General Bank did business with him: Henry Price was known as a good man of good stock, from a distinguished family in the best American tradition. He was not just another one of those Jew peddlers from New York who forever pestered them, trying to ingratiate themselves with a Swiss bank, hoping that some of the polish might rub off on them. Henry Price didn't need that.

"We checked him out. Came up with a few rather fishy-smelling episodes in his past. But they occurred a long time ago, and nobody in the present government seems to be even remotely concerned with them or him. He's got a real money machine there, Lothar."

Again the Swiss banker nodded. They liked money machines in Switzerland. "Tell me," he then said, "is it all true what they say about Las Vegas? That you can get almost any kind of amusement you want?"

"Probably so."

"I hear there are girls there from all over the world."

"I'm sure you've heard correctly."

"You know, Henry, I can tell you now: we'll take thirty-five million of those notes. You have our word on that, and, as you know, that's all you ever have needed from us." Price made a mental note to have somebody from his office in either London or Paris obtain confirmation in writing the next day; or if not really in writing, at least by telex. Winterthur took a few more glances at the memorandum and then tossed it back on top of the coffee table.

"I see you'll be going on the board, Henry."

"Indeed yes."

"I wonder if you might do me a slight favor in that capacity."

"Certainly. Just name it, Lothar."

"Well, I must go to Mexico City next month. To look into the Alfa Industries situation. Doesn't look good. In fact, the whole Mexican situation bothers us. They've borrowed much too much. Or perhaps, more accurately, you Americans have lent them much too much."

"Better Mexico than Poland, Lothar."

"*Ja.* Something to that. But don't exaggerate our position in Poland. It's the Germans who have their necks stuck out there, not us Swiss."

Henry Price decided to remain silent.

"Anyway, I will be going to Guadalajara, and since Las Vegas is more or less on the way, and since we are going to encourage some of our very best clients to invest in your deal, do you think that perhaps . . ."

"Everything will be arranged. They have a system there where they put people like you on a basis they term 'comp,' " Price said.

" 'Comp'?"

"Short for 'complimentary.' The Palace will take care of everything, Lothar."

"That was not the reason I asked."

"Of course not. But I am sure that Mr. Lehman would certainly have it no other way. Please let me know now the exact dates; Mr. Lehman will certainly want to meet you. One more thing: how many will there be in your party?"

"Two."

"I'm sure your wife will enjoy it. Nothing like it in Zurich, you know."

True, although Winterthur's wife was not to find out. Lothar was planning on taking along a girl from the typing pool.

Business done, Winterthur rose. "Where to now, Henry?" he asked.

"Luxembourg. I'm taking the Trans-Europ-Express, in about an hour."

"Eat at the Grand Hotel Cravat. Best veal in Europe."

◆   ◆

Price discovered that the veal was truly magnificent. Then he moved on to Brussels, where the stuffed quail at the Villa Lorraine was also superb. So, too, was the Indonesian Rijsttafel at the Bali in Amsterdam. By the time he arrived in London a week later to complete his rounds, he had placed the entire $250 million of five-year convertible notes, and had gained six pounds in weight. Usually he gained only five on such trips.

Getting older, I guess, he thought to himself during the Pan Am flight back to New York. He never took the Concorde; it made him feel claustrophobic. Furthermore, Pan Am was a client. He spent two hours in the office, bringing his partners up to date, and then flew commercial down to Washington, D.C. He had phoned Natalie Simmons in advance, and she was at the gate at National Airport, waiting for him.

"Nice trip?"

"The usual."

They drove to her town house in Georgetown for a drink. He suggested that they drive out to the farm; she said she would prefer to stay in town, so they did. Neither thought it necessary to even mention casinos, Atlantic City, Danny Lehman, or the quarter of a billion dollars that Price had put together in ten days flat. In their scheme of things, neither the amount, nor the client, nor the business of the client really held any more significance for them. It had all just amounted to a mildly distracting, and appropriately short, interlude that was now over.

PART FOUR

1982

19

The problem started with the Abscam investigation at the beginning of the 1980s. A man by the name of Alvin Malnik was videotaped by the FBI. *The Wall Street Journal*'s description was: "The tape discloses Mr. Malnik trying to convince a federal agent posing as an Arab sheik to secretly buy into the Aladdin Hotel and Casino."

When Eddie Cordoba saw the article, all this hardly came as a surprise; he had already known about the unsettled conditions at the Aladdin casino when trying to divert the attention of Mort Granville and his "principals" away from Danny Lehman and The Palace a year earlier.

There was more: "A portion of the tape seems to offer additional evidence of Mr. Malnik's role with Caesars World. The Malnik-FBI conversation has Mr. Malnik saying, 'What you do is, like what we did with Caesars. We always, we never made any money, because we bet it back into the joint. We always added 200 rooms here, 300 rooms there, so that we have 2,000 rooms there.' "

The *Journal* pointed out the significance of all this by describing Mr. Malnik as a "Miami Beach businessman" who was "an alleged associate of organized crime figure Meyer Lansky."

There was another tape that the *Journal* had not gotten hold of from the FBI. Again it featured two men: this time a certain Saul Meyers of Miami, Miami Beach, Fort Lauderdale—it was never quite clear exactly where he was domiciled—and a certain Daniel Lehman, formerly of Philadelphia, now of Las Vegas and Atlantic City. The tape showed Mr. Meyers picking up Mr. Lehman at the airport in Fort Lauderdale. The same car was then shown as it approached a guardhouse in front of a very imposing building, either a hotel or a condominium complex. In any case, you could see the ocean in the immediate background. Armed guards checked them out and waved them in. A digital clock was running across the tape. Six hours almost to the minute later, according to the digital information on the tape, the same car was driven out of the same gate and went back to the same airport, where Mr. Lehman was seen disappearing into the terminal.

On a second tape, taken exactly eighteen days later, the same scene was repeated; there was never any sound, just black-and-white pictures.

There was a third tape and one that had been secured in a totally illegal manner; to get it, the FBI had blatantly infringed the sovereign rights of the Commonwealth of the Bahamas. There was no sound and in fact there was not much video either; it had a running time of no more than fifteen seconds. It showed a yacht docked at Paradise Island. Mr. Meyer Lansky and Mr. Saul Meyers were seen standing facing each other as they leaned over the railing. Both held glasses, both wore sunglasses. They appeared to be in discussion.

The date of the first taped Lehman/Meyers meeting was

May 9, 1969. The date of the second Lehman/Meyers meeting was dated May 27, 1969. The tape of the Lansky/Meyers rendezvous in the Bahamas was dated June 4, 1969. The FBI had obviously had Meyers and Lansky under close surveillance for well over a decade.

The hearings in Atlantic City to determine whether or not The Palace East would receive a permanent gaming license from the state of New Jersey began on June 23, 1982. At these hearings the Casino Control Commission would also make a determination regarding the "fitness" of each and every member of the board of directors and senior management.

The first week involved skirmishing by the lawyers. But on July 1, it quickly became serious; dead serious. Danny Lehman was scheduled to appear and, according to commission practice, would continue to appear until he had completed his exhaustive testimony. Danny Lehman was nervous, very nervous, about the whole thing, in spite of having sailed through the initial hearings, over two years earlier, cursory hearings that had resulted in The Palace East's being granted its temporary operating license. The difference was that *then* the chairman of the Casino Control Commission and his sponsor in the governor's office had dined at The Palace East the night before the hearings, as well as the night after, and the night after that. Unfortunately, that same chairman had been caught up in the Abscam web and had dragged the governor's man down with him.

That was bad enough. Worse still was the fact that The Palace East was the first casino to come up for permanent licensing in New Jersey, and Danny Lehman was the first principal of such a casino who would be judged as regards his "fitness" to run such an establishment. They would set the precedent for the state's future rulings on such impeccable establishments as Bally's Park Place, Inc., a subsidiary of Bally Manufacturing Corporation, a maker of slot machines, or the Playboy Casino, a sister casino of a string of gambling establishments in England that had just been ordered either closed or sold by the British gaming authority.

Worst of all was the appearance of Randolph Stinson on the New Jersey scene as attorney general. He was in the process of preparing his run for the governorship as Mr. Clean, using the same ploy that had been successfully used in Illinois and Maryland by the successors to Otto Kerner and Marvin Mandel, governors who had both been convicted on criminal charges related to their granting of horse-racing licenses to the wrong people in the wrong way.

Randolph Stinson, unlike *The Wall Street Journal,* was in possession of both Meyers/Lehman tapes as well as the fifteen-second study of that same Mr. Meyers in the company of Mr. Lansky on that boat in the Bahamas. Some of his people said he should have introduced that evidence at the very beginning of the hearings. Others had serious misgivings concerning the illegality of the FBI tapes. Mr. Stinson, thirty-six years old, a Mormon, and a wearer of brown shoes, flatly refused even to listen to them.

And it was he who personally confronted Danny Lehman in the hearing room of the New Jersey Casino Control Commission. Even the room—small, dingy, the paint peeling from the walls—was depressing. Danny entered it flanked by four attorneys, the best that money could buy in New York, including a senior member of the law firm who personally represented such clients as the Kingdom of Sweden, the European Coal and Steel Community, even South Korea. Neither he, nor his three flunkies, seemed especially to like their latest money-maker, Danny Lehman.

◆　　　　　◆

At eight o'clock precisely, it all began.

The chairman: Would the clerk please call the roll.

He called it. All five commissioners were present.

The chairman: Would the clerk please swear in the witness, Mr. Daniel Lehman.

The witness was sworn in.

The chairman: Good morning, sir.

The witness: Good morning, sir.

The chairman: We want to make this as easy as possible for you, Mr. Lehman. We shall also seek to make it brief.

The witness: I certainly appreciate your courtesy.

The chairman: Mr. Randolph Stinson, who is our state's attorney general, will now please proceed with his questions.

Q. Good morning, Mr. Lehman.

A. Good morning, Mr. Stinson.

Q. Mr. Lehman, when did you acquire The Palace?

A. In 1969. It was then known as Raffles.

Q. When did you join the board of directors of that casino in New Jersey, sir?

A. In the summer of that year.

Q. At that time, did you become familiar with the regulations applicable to the operation of a casino in Las Vegas?

A. I would say I did, yes.

Q. Mr. Lehman, are you familiar with regulation 5.011 of the Nevada Casino Control Commission?

A. I am not familiar with it by number, no, sir.

Q. Are you familiar with a regulation that makes it grounds for a disciplinary action against a licensee should the licensee be found "catering to, assisting, employing, or associating with, either socially or in business affairs, persons of notorious or unsavory reputation, or who have extensive police records, or persons who have defied congressional investigative committees, or other officially constituted bodies acting on behalf of the United States, or any states, or persons who are associated with, or support subversive movements through the country, either directly or through a contract of any means with any firm, or any individual in any capacity where the state of Nevada or the gaming industry is liable to be damaged because of the unsuitability of the firm or individual, or because of the unethical methods of operations of the firm or the individual?" Are you familiar with that regulation?

A. Yes, in general I am.

Q. Would you then say that in 1969 you were familiar with that regulation?

A. I would say so.

Q. Do you know a Mr. Saul Meyers?

A. I've met him.

Q. In what capacity?

A. Socially.

Q. Do you know where Mr. Saul Meyers resides?

The attorney for The Palace: Mr. Stinson, I object, in fact I strenuously object to the line of questioning, since in no way has its relevance to the purpose of this hearing been established.

The chairman: Mr. Stinson—why don't you include the relevance in your questioning?

Q. When you met Mr. Saul Meyers socially, were you aware of the fact that he might, in many ways, come under the category of persons described by regulation 5.011 of the Nevada Casino Control Commission?

A. Emphatically, no.

Q. In addition to your social contacts, did you also have business dealings with Mr. Saul Meyers in 1969?

A. Absolutely not.

Q. Perhaps before or after 1969?

The attorney for The Palace: I object. The question is too vague. I don't think that we want this to turn into a fishing expedition.

The chairman: Rephrase that question, please, Mr. Stinson.

Q. Did you have business dealings with Mr. Meyers in 1968?

A. No.

Q. Did you have business dealings with Mr. Meyers in 1970?

A. No.

Q. Did you have business dealings with Mr. Meyers in 1971?

A. No.

Q. Did you—

The attorney for The Palace: Mr. Chairman, I—

The chairman: I think, Mr. Stinson, that you should perhaps proceed to your next—

Q. Mr. Chairman, with your permission I would now like to show some videotapes, and at the same time introduce them into evidence.

The attorney for The Palace: I object. What is their relevance?

The attorney general: Could I and the counsel for The Palace have a word with the chairman?

The chairman nodded his approval, which was duly recorded, and the two attorneys then approached the table where the five commissioners were seated and, after a very quick discussion, returned to their places. Daniel Lehman's five-hundred-dollars-an-hour lawyer did not look happy. The lights were dimmed. The video projection began: first, the two tapes of Lehman and Meyers in Fort Lauderdale.

Danny watched them with equanimity. He had, after all, told the truth about knowing Saul Meyers.

Then the third tape came on. It was Meyers on a boat with some old guy. So what? The lights went back on.

Q. Those first two tapes were of you and Mr. Meyers in Fort Lauderdale, were they not?

A. Yes.

Q. The dates and times indicated on the videotapes, were they correct?

A. Probably, yes. I don't recall exactly.

Q. Would you tell us about the place you visited after Mr. Meyers picked you up at what, I believe, was the Fort Lauderdale airport?

A. It was the Fort Lauderdale airport. The place we went to

is a club south of Fort Lauderdale, on the ocean, as you could see. Mr. Meyers is a member there.

Q. Both times it was the same airport and same club?

A. Yes.

Q. In each instance you spent approximately six hours with Mr. Meyers?

A. I don't recall. That was thirteen years ago.

Q. Are you familiar with any of Mr. Meyers's associates?

A. No.

Q. Do you mean that you never met with Mr. Meyers when he had some of his associates with him?

A. Not that I was aware of, no.

Q. Do you recall ever having met a Mr. Montague Davies?

A. Of course.

Q. Would you tell us who he is?

A. He was one of the directors of the First Charter Bank of the Bahamas in Nassau.

Q. Have you done business with that bank?

A. Yes, sir. A long time ago when I was in the coin business. They used to handle my overseas transactions.

Q. When did that association cease, if it has ceased?

A. It has definitely ceased. In the spring of 1969.

The attorney for The Palace: Mr. Chairman, I again must ask that the relevance of this line be either established or discontinued.

The chairman: I agree. Mr. Stinson?

Q. Did you recognize the man on that yacht with Mr. Saul Meyers?

A. No.

Q. Did you recognize the yacht?

A. No.

Q. The man, Mr. Lehman, was Meyer Lansky. The yacht was that of Mr. Montague Davies.

Danny was visibly stunned and confused. His attorney asked that the proceedings be adjourned for the remainder of the day. The request was granted.

THE PALACE

◆   ◆

Minutes later, one of Eddie Cordoba's henchmen who had been in the hearing room went to a phone booth and, though it was not even 6 A.M. in Nevada, telephoned him at his home outside of Las Vegas, giving the man who ran The Palace's casino operations a full report.

20

After he had hung up the phone, Eddie Cordoba stayed in bed for a full half hour, mulling over the possible implications of this sensational turn of events. At first he could see only positives. They had nailed Lehman. But the more he mulled, the more he started to worry. What if they continued rummaging around in the past and decided to take a closer look at some other events that had occurred in 1969? That could have even deadlier consequences than their establishing the Lansky connection . . . but not just for Danny Lehman. Cordoba quickly dressed and headed for The Palace. There he waited impatiently, smoking cigarette after cigarette, until it was finally eight o'clock, when the office staff began to work. Then he called down to the payroll department and made a simple request: he wanted the names of everybody on the Palace staff who was already working at the casino at the beginning of 1969, when the place was still called Raffles.

The information arrived half an hour later and the list was not long: seventeen names in all. Apparently Danny Lehman had *really* cleaned house when he took over from that gang of three that had been robbing the place blind: Salgo, De Niro, and . . . the auditor. Cordoba picked up the phone and called down to payroll again. "Say, can you try and find the name of that guy who was the chief auditor around here in 1969?"

"Don't have to look it up," answered the payroll chief, who was one of the men whose names were on the computer printout. "His name was Rupert Downey."

"What happened to him?"

"I don't have any idea, sir. Mr. Kelly might know, though. If I recall correctly, he was working in the audit department when Downey was still in charge."

"Thanks."

He checked the list again. Kelly, now the Palace's chief accountant, was on it. "Get Kelly and tell him to come up here," he told his secretary.

The accountant arrived in less than five minutes.

"Good to see you, Matthew," Cordoba said as he rose to greet the man, giving him a little reassuring pat on the back in the process. "Do you smoke cigars?" He did occasionally.

"They're Cuban," Cordoba said, reaching back to his desk for the open box, which he then extended to the accountant. "Let's both sit over there," he continued, walking to the sofa in the corner. "Tell me, how's the month looking?"

"We should be up maybe 45 percent from the same month last year."

"Great," Cordoba said, and then changed the subject. "Say, tell me something: some of us were, you know, reminiscing about the old days last night, and the names of Salgo, De Niro, and Rupert Downey came up. Nobody seemed to remember what happened to them. I mean, everybody assumes that Salgo must have been killed. But what happened to the other two guys? Especially De Niro?"

"Sir, I have no idea about Mr. De Niro. I don't recall ever having even spoken to him. He was not the type of person to take much notice of clerks in the audit office."

So maybe the way to De Niro was through the other guy. "But you must have known Downey."

"Of course. I worked in his department, but I had absolutely nothing to do with—"

Cordoba quickly interrupted, "No, no, Matthew. Nobody has suggested, even in the remotest sense, that you were involved with any of that stuff. They were just wondering what happened to him."

"Well, he turned state's evidence—provided the authorities with everything they needed to nail Salgo and De Niro, I guess. They let him off in the end. Still, he had to give the tax people everything he had."

"Really."

"Yes. They even took his new house. It was a *big* place. Off the old golf course. Must have been worth a million dollars."

"Probably paid for it in cash from all that skimming those guys were doing."

"No. He'd just gotten a big loan."

Who gave a happy fuck about Downey's problems or his house or his goddamn loan? The fox he was chasing was De Niro. With those crazies in New Jersey rummaging around in the past, they might very well try to find De Niro in the hope of getting even more on Lehman, and if they found him, for sure he'd start yapping about Salgo, and maybe by now he might have put together Beirut where Salgo had been heading for when he "disappeared" and Beirut where he, Cordoba, happened to be at that same time. Then, just when it was finally starting to happen to Lehman, just when this whole fucking place was about to fall into his lap, who knows what they might discover. If they could nail Lehman, the world's most careful guy, for hanging around with Lansky and his pals, for Chrissake, how safe was anybody? So if De Niro knew anything, or had found out anything in the last thirteen years, he, Cordoba, wanted to be the first to hear. Right?

But first he had to find the fucker, if he was still alive.

The accountant was still blabbering about his old boss. "In fact, all three of them had big loans. From the same bank in the Caribbean. Mr. Downey used to have me send the payments every quarter."

A bell rang in Cordoba's head. "Where in the Caribbean?"

"Some place in the Cayman Islands."

"What happened when the IRS grabbed Downey's house? Or did they grab everybody's house?"

"I really don't know, sir."

"I mean, they still owed the bank the money, right?"

"Yes."

"How big were those loans?"

"All in all, at least a couple of million."

"Banks don't walk away from that sort of money, do they, Kelly?" And Cordoba let out a laugh.

"I'll bet they know where De Niro is if anybody does, right, Kelly?"

"I really can't say, sir."

"Just for the hell of it, do you think you could check back and find out the name of that bank?"

"That might be possible. I arranged the transfers through the First National right here on the Strip. Banks, as you know, have to keep records forever. But to go this far back will take them a couple of days."

"Matthew, let's do this: first, all this is between you and me. Right? Second, we keep more money at the First National than anybody. Tell them to get the name of that bank real quick. Like in a couple of hours, not days. Okay?"

"Yes, sir."

Cordoba rose. "Say, if you like cigars why don't you take the box. I've got plenty more."

◆   ◆

Exactly one hour later, Kelly came up to Cordoba's office with a slip of paper. On it was a name and an address: The First Charter Bank, 38 Front Street, Grand Cayman. Cordoba took one look at it and, after thanking the man, told him he could leave.

"Unbelievable!" Cordoba exclaimed when the accountant had closed the door behind him. He picked up the phone and called The Palace East in Atlantic City. His man would have just returned from the hearings. "What was the name of that bank in the Bahamas again, the one that came up in the hearings?" Cordoba asked. "Was it by any chance the First Charter Bank?"

It was. He hung up. It *had* to be the same outfit. Now think, he told himself. Lehman had dealt with that bank before 1969. That bank dealt with Salgo and De Niro before 1969. Coincidence that simply could not be. Somehow Lehman, as absolutely crazy as it might sound, had to be linked to that skimming operation. If that could be determined, there was no way that Lehman could stay on. It would make the Lansky business look like nothing. All they had there was circumstantial evidence. This was the real thing. Lehman could go to jail!

This would provide the basis for a very serious negotiation between Daniel Lehman and Eduardo Cordoba. Why bring in those crazies in New Jersey? If Lehman simply stepped down voluntarily, that would be the end of the investigation, wouldn't it? And De Niro, wherever he was, could further rest in peace. Then another thought occurred to him: why risk dealing with Lehman? He might just fire him on the spot, and once out, it might be goddamn hard to get back in. That fucker Granville would want it all for himself. So maybe the guy to engineer the whole thing—the graceful retirement of Mr. Lehman and the natural ascendency of Mr. Cordoba—would be that outside money man from Mercier Frères. Hell, they had two hundred and fifty million dollars to protect and they would hardly want it to go down the drain with Danny Lehman. That guy Price must be worried sick by now.

◆   ◆

He was. When Henry Price had arrived at his office in New York that day at his usual hour of eleven, no less than three of Mercier Frères' partners were waiting for him. "You've heard about your Mr. Lehman?" was the initial question one of them put very quietly to him.

"No. What has happened?"

"He's been tied in with Meyer Lansky."

"Perhaps we should all go to my office to discuss the situation," Price suggested.

After all had taken their seats around the coffee table in his office, Price asked the next question. "When did this happen?"

"It came out in the licensing hearings over two hours ago. The wire services picked it up right away."

"And what's the nature of the tie-in?"

"Through a mutual acquaintance."

"Who?"

"I forget. Sid knows. He's been on the phone all morning." All eyes focused on one of the other partners.

"Yes. His name is Montague Davies. He was with the First Charter Bank of the Bahamas. They closed up shop about five years ago. That was when a lot of banks down there did the same, if you recall."

Henry Price recalled. For many years almost anybody could, and did, form a bank in Nassau. The local rules made it absolutely impossible to get any information about a bank there, especially its officers and directors. This handy arrangement was the brainchild of the Bay Street Boys, that small group of whites who had run the island for decades almost exclusively for the benefit of themselves, and for the very large number of crooks who hid under their wings and their laws. In 1967 a reformist black government came in, and in the 1970s it passed a very simple new bank law. It merely required annual publication of the names of the officers and directors of any bank on the islands, and the amount of resources that the bank had. Rather than do that, almost one hundred banks went out of business voluntarily, First Charter among them, even though it had been generally regarded as one of the more reputable institutions in the Bahamas.

"Any idea where this Davies chap is now?" Price asked.

"Nobody I talked to seems to know."

"I'll find out," Price stated.

"The exchange suspended trading in The Palace's shares at the opening this morning, and it was just resumed half an hour ago. The stock's down six dollars, Henry."

"Doesn't surprise me."

"There are quite a few unhappy people around here."

"Understandable."

"This is going to hurt the firm very badly in Europe, Henry," continued the most senior of the Mercier Frères partners, pouring it on.

Price said nothing.

"What are you going to do?"

Price took his time before answering and then said, "I'll take care of it, and then you can have my resignation. Or perhaps you would prefer it the other way around."

"We'll leave it up to you, Henry."

Immediately after his three partners had left his office, Price went to his desk, picked up the phone, and dialed the thirteen digits that would connect him with Lothar Winterthur's home on the Kilchberg above the Zürichsee. "It's Henry Price," he said. "Hope I'm not interrupting."

"You are, but go ahead," was the gracious response.

"We've got trouble with The Palace. They stopped trading it on the New York exchange at the opening this morning, and when it was reopened, the stock was down six dollars. Apparently that man Lehman who runs the place may be denied a license to continue operating in Atlantic City. It seems he might have indirect links to Meyer Lansky, through a man by the name of Davies, Montague Davies. Davies apparently used to be with the First Charter in Nassau before it closed down."

"*Ja.* We know him. He's in Holland now. With the Zuider Handelsbank in Amsterdam. What do you want to know from him?"

"The Zuider Handelsbank? I know them, too: they came in on the syndication of The Palace."

"I'm fully aware of that. No doubt one of the reasons they did was because we already had our name on it."

Price remained silent. Then the Swiss banker spoke again. "What do you want to know?"

"What kind of business this Davies fellow had with Lehman. Especially before and during 1969."

"How badly do you need it?"

"Very."

"Well then, I'll have to talk to Kordaat. He's their chairman."

"I'd appreciate it, Lothar."

"What are you going to do about all this?"

"It depends upon what we learn from Mr. Davies. That will tell us whether or not we can save Mr. Lehman's neck."

"And if we can't?"

"I'll take care of it."

"I certainly hope so. We came into this because of you, Henry. I think everybody over here will look to you to get us out," said Winterthur. "In one piece," he then added.

Twenty-one hours later, Winterthur called back and Price had his answers. Davies had "brokered" an arrangement whereby a group of individuals, Americans, had provided the funds, thirty-one million dollars to be precise, that had been lent to Lehman via the Zuider Handelsbank and one of its companies in the Netherlands Antilles—because of the tax treaty it has with the United States—to finance the takeover of that casino in Las Vegas in 1969.

"Who were these 'individuals'?"

"Well, let's put it this way, Price: we wouldn't accept a

dollar from any of them even if they paid *us* interest." Strong words from a Swiss banker.

"Anything else?"

"Not only would they not pay any American taxes on the money they hid in the Caymans, but they would be able to *deduct* the interest costs arising from the loan-back of their own illegal funds from whatever taxable income they did declare in the United States."

"What's that got to do with Lehman?"

"Lehman was the man who asked Davies to set it up."

"Who knows all this, Lothar?"

"So far the chairman of the Zuider Handelsbank, Montague Davies, you, and I. But who knows for how long? Of course, the whole Bahamanian–Cayman Islands' operation was shut down years ago, but the Bahamas is not Switzerland. If some of your American authorities find the right ex-employee of that bank and pay him enough, for sure they will nail Lehman right down the line, from beginning to end. Then they'd probably close the casino in New Jersey the next day, and no doubt they'd have to do the same in Nevada the day after. Let me tell you Henry, we're worried. Look what the British did to Playboy. Those casinos that they had in England were valued at a quarter of a billion dollars. When the authorities moved in and forced either a closure or a quick sale, do you know how much they got for them?"

"No," Price answered, and he truly did not.

"Thirty million dollars. They got back just over one *tenth.*"

Price made no comment.

"My advice to you is very simple," the Swiss banker continued. "Get rid of Lehman! Right now."

"The board's meeting the day after tomorrow in Las Vegas. I'll keep you informed."

As soon as he had hung up on the Swiss banker, Price's secretary came in to tell him that Natalie Simmons was waiting on another line. He immediately picked up the phone and punched the blinking light.

"Natalie. What a surprise!" he said.

"Hope I'm not disturbing you," she said, "but I was wondering if you're coming down this weekend."

He said nothing for a few seconds and then answered hesitantly. "I'm not sure."

"You sound funny," Natalie said. "Is something wrong?"

"To put it mildly," Price replied. "I feel like Hugh Lord le Despenser."

"Who, pray tell, was he?"

"He was accused of some dastardly crime and then hung, drawn, and quartered. All this happened way back in 1326."

"What reminded you of him?"

"My last phone call. Hugh Lord le Despenser, you see, having already had his balls cut off, then felt the executioner rummaging around among his kidneys, and thereupon spoke his famous last words: 'Jesus, yet more trouble.' "

"Danny Lehman?"

"You guessed. And I know . . . you warned me."

◆   ◆

The chairman called the board meeting to order at two o'clock. All members were present, as well as the general manager of The Palace's casino operations, Mr. Eduardo Cordoba.

The meeting was held in the room designated for such occasions, located in the executive wing of The Palace—no peeling paint there; the walls were all oak, and well-oiled and polished oak, too. The atmosphere in that boardroom in Nevada was, however, no less poisonous than that of the hearing room in New Jersey just a few days earlier. You could almost hear the hum of the knife sharpener. You could almost sense the board members flinch when Danny Lehman entered the room last, bringing with him an aura of contamination, bearing a plague that would infect all unless the disease was cut off at the source.

These board members, each and every one, had been appointed by Danny Lehman. And except for Mort Granville, who, after all, had brokered the deal that had enabled Danny to take over the casino, and Benjamin Shea, Danny's personal attorney and vice-chairman of The Palace, who had arranged for Danny to be licensed in Nevada, not one of them had ever contributed a thing to the success of The Palace. To a man, however, they had accepted perks on a scale that was mind-boggling, almost amounting to a full-time occupation in some cases, and all at Danny Lehman's company's expense: the yachts, the villa in Palm Springs, the jets, the girls, the fights, the races—all on top of board fees that ranked right up there with those of General Motors.

Then just one whiff of trouble and, boom, Danny Lehman was suddenly a stranger who had entered their casino, uninvited and bearing the stench of the unclean. "All right," Danny began, "you all know why we're here. Let me tell you here and now, once and for all, that the story about Lansky and me is an absolute, total fabrication. I have never done business with the man; I have never met the man; I have never spoken to the man. I don't even know what he looks like, for Chrissake."

Silence in the room.

"You've got a classic case here of guilt by association, but not even direct association: association twice, three times, ten times removed. It's a goddamn outrage! Not only that, but on top of it you have the FBI, who are supposed to be protecting our liberties, sneaking around trying to entrap anybody who is in a business they don't approve of."

Still silence.

"What do you intend to do, Danny?" The question came from Mort Granville.

"Do? What should I do? I've not been either charged or convicted of anything in my entire *life*. Not even a speeding ticket! Now, because of some young jerk who's running for governer of the most corrupt state in the union at my expense, who intends to win by ruining me, by spreading malicious out-and-out lies about me, I'm supposed to *do* something? Like what?"

"Maybe take a leave of absence, Danny," Granville replied. "Some of us have been discussing the matter prior to the meeting and concluded that we simply cannot escape our moral responsibilities. After all, the jobs of 3,600 employees in our casino in New Jersey are threatened. Ultimately the jobs of the 5,200 here might be. Then there are the outside shareholders. The price of our stock is now down to eighteen dollars, as you know. That's down eight from where it was before trading was suspended and then reopened. If the gaming commission closes down The Palace East—and we've heard that this decision might be imminent in spite of the fact that the hearings have been suspended for the moment—it will drop at least another eight dollars. You're the biggest shareholder, Danny: you'd be hurt worst." All heads in the room seemed to be nodding collectively at every word.

"What we thought is that you might take a leave of absence, Danny, and, as a matter of further goodwill, put your stock in the corporation in trust, maybe allowing some member or members of this board to exercise the voting rights in your absence to make sure that no outsider could come in and take advantage of this crisis situation. I think that I, as company secretary, would probably be in the best position to guard your interests in that respect."

Danny just sat there.

"I think, too, that it would be to your advantage if the impression was given to both the authorities in New Jersey and the financial people in New York that you'd decided to completely neutralize yourself as an operating officer. So I think it would be prudent if this board were to designate Eddie Cordoba as acting chief executive officer."

There it was: the blatant, greedy, open takeover-from-inside laid right out in front of Danny in the full knowledge that he had no choice but to accept. For thirteen years he had coddled this group of men, made some of them rich, asked nothing in return. Now, within just three days, they—and collectively they owned less than 1 percent of the fucking place—were throwing him out. *He,* who owned over ten million shares, was being forced off his own board and out of his own management. Next they would throw him out of his office.

That was, in fact, the next suggestion that Mort Granville was about to make. But Price, the outside money man from New York, the person who thus far had never said a single word, not one word, at a board meeting of The Palace even though he had been a member now for almost two years, suddenly intervened. "I beg to disagree with the suggestions that have been made by the company secretary. Matters are anything but clear. This is no time to make precipitous decisions. I would like to move that this meeting be adjourned."

Danny Lehman was almost as stunned by these words as he had been by the revelation that the old man in the picture with Meyers on that yacht had been Lansky. But not Mort Granville. "I'm afraid, that this board's primary responsibility is to the shareholders, not the debtholders. I'm afraid you and your group fall into the latter category."

"Wrong," said Price. "I can convert our debt into equity tomorrow morning. That would give us, with or without Mr. Lehman, absolute control of this company. I can guarantee you that if we are forced to take that route, all of you, with the exception of the chairman and vice-chairman, gentlemen, all of you will be off this board just as soon as I can convene an extraordinary shareholders' meeting."

All completely stunned, none of the board members now chose to challenge the investment banker.

He continued, "I would like it to be on the record that the chairman continues to have the confidence of this board. Now I again propose adjournment." Vice-Chairman Benjamin Shea seconded it, and the meeting ended.

"Mr. Price," said Eddie Cordoba, in a low voice, as they all filed out of the door and toward the elevators. "Could I have a word with you? In private?"

"Certainly."

Cordoba led him down the stairs rather than wait for the

elevator and then directly to hookers' corner, which, since it was only two-thirty in the afternoon, was almost deserted. Both the Argentinian and the New York banker ordered scotch. Then: "You made a big mistake up there, Mr. Price."

"Really? Why?"

"Danny Lehman has been involved with the mob from the very beginning."

"Oh?"

"I've got the proof. And it goes far, far beyond what came out in the hearings."

"If you think you know something, spell it out."

Cordoba proceeded to do just that. Lehman's pre-1969 association with the First Charter Bank in Nassau. The subsequent association of Salgo, a felon who never went to jail because no doubt somebody had murdered him first, with the same bank. Then there were De Niro and Downey—both guilty of skimming millions from this very casino—receiving loans from the First Charter Bank in the Cayman Islands. What the exact connection was, what scheme they were all involved in, that had yet to be determined, but there could be no doubt that Lehman was working in collusion with the mob from day one. Lansky had probably masterminded the whole thing, using a coin dealer from Philadelphia as his front.

"That's all you've got?" asked Price.

"All? All?" countered Cordoba, incredulously. "What more do you want? Canceled checks proving Lehman bought this place with mob money? Well, you'll get them. I'll make sure of that."

"Hardly. The First Charter Bank in Nassau, you see, as well as the First Charter Bank in the Cayman Islands—both were liquidated in 1977. Not a scrap of paper was left behind."

Cordoba looked as if he were on the edge of cardiac arrest.

Price looked at his watch. "I've changed my mind about that drink. I must be off. Goodbye, Mr. Cordoba."

21

One hour later, Henry Price took a Delta flight to Dallas. He stayed there for three days. On Friday he flew directly from Dallas to London and connected there with a flight to Amsterdam. Kordaat, chairman of the Zuider Handelsbank, came to Schiphol Airport on Saturday to meet him. From there they went to the Amstel Hotel in the center of the city. Lothar Winterthur of the General Bank of Switzerland was sitting in the lobby when they arrived, drinking an Amstel beer. Kordaat suggested they go over to his bank, which was just across the canal; Winterthur preferred to stay in the hotel, where he knew he would be able to get another beer or two should he feel the need. So after Price had checked in, the three of them rendezvoused in the Swiss banker's suite.

"The reason I imposed on both of you on a weekend," said Price, starting the proceedings, "is that time is of the essence. It always is, I know, but in this instance there is an acute urgency. We and our colleagues mutually have two hundred and fifty million dollars in that casino, and if we do not act swiftly and decisively we are at the very least going to lose all semblance of influence or control over that operation; at the very worst we are going to see the New Jersey casino closed down, and have the parent company default on its interest payments immediately.

"After what you people have gone through with Poland, and with AEG in Germany, and with Alfa Industries in Mexico, I don't think you need any more trouble."

"*Ja,*" said Kordaat, "these are sad times. Our Dutch banking commission is getting very nervous about these things."

"So is ours," admitted Winterthur. "I'm surprised we're in this mess, frankly." He then faced Price square on. "You told me this man Lehman was clean. I specifically asked you, Henry."

"I told you he was as clean as anybody in that business. I stand by that. The problem is that he is getting hanged for a crime he didn't commit. All that stuff about him and Lansky being partners simply does not wash."

"Don't be too sure," said the Dutch banker. "As you know, I have been having a few words with Montague Davies about this matter. It might not be Lansky, but suffice it to say that the less said about Lehman and some of his transactions, the better."

"Where is Davies, by the way?" asked Price.

"We sent him to Indonesia at the beginning of the week. He will stay there for a while, and not in Jakarta either."

Both of the other bankers grunted their approval.

"Well," continued Kordaat, addressing the American, "what have you come to offer?"

"First, let me tell you what we *don't* want to do. On the one hand, we don't under any circumstances want to let the Palace board dump Lehman. If they do, some of the insiders are going to take over, rob the place blind, and laugh at us when we try to get our money back. On the other hand, we don't want to exercise our rights to convert the debt into equity in order to take control because under the terms of the convertible note that's only possible at twenty-four dollars a share, and the price is already down to sixteen dollars and my guess is that, since it's been sinking steadily, the price is going to be down in the range of ten to twelve dollars within a week."

"*Klar,*" said Winterthur.

"Second, what we don't want to do is let the gaming commission in New Jersey continue with its investigation of Lehman. If it does, it is going to turn up some very nasty things—leaving Lansky aside—that could force it to close down the Atlantic City casino. With that cash flow gone, once again the parent company is going to be forced to default on the note, as I mentioned at the outset."

"Damned if we do something and damned if we do nothing," the Dutchman said.

"Precisely," Price commented, "except . . ."

"Except what, Henry?" asked Winterthur, now alert.

"Except if I arrange for a leveraged buyout of the whole company, which would include its casinos in Atlantic City and Las Vegas, as well as the other properties it owns."

"You mean Mercier Frères," the Swiss banker stated.

"No, I mean Henry Price, and, if you are interested, also Jan Kordaat and Lothar Winterthur. As silent minority partners."

Now the Swiss banker was *really* interested. "What does 'leveraged buyout' mean precisely?" he asked.

"Very simply, it is a method allowing a few individuals to buy a company with other people's money, Lothar. Normally, institutional money. It involves almost all debt and virtually no equity."

"I still don't get it."

"Okay, let's say somebody—somebody you don't like—tries to take over your bank, Lothar, by offering the bank's shareholders 25 percent more than the price they can get on the stock exchange. To stave off this hostile takeover, you and some of your colleagues in the bank's senior management get together, set up a shell company, capitalize it by putting in whatever cash you can scrape together, and then borrow whatever is needed to finance a more attractive offer to the shareholders. If it works, you and your colleagues end up both managing and owning the bank. In the United States, that's the new hot thing, since everybody makes out like bandits. The shareholders get a windfall gain. The managers end up owners. And the investment bank which put the deal together, and arranges the financing, makes millions of dollars in advisory fees. Sometimes tens of millions. That's why American investment bankers are the highest-paid people on earth. And that's also why every other kid who graduates from Harvard Business School wants to become an investment banker."

"All right," the Swiss banker then said, "I understand everything, except for one point. Why should anybody lend all that money to what is essentially a shell company with almost no equity?"

"The lenders don't look at the acquiring company. They look exclusively at the assets and the cash flow of the company being taken over. Its earnings stream goes right through the shell company directly to the lenders."

"Big cash flow, big loans," said the Dutchman.

"Precisely, plus big interest: normally 15 percent," said Price.

"How much more debt can The Palace operations support at that rate?" asked Winterthur.

"Lothar, as usual, you asked *the* question. I checked with the chief accountant at The Palace. He said that receipts at Atlantic City are a full 45 percent ahead of their projections, and gaining every month. In Nevada, revenues are up 17 percent. Tahoe was plagued by bad weather last year, but this year all the casinos on the lake are on the rebound and The Palace's Tahoe revenues are expected to set a record."

"So how much more debt can it reasonably support?"

"Another hundred and fifty million. If one stretches things a bit, two hundred million."

"How many shares does The Palace Inc. have outstanding?"

"Twenty million."

"Ten dollars a share, then," said Winterthur. "You said it is sixteen dollars now. That's out of the question. Nobody's going to accept a tender bid *below* market."

"I said it will probably head toward the ten-to-twelve dollar range within a week in any case. A little judicious short selling could nudge it down to eight dollars easily. Look what happened to Caesars stock when those New Jersey people started to go after them: from thirty-six dollars down to four dollars. Everybody on the Street knows that. So when they realize what's coming down on The Palace, a lot of people are going to want out, and quickly."

"Where would you seek the loans?"

"Here in Europe."

"That's not going to be easy. People here don't understand what you call 'leveraged buyouts,' " Kordaat said.

"It depends on who puts their name on it first," Price countered.

"We put our name on that convertible note," said Winterthur, "and look at the mess we are all in now."

"You're going to be in a bigger mess if they default on it," answered Price. That made an impression. "But there's more," continued Price. "I came directly from Dallas, Texas. There is a family there that Mercier Frères has been very close to for years. They would be willing to pay fourteen dollars a share for The Palace stock, provided they would not have to deal in any way, shape, or form with Lehman and his crowd, provided I can get The Palace East its permanent license, and provided they could get the whole thing."

"Wouldn't it bother them if we got it for ten dollars?" asked Winterthur.

"That's what investment bankers are for, Lothar. We put

ourselves, our good names, and our money, at risk on this takeover. Nobody can really argue about four dollars a share differential for our trouble. They recognize that in Texas just as they do in New York."

"How sure is their offer?"

"As sure as such things ever get," answered Price. "I'd say if we can produce one hundred percent of The Palace, with the permanent license, there is at least a 99 percent chance of their coming through. They shook hands with me on it. That still means more than a written contract in Texas."

"What's your Mr. Lehman going to think about all this?" Kordaat then asked.

"He'll get a hundred million dollars for his ten million–odd shares instead of maybe nothing if he tries to tough it out alone. I suspect he will need very little convincing."

For a minute they just sat there in silence. The Dutchman then looked at the Swiss banker. Winterthur nodded his assent.

"Let's discuss the terms and conditions of the borrowings, Henry," Kordaat then said. "Don't you think we could cut back a bit on the interest rate? I have the feeling that 14 percent, maybe even 13½, should do it."

The three bankers stayed in that suite at the Amstel Hotel until the early hours of Sunday morning. When they parted, they had a deal.

◆   ◆

Short selling of The Palace stock, most of it apparently originating in Europe, started in earnest at the first uptick on the New York Stock Exchange on Monday morning. The downward movement of the stock was accelerated on Monday afternoon when the Enforcement Division of the New Jersey Casino Control Commission announced that it was going to resume hearings in two weeks to determine whether or not Daniel Lehman and The Palace East would be licensed to continue operations. By Wednesday the stock was down to 10½. At closing on Friday it was 7⅞. On Saturday morning Henry Price called Danny Lehman and suggested that they meet in private in New York the following day. They met at the Carlyle at noon. By three o'clock Danny Lehman had agreed to sell his complete holdings in The Palace Inc., exactly 10,312,000 shares, for exactly $103,120,000 to Casino Properties Inc., a Delaware Corporation, the chairman of which was Henry Price. The sale was conditional, however, upon the success of the tender offer that Casino Properties Inc. intended to make for the rest of the outstanding shares of The Palace Inc. residing with the general public. It was lock, stock, and barrel or nothing. Such is the nature of a leveraged buyout.

On Monday morning very shortly after nine, Henry Price phoned the attorney general of the state of New Jersey and sought to arrange an appointment with him in Trenton at his earliest possible convenience.

The meeting was fixed in the late afternoon of that same day. Initially, the attorney general was extremely disappointed when he was told that the unsavory situation at The Palace East, particularly that relating to its chairman, was in the process of being taken care of by Mr. Price. His disappointment was assuaged when he was led to believe that substantial financial support would be forthcoming from Mr. Price and his friends should he seek higher political office, for instance the governorship, in 1985. Mr. Price requested two things: first, that the hearings of the Casino Control Commission related to The Palace East be postponed a further two weeks, but that this decision be made public only at the last moment. Second, that the attorney general immediately initiate his investigation of Mr. Price to determine his acceptability as far as the state of New Jersey was concerned. With the results of the attorney general's investigation to hand, it seemed likely, he thought, that the commission would be in a position to grant a permanent license to The Palace East and its new owner in rather short order. The attorney general agreed, both with his logic and to the requests.

The tender offer for the ten million–odd shares of The

Palace Inc. in the hands of the general investing public was handled by Mercier Frères. The organization of the syndicated loan which would provide the two hundred million dollars which would be needed to buy out both Daniel Lehman and the public was organized jointly by the General Bank of Switzerland and the Zuider Handelsbank of Amsterdam. The acquiring company, Casino Properties Inc. of Delaware, had been capitalized at ten million dollars, and the financial communities of New York and Europe were given to believe that the funds involved were the personal assets of Mr. Henry Price. The financial communities of New York and of Europe were also led to believe, and they believed it, that if this buyout of The Palace Inc. was not successful, The Palace East would be closed by the New Jersey authorities within a month, which would, inevitably, lead to a default on its outstanding convertible debentures in Europe, and a total collapse of its parent company's stock on the New York Exchange.

Thus the eager American sellers of The Palace Inc.'s shares (ten dollars a share was a hell of a lot better than next to nothing), and the equally eager European lenders of the funds necessary to provide the ten dollars each for such shares (lending another $200 million at 13.5 percent, secured by the complete underlying assets of The Palace Inc. in both Nevada and New Jersey, and the name of Henry Price, was a hell of a lot better than seeing their first loan in the form of those convertible debentures going into default), were quickly and easily lined up for the closing that was scheduled for August 1.

On July 25 Eddie Cordoba telephoned Henry Price in the offices of Mercier Frères in New York. He said he had to see Price immediately.

They met an hour later in the coffee shop next to the St. Moritz Hotel, where Cordoba was staying. Eddie Cordoba came right to the point. Through his Corsican friends in Paris, he had determined that the infamous Mr. Montague Davies, owner of that yacht upon the deck of which Mr. Lansky had been standing as he chatted with that man from Fort Lauderdale, the one who was also seen so regularly in the company of Danny Lehman, was still alive and well. He apparently now worked for the Zuider Handelsbank in Amsterdam. According to what he had just read the previous day in the New York *Times* it was this same Zuider Handelsbank that was the co-leader of the banking syndicate that was financing the takeover of The Palace by Mr. Price.

This could hardly be yet another coincidence, could it? All that was happening really was that the "friends" of Montague Davies were in essence buying themselves out, this time through Holland instead of the Caribbean. And they were essentially just changing their front man from Daniel Lehman to Henry Price, were they not? *"Plus ça change, plus c'est la même chose, n'est-ce pas, Monsieur Price?"* was Cordoba's flippant way of summing up the situation.

Cordoba's final question was this: when the Enforcement Division of the New Jersey Casino Control Commission found this out, as indeed it would because Eddie Cordoba would see to it *that* it did, did Mr. Price think it likely they would grant him the permanent license that they were certainly going to deny to Danny Lehman because of his links with the same Montague Davies and his pals?

Henry Price was faced with an impeccable line of reasoning based on irrefutable facts. So he asked, "What exactly do you want of me, Mr. Cordoba?"

Cordoba, of course, had his answer ready: "Not that much, considering; say five million dollars in an account at that Dutch bank, plus job security at The Palace, permanent job security . . . I believe they term it 'tenure' in university circles."

It *wasn't* that much, considering the personal and financial devastation Price would face if the Casino Control Commission decided not to license him. "How soon do you require an answer?"

Cordoba parried, "How soon are the hearings scheduled to resume?"

Price did not reply.

"The day before, whenever that may be, should do," the Argentinian said, after which he simply got up and walked out.

◆   ◆

On exactly the same day, although it was evening and three time zones to the west of New York, Lenny De Niro walked into The Palace in Las Vegas. After just a few steps he stopped in his tracks, stunned, dazed by the opulence of it all. When he resumed walking he constantly scanned the casino floor, looking anxiously for a familiar face. He found none. Small wonder; after all, it had been almost thirteen years ago that he had left not only the casino but also the town. More accurately, he had been *forced* to leave in disgrace, broken by the same man who, it had been reported in the San Francisco *Chronicle,* was now going to get over one hundred million dollars when he sold the casino that thirteen years ago he had taken away from De Niro and his friend Salgo for not one red cent.

Having half circled the main casino floor, he saw a bar on the left that came as close to the bars he frequented in San Francisco as anything he had seen thus far in the place. The maître d' didn't even look in his direction as he mounted the two steps up to the drinking area surrounded by a brass railing. Then he remembered: this was the old hookers' corner. And sure enough, since it was ten o'clock, there was a whole gang of them sitting at the bar. And there was that big baccarat area he'd heard about. What he hadn't heard was they they now had black women, for God's sake, dealing in the casino.

Sandra Lee saw him about thirty seconds after he saw her. She then kept watching him. As the minutes passed she determined with relief that he apparently wasn't watching *her.* Then again, she reasoned, why should he recognize her? She had been twenty-two then, and a hooker. He must now be sixty-five at least, she thought, but he looked as big and tough, and as mean, as he did then.

There was only one reason he would come back: Danny

Lehman. Danny Lehman and his one hundred million dollars that had been plastered all over the papers; the Danny Lehman that had been revealed as an associate of Meyer Lansky and who was thus now fair game, even for De Niro.

Cordoba was out of town, she knew, and from what she had heard wouldn't be back until the following day. So if she went to the bar during her break there was no risk that he would come and make another big scene over her drinking in the company of the hookers. "Do you mind?" she asked, standing beside the small table where De Niro was seated.

"Naw, sit down." She did.

"What'll you have?" he asked.

"Just a glass of white wine" was her answer. Then: "Don't you remember me, Lenny?"

His head moved back as if she'd threatened to slap him. "Who are you?" he demanded in a low voice.

"Sandra Lee. You must remember me. We all used to sit here in the good old days—you and Roberto Salgo and Mort Granville and all the guys from the counting room and the dealers. We were all here almost every night before everybody went home."

It dawned on him. "You were that gorgeous black whore!"

"That's right."

"I'll be goddamned!"

"What're you doing here?"

"Nothing. Just looking around."

"Looking for the boss?"

"Maybe."

"Better be careful."

"Of Lehman, that crooked little bastard from Philadelphia?" He actually sneered.

"No. Of Eddie Cordoba."

"Who's he?"

"He manages this place. Lehman is out. Cordoba's also the guy who killed your partner, Salgo."

De Niro sat there as if poleaxed. "How do you know somebody killed him? How do you know he's even dead?" he finally asked.

"I know. Look, you better get out of here fast. Meet me at my place in an hour." She wrote her address on the drink napkin and handed it to him. "Go on," she said, "I'll take care of the drinks."

He left immediately. She pleaded illness and left the casino half an hour later. When she got home, she began explaining what had happened thirteen years ago in Beirut, revising history as she told it. Danny, according to her, had intended to be as conciliatory toward Salgo as he had ultimately been toward De Niro. But Cordoba, driven by ambition, had instead eliminated Salgo. Three months later the Argentinian was running The Palace. De Niro listened in silence, downing one scotch after another. Finally she went to bed, leaving De Niro to sleep in the living room. At eight the next morning she woke him, and an hour later sent him on his way. But before he left he returned to the subject of Danny and Cordoba, asking her advice about where and when he might have a word with them. Again she stressed that he was wasting his time where Lehman was concerned. He was out of the casino business, and, for all she knew, had already left town. As to Cordoba, she knew he was coming back from New York that afternoon, and there was no reason to doubt he would then proceed to do what he always did: spend a few hours walking the casino floor, checking everything out. When did he usually show up? Around seven o'clock, before the big action started, allowing him the opportunity to talk to his employees before they got too busy. She suggested that he stay away from The Palace in the meantime. He said he would, but she didn't believe him.

So as soon as he had left, she called Danny Lehman in the executive wing of The Palace. He was cleaning out his office in preparation for leaving later that day for the East Coast. Henry Price had invited him to stay at his country place in Virginia overnight. Then on Friday morning the two of them would proceed together up to New York for the closing.

The phone call prompted a radical change of plan. Lehman left the casino just an hour later. And he did not leave alone; he was accompanied by William Smith, who, since 1969, had been the head of security at The Palace. When they got to the airport, the other originally unscheduled traveling companion, Sandra Lee, was outside waiting for them.

"Thanks," he said to her while Smith was helping to stow their luggage into the Gulfstream III.

"I'm glad you've got Smith with you."

"So am I. And both of you are going to stay with me for a while. At least for as long as that maniac is on the loose."

On the flight east neither of them even remotely referred to the subject of Lenny De Niro, even though he was the man who had brought them together so unexpectedly.

◆   ◆

Three hours before The Palace's Gulfstream III came in to land at National Airport in Washington, D.C., United Flight 297 took off from Dulles International Airport, bound for Los Angeles, with a stopover in Las Vegas. Eddie Cordoba was sitting in first class, and by the time the 747 had reached cruising altitude, he was already into his second bloody Mary. He was celebrating.

Cordoba had spent most of the day in New York looking at properties. He had always wanted to have a place in that city, but had never been able to afford what he wanted there: a luxury apartment in midtown. Now, with a windfall gain of five million dollars a sure thing, compliments of Henry Price, nothing stood in his way. In the early afternoon he'd been shown a co-op on Park Avenue that he had liked, really liked, and had spent so much time looking and relooking at it that he'd almost missed his flight. Cordoba's last act had been to write out a check for seventy-five thousand dollars. That would hold it for at least thirty days.

After dinner, Cordoba decided to take a nap. He needed some rest, since he intended to spend a lot of time walking around the casino in Vegas that evening for a special reason: to assure his key employees that, despite everything they had been reading in the papers, The Palace was definitely going to stay in business and he, and he alone, was going to remain in charge.

The United flight pulled up to the gate at McCarran Airport at 6:15 P.M. West Coast time, and twenty minutes later Cordoba stepped out of his limousine in front of The Palace. He went directly to the elevator that took him up to his office. There he checked out the messages his secretary had neatly lined up on his desk, and then took the elevator back down to the casino floor. He proceeded directly to the baccarat tables, for it was at these tables that the big action always took place in The Palace, and here were where the big profits were made; consequently, it was extremely important that the baccarat pit bosses and dealers be the first to receive his reassurance that all was now well.

Except for *one* dealer. He was going to fire *her* on the spot. With Lehman gone, she'd finally lost her protection. And she was the only link left to Salgo.

"I want to have a word with all of you," he said to the man in charge of the baccarat area during the early evening shift, "but first, tell me: where's Sandra Lee?"

"She went home sick about ten-thirty last night."

"When was she due in tonight?"

"About an hour ago."

"Go call her. And tell her to get her lazy ass over here. *Now!* Then come back here. Like I said, I want to talk to you about something else." The dealer immediately plunged into the crowd, and headed across the casino floor in the direction of the nearest phone with an outside line.

Out of that same crowd at exactly that same time appeared a bull of a man in his late sixties. He went directly to the bar that overlooked the baccarat area, approached the maître d', and asked a question. The maître d' nodded his head affirmatively and pointed in the direction of Eddie Cordoba. Then the stranger left the bar and approached Cordoba, who was still standing on the edge of the baccarat area, watching the action on the casino floor. The old man stopped about six feet in front of him, and just stood there. Soon Cordoba noticed him.

"Are you Eddie Cordoba?" the old man asked.

"Yes."

"I've got something for you, compliments of Roberto Salgo."

As De Niro reached inside his jacket, Cordoba started to recoil. A second later two shots hit Cordoba's chest and slammed him to the carpet. De Niro then stepped forward, put his .44 magnum a foot from Cordoba's skull, and fired four bullets into it.

The time was exactly 7:16, West Coast time, on Thursday, August 12, 1982. The chief baccarat dealer never got to deliver his message to his boss, which was that nobody had answered the phone at Sandra Lee's home.

◆　　　◆

After a late dinner in the main house of Henry Price's farm, he and his guests had retired to the living room for coffee. Price, Danny Lehman, and Sandra Lee sat together on the sofa. Natalie Simmons sat off to the side, chatting with Bill Smith. Since investment bankers like to keep in touch the TV set had, as usual, been turned on in the far corner, tuned to the NBC affiliate in Washington. At eleven they aired the late news. The Israeli siege of Beirut was at its peak, and everybody stopped talking for a moment to watch the carnage. Before the commercial break, a scene flashed onto the screen showing still more bodies, though this time only two, and this time in black-and-white. It was accompanied by a voice-over teaser by the anchorman, suggesting that the audience stay tuned, since after the break they were going to see some remarkable footage taken just one hour ago inside the famed Palace casino in Las Vegas.

The effect of these words on the select audience of five in

the Virginia farmhouse was immediate. All conversation stopped. Then, after the commercial, the anchorman came back on camera with these words: "We are about to show you some scenes shot in rather primitive black-and-white video. Please do not attempt to adjust your sets for the next couple of minutes. The tape was taken by cameras which, for security reasons, are installed behind mirrors in the ceiling above the casino floor at The Palace in Las Vegas. They constantly monitor all activity taking place below."

The tension in the living room of the farmhouse was almost palpable as the grainy black-and-white tape began to run, and the action was shown once at normal speed and then in agonizing slow motion, not unlike the scenes when President Reagan was shot. The first sequence depicted Eduardo Cordoba having words with someone who appeared to be one of The Palace's dealers. The dealer then disappeared, and another figure came on camera, Lenny De Niro. Slowly, calmly, and with terrifying deliberateness, he approached Cordoba and then paused about six feet away from him. When Cordoba turned toward him, the two appeared to exchange a few cordial words. Then in one swift motion De Niro drew a gun from inside his jacket and, judging by the little puffs of smoke which were the viewers' only evidence of gunfire, since, of course, the video security film has no sound, fired two shots at Cordoba's heart, slamming him backward and down. As he lay there, Cordoba's face was turned up toward the camera, and he was obviously still alive. De Niro then stepped forward, put his gun to Cordoba's head, and fired four more rounds which, at that range, caused Cordoba's skull to disintegrate and, in appalling slow motion, literally to fly apart.

In the same eerie silence, a posse of The Palace's security guards suddenly appeared on camera and, as De Niro started to flee, opened fire and seemed almost to cut the man in half.

When this grisly video clip had ended, the anchorman calmly identified Cordoba as an Argentine national and a key figure in the management of The Palace. He went on to report that his slain assailant, reported to have no less than eighteen bullets in him, was one Lenny De Niro, apparently a resident of San Francisco and believed to have been a disgruntled former employee of the casino.

Natalie Simmons was the first to find her voice. "How absolutely horrible! That poor man Cordoba! And who was that man who killed him?"

Henry Price, who had gotten up to turn off the TV set, said nothing. Danny Lehman and Sandra Lee just looked at each other. So it was William Smith who answered the question. "He was an evil man, believe me, Miss Simmons. But no doubt he had a reason for doing what he did. As crass as this might sound, I say good riddance to both of them."

If it had been a church instead of a farmhouse, this statement would have been greeted by a chorus of "Amens." But instead Henry Price finally said, "I don't think any good will come from us dwelling on this any further this evening. Danny and I have to leave for New York in about seven hours. The closing is scheduled for 9 A.M. So let's all go to bed."

Bill Smith went to bed in his cottage, but it took a while before he found sleep. His mind kept circling the events not only of that evening, but also of the last thirteen years, and it kept coming back to Sandra Lee. It had been a woman who had placed that call thirteen years ago, that anonymous phone call, which had put the FBI onto Salgo and De Niro in the first place. He had no doubt now that it had been Sandra Lee. And who had been in Beirut with Danny when Salgo had disappeared? Sandra Lee, of course. And now here she was with Danny in Virginia, three thousand miles away from where the killings had just taken place, as a result of a hurried departure from Las Vegas which, no doubt, she had engineered.

His final thought was that he had better get back to Las Vegas first thing in the morning. There would be a lot of questions asked by the authorities, including his former colleagues at the FBI. It was better that he be there personally to give the answers, and to ensure that this entire matter was closed as quickly and quietly as possible.

◆   ◆

The next morning, Natalie Simmons was up at dawn to prepare breakfast for Henry Price. Despite all that had happened, the look of worried preoccupation that she'd noticed on the investment banker's face the day before seemed to have disappeared. And when Sandra Lee arrived with Danny Lehman—who had also been so uncharacteristically subdued the previous evening—he was once again his usual cheerful self. William Smith was the last man to show up for breakfast. He immediately had a private word with Danny, who subsequently asked Price if Smith could ride with them to National Airport. Smith now planned to return immediately to Las Vegas on the first available commercial flight. Sandra Lee, upon hearing that, said that she'd return to Vegas with Smith. She wanted to get back to her job.

The sale of The Palace Inc. was consummated at the New York office of Mercier Frères at 9:15 A.M. that day. Danny Lehman received a cashier's check in the amount of $103,120,000.00 in exchange for his delivering 10,312,000 shares of The Palace Inc. The remaining ten million–odd shares that had been in the hands of the public had also been tendered, to Casino Properties Inc., so that when the closing was completed, Henry Price and his two European silent partners now controlled 100 percent of The Palace.

When Danny walked out onto Wall Street fifteen minutes later, alone, he had absolutely no idea what he would do next, not even where he would live. He decided to stay in New York for a while and checked into the Pierre. After a week of moping around, while eating a room service hamburger all alone in his suite, he suddenly came to the conclusion that what had happened to him was probably inevitable.

Furthermore, Danny realized that although he had lost the battle, he might still be in a position to win the war. After all, when he added up everything he had stashed away during his thirteen years as the proprietor of the greatest casino on earth, he now had a net worth well in excess of one hundred million dollars. And how had he done it? By emulating success. By following the example of Cliff Perlman and Caesars Palace, and by going one better. Who was setting the pace now? It was obvious: the Wall Street gang and investment bankers like Henry Price. That was the route to go.

But how? After everything that had been in the papers, he was a pariah on Wall Street, and probably everywhere else in America. Then he had an idea. Last time around thirteen years ago, it had all started to come together when he and Sandra Lee had gone to Europe. Maybe it would work again. In Europe nobody would know, or if, by chance, they knew, care about the notorious past of Danny Lehman.

He picked up the phone and called Sandra Lee.

22

When they checked into Claridge's this time, they were shown the utmost courtesy and given one of the finest suites in the hotel. Sandra Lee immediately went apartment-hunting, and, early on the second day, found what she wanted just around the corner from the hotel: a seven-room flat, fully and exquisitely furnished, with maid service and, most important, ready for immediate occupancy.

They found out right away that what Danny had suspected was correct: when he mentioned his name in a restaurant, nobody knew him from Adam. And nobody gave Sandra Lee any funny glances either. He finally began to grasp what it meant when London was described as the most civilized city on earth.

As to his plans to get back into business, well, as Sandra Lee had pointed out, he didn't have to do a thing. When the realization sank in that he had, at a very minimum, $100 million of idle cash, and when the vultures found out where he was, the phone would start ringing of its own accord. It did, with proposals for investments in everything from gold mines in Peru to perpetual-motion machines in Texas. He ignored them all, since he had decided to put most of his money into tax-free municipal bonds, and then to move it progressively into real estate. So Sandra Lee took over the phone, screening the calls, and, more often than not, just telling the parties on the other end to forget about it without even asking Danny.

One morning in late September 1982 when a call came through from Amsterdam, she wasn't quite sure what to do, so she asked the caller to hold and went into the bedroom, where Danny was eating breakfast and reading the papers. "There's a man on the phone who says he's with the Zuider Handelsbank. He refused to give his name to me. Do you want to talk to him?"

Danny hesitated. Was the tranquillity he had found in London about to go down the drain? If so, he might as well find out right away. He picked up the phone on the bedside table.

"This is Danny Lehman," he said tentatively.

"Mr. Lehman, this is Montague Davies. Remember me?"

Although it had been over thirteen years ago, he could hardly forget *that* name.

"I do indeed," he said warily.

"I hear you've become a man of leisure," said Davies. He then continued, "However, something came up yesterday here at the bank that I thought might interest you. It has to do with making money, using a new approach."

Danny immediately relaxed: this was apparently not to be an attempt at blackmail.

"I'd like to explain further, but would prefer to do it in person," Davies went on. He had obviously maintained the aversion to talking business on an unsecured telephone line that he had developed when working for the now defunct First Charter Bank of the Bahamas. "I could manage to fly over to London today, if it is convenient."

At 3 P.M. Montague Davies was shown into the drawing room of Danny's London flat. He was deeply tanned and immaculately dressed and had entered carrying an attaché case and an umbrella—every inch the merchant banker. Danny suggested tea, and after the maid had served them, Davies got right down to business.

"Perhaps I should first explain how I got your number here in London. It was through the chairman of our bank, Jan Kordaat. He and Henry Price are very close, and Price got the number for him."

Price had no doubt gotten it from Natalie Simmons, Danny thought. For he knew that Sandra Lee and Natalie had stayed in contact.

Davies then continued, "While I'm on that subject, you might be interested to hear that Mr. Price has already resold The Palace."

That surprised Danny. "I've read nothing about it in the papers," he said, rather skeptically.

"No, you wouldn't have. It was a strictly private transaction. Price sold out to a family partnership, some people from Dallas, it seems. Furthermore, from what I hear, he got a rather good price: the equivalent of fourteen dollars a share of the old Palace common stock."

Now Danny was really shocked. He'd sold out at ten dollars just six weeks earlier!

Montague Davies could not help but sense what was going through Lehman's mind. "That works out to Price making about forty million dollars on the deal. Not bad, eh? Then do you know what he did? Resigned from Mercier Frères. Apparently his whole involvement with you had resulted in an estrangement between Price and the other senior partners at his bank, so now that he'd taken them completely off the hook, and made forty million in the process for himself, he took the opportunity to tell them to bugger off."

The maid returned to refill their teacups. After she had left again, Montague Davies resumed. "I learned all this the day before yesterday, when one of the partners at Mercier Frères paid us a visit in Amsterdam. He was delegated to replace Price as their contact man with our bank and came over to introduce himself. I take it the bank gave him this job more as a perk than anything else. Which brings me to the purpose of this visit. His main activity is heading up the M&A—mergers and acquisitions—activities at Mercier Frères, advising clients, for a fee, on leveraged buyouts, like the one Price pulled on you, as well as straight takeovers of one corporation by another, including hostile takeovers.

"Now to the point," continued Davies. "He advised me that Litton Industries is going to buy Itek Corporation. He thinks it will take about three, maybe four, months before the deal is finalized. The takeover price will be forty-eight dollars a share. Itek is now selling at twenty-one dollars a share. Follow me?"

Danny followed him completely. "Is Mercier Frères handling the deal?"

"No. Another investment bank in New York. But apparently these M&A types, at least some of them, talk to each other."

"And why has this successor to Henry Price talked to you?" Danny asked, and then added, "By the way, what's his name?"

"I think it best we leave names out of this. As to the reason, I think it probably best you also don't ask me that."

"But I prefer to."

"All right. He'll get a commission of 5 percent of all profits generated by his 'suggestions,' which I guess is standard practice. But do you know how he insists upon being paid? Would you believe an attaché case full of cash to be hand-delivered on a yet-to-be-determined street corner in New York?"

Danny was stunned. "But why would a partner of a prestigious bank like Mercier Frères . . . ?"

Davies just shrugged. Then he said, "Greed. Tax-free greed." He immediately continued. "I'm putting together a little investment syndicate for the purpose of investing in such situations. I was assured that there are more, many more, to come. I've set up an Anstalt in Liechtenstein, which will trade through the General Bank of Switzerland, through either their Zurich office or one of their branches in the Bahamas or the Cayman Islands." Obviously Montague Davies was going full circle.

"But isn't that illegal?" Danny asked.

"Merger arbitrage? Hardly, it's the new art form on Wall Street."

"That's not what I meant? I'm referring to the use of insider information."

"It's not illegal in Switzerland. They do it all the time. Same goes for Liechtenstein. But it always pays to be as prudent as possible. If you decide to participate in our syndicate, my suggestion would be that you transfer the funds to our bank in Amsterdam. We will treat it as a time deposit, giving you full documentation to that effect. The funds will, however, go immediately to Liechtenstein, and proceed from there to accounts at the General Bank of Switzerland either in Zurich or at their branch in the Bahamas. The trades will be made in the bank's name. Follow me?"

"Perfectly."

"Are you interested?"

"Maybe. Can I call you tomorrow?"

"Yes. I'll be back in Amsterdam. Here's my number." Davies handed him his card. Then he reached down into his briefcase and pulled out a long yellow envelope. "And here are the papers you would have to execute to open up that deposit account at the Zuider Handelsbank. I don't think you would want to get involved in any paperwork beyond that point. I'm afraid you will simply have to trust me. But I can assure you that when we take our profits in Liechtenstein, your share will be immediately deposited back to your account in Amsterdam."

Montague Davies rose and said, "Then I'll be hearing from you. By the way, the minimum amount will be one million dollars. If I were you, however, I would make it two. After all, it's about as close to a sure thing as you will probably ever get." Montague Davies then shook hands with Danny, picked up his attaché case, collected his brolly at the door, and left.

The next day Danny deposited two million dollars with the Zuider Handelsbank in Amsterdam. On January 17, 1983, Litton Industries bought Itek for exactly forty-eight dollars a share. Danny had turned his two million dollars into almost five million. It had taken just over four months. He hadn't done as well as Henry Price had done with The Palace. But he was learning.

◆   ◆

In the three years that followed, it was one sure thing after the other. Gulf + Western bought out Esquire at $23.50 a share in 1984. The Liechtenstein "syndicate" prebought the stock at $17. American Stores made a tender offer for Jewel Companies at $70 a share. Four months earlier Montague Davies had inaugurated a purchase program for Jewel's shares: his average price was $49.50. And so it went.

Davies always kept Danny informed, via cryptic phone calls from Amsterdam, Zurich, and Nassau. Cryptic, because Davies adhered strictly to a "need to know" policy. From that point of view, all that Danny needed to know was the results—not how they were made possible, nor who else was sharing them. Davies's mode of operation was very simple: every informant and every client was put into a separate compartment, and the only person who was able to connect them up was Davies himself. In this sense, the M.O. of the world of international investment banking closely resembles that employed by the world's intelligence services. In fact, they often complement each other. Thus the CIA is a regular customer of the Crédit Suisse, using the Swiss bank's airtight compartments, known as numbered accounts, to protect its informers and clients when money changes hands.

Danny understood this and had no problem with it . . .

as long as the profits kept rolling in. And they did, month after month, year after year. But he did try to keep himself more fully informed independently. He had bought a country home in the Buckinghamshire village of Chalfont St. Giles, and had installed his office there.

Then in February of 1986 Danny received an urgent call from Montague Davies at his country place. Davies was phoning from a public booth at Schiphol Airport and about to board a plane for London. He told Danny that it was imperative that they meet at Danny's flat in three hours. Danny assured Davies that he would be there, and, accompanied by Sandra Lee, left for London an hour later.

Davies's first words were "I'm afraid that the last transaction, which involved FMC, is going to be just that: our last transaction, Danny. Our friends at the General Bank of Switzerland have heard that somebody else has been playing the same game, using the facilities of the Bank Leu in the Bahamas. The SEC is onto them, and the Bank Leu, to protect itself, is in the process of finking on the guy. His name's Dennis Levine. He does M&A for Drexel Burnham Lambert and he's a friend of our friend at Mercier Frères. This thing is bound to spread, with everybody finking on everybody else. I've heard that Ivan Boesky, the king of arbitrage, is the next guy in line to get it. Who knows where it will all end? I've already closed down all of our trading accounts in Zurich and Nassau and liquidated the Anstalt in Liechtenstein. I've also suggested to our friend in New York that he take a long vacation, preferably on a beach in Romania." He paused. "Here's the final accounting where you're concerned."

He handed Danny a single piece of paper. The accounting was handwritten. It showed that between September 1982 and February 1986 Danny Lehman had made $47 million dollars, plus change. This did not include the interest he had earned on his deposit account at the Zuider Handelsbank in Amsterdam on the profits that had accumulated there over time.

After Danny was done reading it, Davies held out his hand.

"If you don't mind," he said. Danny returned the piece of paper. Davies then walked over to the fireplace in Danny's drawing room, reached into his vest pocket for some matches, lit the piece of paper, threw it into the fireplace, and watched it burn.

"That's the end of the paper trail," he said, with a grin. "Now, Danny, I've got a new idea. Japan. The Japanese are in the process of accumulating the biggest cash hoard in history. As a result, the stock market over there is about to explode on the upside, and the yen is bound to go through the roof. A double whammy. And strictly legit, meaning that it is no longer a sure thing, but close to it. What do you say?"

"I'll think it over."

"No hurry on this one. But in the meantime, I do suggest that you shift from dollars to either yen or DM in your account at our bank. We would put the funds on three-month deposit and keep rolling them over until you tell us differently. The dollar's a goner, you know."

"Sure. Go ahead, Montague. But I think I'll pass on Japan for the moment."

An hour later, Sandra Lee returned from her shopping. "Is that man gone?" she asked.

"Yes. Permanently."

"Good, I'm glad. Now what are you going to do?"

"I really don't know. Any ideas?"

"No."

"Remember all the times we've been in Seabys and in Spinks?" he asked, referring to the two most prestigious coin stores in the world, where he often made purchases, adding to his personal collection. "What if we opened our own store?"

"That's pretty stiff competition. Anyway, where would you put the store?"

Danny didn't hesitate. "Bond Street. And by the time we're done they'll look paltry by comparison."

"You're always saying 'we.' Do you mean it?"

"Are you willing to take on a partner?" he asked.

"Anytime, anywhere, Danny," she replied.

"All right. It's a deal. I'll be back at one and we'll head over the the Savoy to celebrate."

An hour later, they walked into the Savoy Grill. All eyes turned to Sandra Lee. Nobody noticed the middle-aged, short, pudgy man at her side. Danny didn't mind. He had just retired from a brief but highly successful career as a merger arbitrageur. Now he was about to go back, full circle, to the business he had always liked best: the coin business. Now that all the world was becoming a gambling casino, it felt good to be back in real money again. As coffee was being served, he reached into his pocket and handed a small velvet-covered box to Sandra Lee. She opened it and took out an American double eagle gold coin. "Thanks, partner," she said and leaned over the table to kiss Danny.

Every man in the Savoy Grill who watched them felt the same thing: pure envy.

Life *was* fair.